Groundwater Chemistry and Quality in Coastal Aquifers

Groundwater Chemistry and Quality in Coastal Aquifers

Editors

Guanxing Huang
Liangping Li

MDPI

Basel • Beijing • Wuhan • Barcelona • Belgrade • Novi Sad • Cluj • Manchester

Editors
Guanxing Huang
Institute of Hydrogeology
and Environmental Geology
Chinese Academy of
Geological Sciences
Shijiazhuang
China

Liangping Li
Department of Geology
and Geological Engineering
South Dakota School of Mines
and Technology
Rapid City
United States

Editorial Office
MDPI AG
Grosspeteranlage 5
4052 Basel, Switzerland

This is a reprint of articles from the Special Issue published online in the open access journal *Water* (ISSN 2073-4441) (available at: https://www.mdpi.com/journal/water/special_issues/groundwater_chemistry_quality).

For citation purposes, cite each article independently as indicated on the article page online and as indicated below:

Lastname, A.A.; Lastname, B.B. Article Title. *Journal Name* **Year**, *Volume Number*, Page Range.

ISBN 978-3-7258-1798-6 (Hbk)
ISBN 978-3-7258-1797-9 (PDF)
doi.org/10.3390/books978-3-7258-1797-9

Contents

About the Editors

Guanxing Huang

Dr. Guanxing Huang is a Professor within the Institute of Hydrogeology and Environmental Geology, Chinese Academy of Geological Sciences, China. He received his Ph.D. degree from the Chinese Academy of Geological Sciences, China, in 2014. He worked as a Visiting Scholar at the University of Massachusetts, Amherst, in 2015–2016. He has published over 70 refereed journal articles, with a total citation exceeding 2,000. Dr. Huang is a Guest Editor of the journal *Water*, and he is a Youth Editorial Board Member of the *Carbon Research* and *China Geology* journals. He served as an Editorial Board Member of *Frontiers in Water* (2021–2022), and as a Guest Editor of the *International Journal of Environmental Research and Public Health* (2021). He is a member of both the International Association of GeoChemistry (IAGC) and International Association of Hydrology (IAH). He has received several awards/honors, including the Leading Talent of Science and Technology of the Ministry of Natural Resources of China in 2021, Excellence in Review Award of Applied Geochemistry in 2020, Young Scientist Award of the Chinese Society of Environmental Sciences in 2020, and Outstanding Youth Science and Technology Talent Cultivation Plan, Ministry of Land and Resources of China in 2015. His current research focuses on regional groundwater contamination, the evolution of groundwater chemistry, groundwater quality assessment, and natural background levels in groundwater.

Liangping Li

Liangping Li is an Associate Professor (since 2022) within the Department of Geology and Geological Engineering, South Dakota School of Mines and Technology, Rapid City, USA, where he has been actively involved in teaching and research since 2015. Dr. Li received his Ph.D. degree from the Technical University of Valencia, Spain, in 2011, and he then worked within post-doc positions at the University of Texas, Austin, USA, during 2012-2015. Dr. Li's current research includes groundwater modeling, contaminant migration, groundwater sampling investigations, groundwater supplies, geochemistry, analytical methods in groundwater, and field methods. With his students and colleagues, he has published over 60 refereed articles, most in top journals such as *Water Resources Research*, *Advances in Water Resources*, and the *Journal of Hydrology*, many of which are frequently cited by his peers. His h-index is currently at 22, and he has a total citation of over 1700. He has been invited to present his research results at several universities and institutions. Dr. Li served as an Associate Editor for *Hydrogeology Journal*, the *Journal of Hydraulic Engineering*, and *Stochastic Environmental Research and Risk Assessment*, as a Guest Editor for *Geofluids*, *Mathematical Geosciences*, and *Water*, and as a Co-Chair of the Organizational Committee for the 20th Annual Meeting of the International Association of Mathematical Geosciences. Dr. Li obtained funds from NSF, USGS, NASA, and the South Dakota Board of Regents over the last five years. His research program is recognized internationally.

Preface

Groundwater is the most abundant freshwater resource available on Earth. Today, coastal aquifers, regarded as the nexus of ocean and terrestrial hydrologic ecosystems, provide an available groundwater resource for one billion people worldwide, as well as social and economic development in coastal areas. However, knowledge on the current status of groundwater chemistry and quality in coastal areas is limited. This constrains the utility of groundwater resources in coastal areas and hinders their social and economic development. Therefore, this reprint presents a vast array of information on regional groundwater chemistry and quality in coastal areas in ocean countries, including China, Pakistan, and Egypt. We hope that it will be a useful reference for related readers, hydrogeologists, and departments of water resources' development and utilization.

Guanxing Huang and Liangping Li
Editors

water

MDPI

Editorial

Groundwater Chemistry and Quality in Coastal Aquifers

Guanxing Huang [1,2,*] and Liangping Li [3]

[1] Institute of Hydrogeology and Environmental Geology, Chinese Academy of Geological Sciences, Shijiazhuang 050061, China
[2] Fujian Provincial Key Laboratory of Water Cycling and Eco-Geological Processes, Xiamen 361024, China
[3] Department of Geology and Geological Engineering, South Dakota School of Mines and Technology, Rapid City, SD 57701, USA; liangping.li@sdsmt.edu
* Correspondence: huangguanxing2004@126.com

1. Introduction

Groundwater is the most abundant freshwater resource available on earth, and it accounts for more than 95% of all liquid freshwater [1,2]. This freshwater resource is important for coastal areas where it is commonly regarded as the center of social and economic development because the availability of water resources constrains socio-economic development in these areas [3]. It is reported that coastal aquifers are a nexus of the ocean, and terrestrial hydrologic ecosystems provide groundwater resources for about one billion people worldwide as well as social and economic development in coastal areas [4]. There is a direct connection between the stores of available freshwater provided by groundwater and their status in terms of quality because the good quality provided by groundwater is critical for sustaining agriculture, industries, and drinking water services [5,6]. We need to improve our knowledge of groundwater quality if we are to use groundwater sustainably and not further burden future generations by limiting resources and/or increasing treatment costs [7]. Therefore, understanding the status and challenges of groundwater quality to support sustainable water supply is fundamental to reaching key Sustainable Development Goals (SDGs) in coastal areas.

Two groups of chemical components threaten groundwater quality. One includes geogenic contaminants, such as arsenic, fluoride, and iodide [8,9]. Another is human-induced pollutants, such as nitrate and organic pollutants [10,11]. Both of them influence groundwater quality and are often complicated because of complex geological settings and anthropogenic factors. In recent decades, land use transformation from agricultural and natural ecosystems to urbanization accompanied by various human activities has intensified the complexity of groundwater chemistry and quality in global coastal areas [12,13]. For instance, large-scale urban agglomerations, such as the Yangtze River Delta (YRD) and Pearl River Delta (PRD) megalopolises of China and the Boston–Washington megalopolis of the United States, are distributed in coastal areas adjacent to the Pacific and Atlantic oceans, respectively. As a result, in order to reflect recent research advances on groundwater chemistry and quality in coastal areas, this Special Issue for the journal *Water*, entitled *Groundwater Chemistry and Quality in Coastal Aquifers*, was developed.

Citation: Huang, G.; Li, L. Groundwater Chemistry and Quality in Coastal Aquifers. *Water* **2024**, *16*, 2041. https://doi.org/10.3390/w16142041

Received: 8 July 2024
Accepted: 12 July 2024
Published: 19 July 2024

2. Advances on Groundwater Chemistry and Quality in Coastal Areas

In order to understand the advances in groundwater chemistry and quality in coastal areas worldwide, we searched related publications in the Web of Science database using the following three keywords: groundwater quality, coastal area, and groundwater chemistry in the coastal area. As shown in Figure 1A, more than two thousand papers related to groundwater chemistry and quality in coastal areas were published in the last twenty years, and the annual number of papers has increased rapidly, from around 20 papers in one year in the early twenty-first century up to nearly 200 papers per year in recent times. All countries with a high number of papers are adjacent to oceans (Figure 1B). For example, the three countries with the highest number of papers are India, the USA, and China, which are adjacent to the Indian Ocean and Pacific Ocean, respectively, and their numbers of papers have nearly doubled compared to other countries (Figure 1B). This indicates that India, the USA, and China pay much more attention to groundwater chemistry and quality in coastal areas than other ocean countries. Correspondingly, this Special Issue mainly focuses on groundwater chemistry and quality in coastal areas in one of the above countries (China). In addition, groundwater chemistry and quality in coastal areas of two other ocean countries, including Pakistan and Egypt, are also mentioned in this Special Issue.

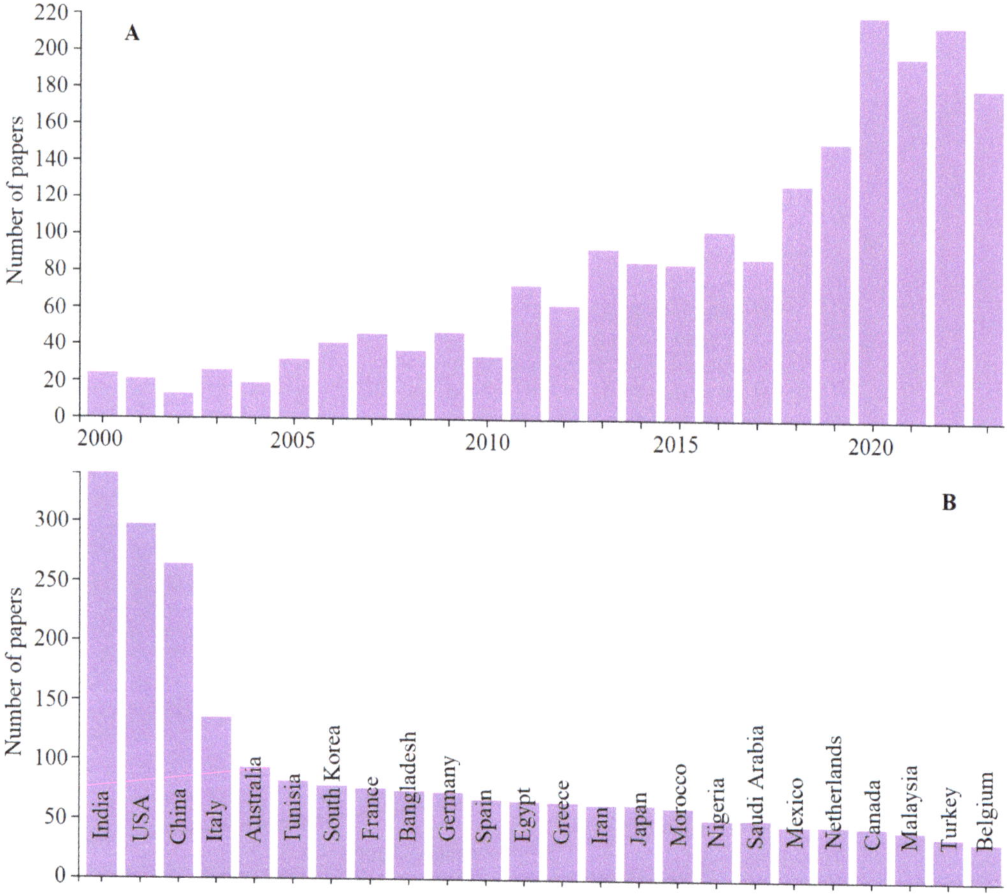

Figure 1. Statistics of number of papers related to groundwater quality, coastal areas, and groundwater chemistry for different years (**A**) and countries (**B**) in the Web of Science database.

2.1. *Groundwater Chemistry and Quality in Coastal Areas of China*

This Special Issue includes 10 papers related to groundwater chemistry and quality in the coastal areas of China from the south to the northeast. Hainan Island, located at the most southern point of China and adjacent to the South China Sea, is a well-known center for tropical agricultural production, but its groundwater quality has received little attention for a long time. In this Special Issue, Liu et al. (contribution 1) paid attention to groundwater quality in a typical coastal area dominated by agriculture on Hainan Island. In order to investigate the occurrence of fluoride in the western coastal area of Hainan Island and discuss factors affecting groundwater F^- contamination in various aquifers and areas with different land-use types, the authors collected a total of 100 groundwater samples in porous and fissured aquifers from this coastal area of Hainan Island in 2022 and analyzed 20 parameters including pH, dissolved oxygen (DO), oxidation-reduction potential (ORP), total dissolved solids (TDSs), bicarbonate, chloride, sulfate, nitrate, nitrite, fluoride, potassium, sodium, calcium, magnesium, arsenic, selenium, aluminum, barium, manganese, and strontium in water samples. Specifically, they depicted the spatial distribution of groundwater fluoride concentrations in this coastal area of Hainan Island using the inverse distance weighting method. They revealed that high levels of fluoride in porous groundwater are mainly attributed to the leaching of fluoride/aluminum-containing minerals such as phlogopite and calcite in the vadose zone using principal component analysis (PCA) and hierarchical cluster analysis (HCA). As a result, the authors recommend that the use of fluoride-containing fertilizers in the study area should be limited to prevent an increase in high-fluoride groundwater. This study is the first time to reveal the co-impacts of anthropogenic and geogenic factors on the occurrence of high-fluoride groundwater in coastal areas dominated by agriculture on Hainan Island. In addition, a recent publication pointed out that nitrate was another major groundwater contaminant in coastal areas dominated by agriculture on Hainan Island [14]. Therefore, in coastal areas of Hainan Island, preventing groundwater contamination of fluoride and nitrate is a major groundwater quality issue.

The PRD is a large-scale urbanized area, also adjacent to the South China Sea and located in the Guangdong Province of South China. Unlike Hainan Island, groundwater chemistry and quality in the PRD has received a lot of attention. During the past two decades, we investigated the characteristics of groundwater chemistry in granular, fissured, and karst aquifers of the PRD at a regional scale [13]. We also put forward contamination patterns of heavy metal(loid)s and organic contaminants in the shallow groundwater of the PRD [10], revealed the origins and driving mechanisms of iodide-rich, iron-rich, and aluminum-rich groundwater in the PRD [8,15,16], and investigated the distribution and origins of groundwater phosphate contamination in the PRD [17]. In addition, Hou et al. reported the origins and driving mechanisms of elevated manganese concentrations in shallow groundwater in the PRD [18]. Zhang et al. investigated the distributions and origins of nitrate, nitrite, and ammonium in various aquifers in the PRD [19]. Zhang et al. evaluated the shallow groundwater quality in granular, fissured, and karst aquifers of the PRD using a fuzzy synthetic evaluation method and discussed the driving forces controlling groundwater quality in these aquifers [5]. However, knowledge of the influence of large-scale land use conversion on groundwater chemistry and quality in this coastal alluvial aquifer is still limited, though shallow groundwater in the coastal alluvial aquifer is an important source of water supply in the PRD. Considering this, in this Special Issue, Liu et al. (contribution 2) investigated the impact of land use on hydrogeochemical characteristics and groundwater quality in a coastal alluvial aquifer of the PRD. They collected a total of 149 groundwater samples at once and analyzed 19 physicochemical parameters. Specifically, 75 samples, 46 samples, 25 samples, and 3 samples were collected from urban areas, peri-urban areas, agricultural areas, as well as the remaining area, respectively. They found that groundwater chemistry in this coastal alluvial aquifer was dominated by Ca-HCO_3 and Ca·Na-HCO_3 facies. They also used the fuzzy synthetic evaluation method with the groundwater quality standards of China to assess groundwater quality in this coastal alluvial aquifer and reported that the occurrence of poor-quality groundwater in urban and agricultural areas was more regular than that in peri-urban areas. In addition, they revealed that groundwater chemistry and quality in this coastal alluvial aquifer of the PRD were mainly controlled by five factors via the PCA. Therefore, they recommend many useful suggestions according to these factors to protect groundwater quality in this coastal alluvial aquifer of the PRD.

On the other hand, we assessed the natural background levels (NBLs) of many contaminants such as arsenic, manganese, nitrate, and chloride in the shallow groundwater of the PRD in recent studies via developing a new preselection method, which consists of the chloride/bromide mass ratio versus chloride concentration, the oxidation capacity, and Grubbs' test [9,11]. However, knowledge of the NBLs of two other important contaminants (e.g., iodide and ammonium) in the groundwater of the PRD is still limited, though the contamination of these two contaminants was widely distributed in the shallow groundwater of the PRD. Therefore, in this Special Issue, Pei et al. (contribution 3) evaluated the NBLs of these two contaminants in shallow groundwater of the PRD via the method established by us and discussed the main factors controlling the NBLs of these two contaminants in groundwater to enhance the knowledge on groundwater NBLs in the PRD. The authors compared the variation in groundwater iodide and ammonium concentrations in areas with different land-use types in original and residual datasets to verify the effectiveness of the used method for assessing the groundwater NBLs of iodide and ammonium in the PRD. They found that NBLs of iodide and ammonium in shallow groundwater of the coastal alluvial aquifer were more than twice those in the other three groundwater units. They revealed the driving forces for controlling higher NBLs of iodide and ammonium in groundwater of the coastal alluvial aquifer compared with other groundwater units via the combination of hydrogeological data, land use data, and the PCA with hydrochemical data.

Fujian Province is located in the southeast of China and is adjacent to the East China Sea. Similar to Hainan Island, groundwater chemistry and quality in Fujian Province has received little attention because of its sufficient surface water resources in comparison with poorer groundwater resources. In this Special Issue, Li et al. (contribution 4) investigated the hydrochemical characteristics of groundwater in a coastal city (Xiamen) of Fujian Province and analyzed its anthropogenic impacts via statistical analysis methods, such as the Mahalanobis distance. Moreover, they compared the current status of hydrochemical characteristics in groundwater in Xiamen City to the historic hydrochemical characteristics of 1993 and found that six new groundwater contaminants, including nitrate, lead, ammonium, aluminum, nitrite, and copper, had increased during the past two decades, and the number of hydrochemical facies in groundwater rose from 19 to 28, such as the occurrence of NO_3 facies and the increase in Cl and SO_4 facies. In addition, the authors highlighted that the Mahalanobis distance method is useful for identifying hydrogeochemical anomalies and determining the intensity of anthropogenic influences on groundwater at regional scales.

The Yangtze River Delta (YRD) is also a large-scale urbanized area and is adjacent to the East China Sea. Unlike the PRD, many concerns over groundwater chemistry and quality in the YRD still remain unclear. In this Special Issue, Chen et al. (contribution 5) investigated the hydrochemical characteristics of groundwater in a pilot promoter region of the YRD (southeast of the Taihu Lake) and used the absolute principal component score–multiple linear regression (APCS-MLR) model to evaluate the groundwater quality in this area. Briefly, the authors collected groundwater samples from 84 sampling sites in June 2020/2021 and analyzed 16 hydrochemical parameters, including pH, DO, TDS, chloride, sulfate, nitrate, nitrite, ammonium, potassium, sodium, calcium, magnesium, manganese, total phosphorus, iodine, and antimony, in these samples. They used the PCA to identify the potential natural and anthropogenic factors impacting groundwater quality. In addition, they used the APCS-MLR model to quantify the contribution of factors for groundwater quality in the pilot promoter region of the YRD. They highlighted that groundwater quality in this region is mainly contaminated by industrial effluent and agricultural activities rather than domestic sewage.

Hebei Province is located in the east of China and is adjacent to the Bohai Sea. Similar to the PRD, groundwater chemistry and quality in Hebei Province has received a lot of attention because of the importance of groundwater resources for water supply in this area. Meanwhile, this Special Issue has four papers related to groundwater chemistry and quality in Hebei Province. For example, Qian et al. (contribution 6) investigated the current status of hydrogeochemical characteristics and groundwater quality in the Hebei Plain because of the strong impact of human activities on groundwater in this area. They collected a total of 54 groundwater samples, including 35 shallow and 19 deep samples, from the Piedmont Plain to the littoral plain and analyzed 31 hydrochemical parameters in these water samples. They found that the quality of deep groundwater was better than the shallow samples according to the fuzzy synthetic evaluation method and suggested that deep groundwater is more suitable for drinking purposes in comparison with shallow groundwater in the Hebei Plain. They revealed the factors controlling groundwater quality using PCA and pointed out that shallow groundwater quality is mainly controlled by five factors, including the water–rock interaction, marine geogenic sources, agricultural pollution, acidification, and the reductive environment, while deep groundwater quality is mainly controlled by three factors, including the water–rock interaction and redox processes, agricultural pollution, and the input of external water. Furthermore, Qian et al. (contribution 7) also investigated the distribution and origins of groundwater total hardness (TH) in various sub-plains and different land-use areas of the Hebei Plain on the basis of analyzing 445 groundwater samples because the current status of TH-rich groundwater in the Hebei Plain is still unclear. They found that TH-rich shallow groundwater is mainly distributed in central and littoral plains rather than the Piedmont Plain, while TH-rich deep groundwater mainly occurs in the central plain. They also found that TH-rich groundwater in agricultural areas in the central plain is higher than that in rural areas, but this is the opposite case in the littoral plain. In addition, the authors revealed the driving factors controlling TH-rich groundwater in both central and littoral plains using the PCA and Gibbs diagram. According to these driving factors, the authors recommended restricting the use of nitrogenous fertilizers to limit groundwater hardness concentrations in the Hebei Plain. Unlike the above two papers, which focus on groundwater chemistry and quality in Hebei Plain at regional scales, Yuan et al. (contribution 8) investigated the distribution of groundwater chemistry and quality in a drinking water area (Hutuo River) of the Piedmont Plain in the Hebei Plain at a site scale with a sampling density of more than 60 groundwater samples/100 km^2 because knowledge on groundwater chemistry and quality at different scales is often distinct. The authors reported that groundwater chemistry in the Hutuo River drinking water area is mainly controlled by the water–rock interaction. They assessed groundwater quality using the entropy-weighted water quality index and pointed out that approximately 99.4% of groundwater samples in the Hutuo River drinking water area are suitable for drinking purposes. This indicates that groundwater quality in the Hutuo River drinking water area is better than that in the entire Hebei Plain. In addition, unlike the above three papers, which focus on groundwater chemistry and quality in the Hebei Plain, Gao et al. (contribution 9) focused on groundwater chemistry and quality in the hilly area of Hebei Province. They investigated the hydrochemical characteristics of groundwater (dominated by karst aquifers) in the Heilongdong Spring Basin in the Taihang Mountain of Hebei Province using two batches of samples (2013 and 2019). They revealed that groundwater nitrate and chloride contamination occurred in runoff and discharge areas in the Heilongdong Spring Basin because of domestic sewage leakage and agricultural activities.

Heilongjiang Province is located in the northeast of China and is near the Japan Sea. Similar to Hebei Province, groundwater chemistry and quality in Heilongjiang Province has also received a lot of attention because it includes many large-scale plains (e.g., Songnen Plain, Muling–Xingkai Plain) with abundant groundwater resources. In this Special Issue, Su et al. (contribution 10) focused on groundwater chemistry in the Muling–Xingkai Plain of Heilongjiang Province. They investigated the characteristics and evolution of groundwater chemistry in the Muling–Xingkai Plain by analyzing 164 groundwater samples collected in 2016–2018. They found that groundwater chemistry in this area is mainly controlled by the dissolution of silicate and carbonate minerals and the infiltration of domestic sewage, and the latter is mainly responsible for the elevated nitrate, chloride, and sulfate concentrations found in shallow groundwater. Thus, they recommended that deep groundwater below 80 m in this area is suitable for water supply.

2.2. Groundwater Chemistry and Quality in Coastal Areas of Pakistan

Sindh is located in the south of Pakistan and is adjacent to the Arabian Sea. In this Special Issue, Landar et al. (contribution 11) focused on groundwater quality in Sindh and evaluated the groundwater quality in the Sanghar district of Sindh by collecting 74 groundwater samples and analyzing 26 chemical parameters in these samples. The authors found that six contaminants, including chromium, lead, nickel, cadmium, arsenic, and fluoride, in groundwater samples in the Sanghar district exceeded the allowable limits recommended by the World Health Organization (WHO). They used the water quality index to assess groundwater quality and found that approximately 77% of groundwater samples had excellent-to-good water quality. They also reported that about 89% of groundwater samples with a low contamination index (<3) of cadmium were suitable for drinking purposes. Moreover, the authors assessed the human health risk of groundwater by calculating chronic daily intake indices (CDIs) and hazard quotient indices (HQIs) and found that only CDI values of lead in a few groundwater samples were above one, and HQ values of lead and cadmium in some groundwater samples were unsafe for human health, while others were safe. In addition, they also used other methods, such as the sodium adsorption ratio (SAR), to assess groundwater quality for irrigation purposes.

2.3. Groundwater Chemistry and Quality in Coastal Areas of Egypt

The West Nile Delta (WND) is located in the northwest of Egypt and is adjacent to the Mediterranean Sea. In this Special Issue, Hasan et al. (contribution 12) focused on groundwater chemistry and quality in the WND using integrated statistical and graphical techniques. They collected 75 groundwater samples from quaternary aquifers in the WND in 2018 and analyzed 16 chemical parameters in water samples. They used the HCA to divide groundwater samples into four groups and revealed that groundwater chemistry in two groups located to the north of the study area near the Mediterranean Sea was mainly controlled by the evaporation process, seawater intrusion, and ion exchange. They also pointed out that groundwater chemistry in the other two groups, relatively far from the Mediterranean Sea, was mainly affected by the weathering of silicate minerals and the dissolution of carbonate minerals. In addition, they highlighted saltwater intrusion in this coastal aquifer.

3. Conclusions

This editorial introduces the foundation of the current Special Issue, *Groundwater Chemistry and Quality in Coastal Aquifers*, which briefly reviewed recent research advances in groundwater chemistry and quality in coastal areas and summarizes the main contribution of published papers in this Special Issue. Specifically, this Special Issue reported the status and factors controlling regional groundwater chemistry and quality in eight coastal areas related to three countries, including China, Pakistan, and Egypt. Most papers in this Special Issue used some multivariate statistical techniques, such as PCA and HCA, to enhance their understanding of the factors controlling groundwater chemistry and quality, and only a few papers used some relatively new methods such as the APCS-MLR model. Most of them revealed that agricultural contamination is a main anthropogenic factor influencing groundwater chemistry and quality in coastal areas, especially in areas dominated by agriculture. The infiltration of domestic sewage and industrial wastewater is another major anthropogenic factor affecting groundwater chemistry and quality in coastal urbanized areas, such as the PRD and YRD. Nitrate contamination and seawater intrusion are also two widespread issues regarding groundwater quality in coastal areas regardless of organic contamination. Therefore, in the future, we should focus on the treatment of nitrate contamination and seawater intrusion in the groundwater of coastal areas. New statistical techniques should be adopted and used to enhance the knowledge of groundwater chemistry and quality in coastal aquifers. Organic contaminants in groundwater in coastal areas, especially in coastal urbanized areas, should not be ignored.

Author Contributions: Conceptualization, G.H. and L.L.; investigation, G.H. and L.L.; methodology, G.H.; writing—original draft preparation, G.H.; writing—review and editing, G.H. and L.L. All authors have read and agreed to the published version of the manuscript.

Funding: This research was supported by the Natural Science Foundation of Xiamen City in China (3502Z202373150) and the Science and Technology Strategy Research Project of Natural Resource in China (2023-ZL-10).

Conflicts of Interest: The authors declare no conflict of interest.

List of Contributions

1. Liu, R.; Li, X.; Yang, X.; Zhang, M. Occurrence of and Factors Affecting Groundwater Fluoride in the Western Coastal Area of Hainan Island, South China. *Water* **2023**, *15*, 3678. https://doi.org/10.3390/w15203678.
2. Liu, C.; Hou, Q.; Chen, Y.; Huang, G. Hydrogeochemical Characteristics and Groundwater Quality in a Coastal Urbanized Area, South China: Impact of Land Use. *Water* **2022**, *14*, 4131. https://doi.org/10.3390/w14244131.
3. Pei, L.; Lu, X.; Li, X.; Zhang, M.; Wu, H. Factors Controlling Natural Background Levels of Ammonium and Iodide in Shallow Groundwater of Coastal Aquifers, South China. *Water* **2022**, *14*, 3737. https://doi.org/10.3390/w14223737.
4. Li, Z.; Li, J.; Zhu, Y.; Li, Y.; Hao, Q. Anthropogenic Influences on the Hydrochemical Characteristics of the Groundwater in Xiamen City, China and Their Evolution. *Water* **2022**, *14*, 3377. https://doi.org/10.3390/w14213377.
5. Chen, Z.; Zhou, Q.; Lv, J.; Jiang, Y.; Yang, H.; Yang, H.; Mei, S.; Jia, Z.; Zhang, H.; Jin, Y.; et al. Assessment of Groundwater Quality Using APCS-MLR Model: A Case Study in the Pilot Promoter Region of Yangtze River Delta Integration Demonstration Zone, China. *Water* **2023**, *15*, 225. https://doi.org/10.3390/w15020225.
6. Qian, Y.; Hou, Q.; Wang, C.; Zhen, S.; Yue, C.; Cui, X.; Guo, C. Hydrogeochemical Characteristics and Groundwater Quality in Phreatic and Confined Aquifers of the Hebei Plain, China. *Water* **2023**, *15*, 3071. https://doi.org/10.3390/w15173071.
7. Qian, Y.; Zhen, S.; Yue, C.; Cui, X. Distribution and Origins of Hardness in Shallow and Deep Groundwaters of the Hebei Plain, China. *Water* **2024**, *16*, 310. https://doi.org/10.3390/w16020310.
8. Yuan, Z.; Jian, Y.; Chen, Z.; Jin, P.; Gao, S.; Wang, Q.; Ding, Z.; Wang, D.; Ma, Z. Distribution of Groundwater Hydrochemistry and Quality Assessment in Hutuo River Drinking Water Source Area of Shijiazhuang (North China Plain). *Water* **2024**, *16*, 175. https://doi.org/10.3390/w16010175.
9. Gao, M.; Li, X.; Qian, J.; Wang, Z.; Hou, X.; Fu, C.; Ma, J.; Zhang, C.; Li, J. Hydrogeochemical Characteristics and Evolution of Karst Groundwater in Heilongdong Spring Basin, Northern China. *Water* **2023**, *15*, 726. https://doi.org/10.3390/w15040726.
10. Su, C.; Li, Z.; Wang, W.; Cheng, Z.; Zheng, Z.; Chen, Z. Key Factors Dominating the Groundwater Chemical Composition in a Grain Production Base: A Case Study of Muling–Xingkai Plain, Northeast China. *Water* **2022**, *14*, 2222. https://doi.org/10.3390/w14142222.
11. Landar, A.Q.; Jahangir, T.M.; Khuhawar, M.Y.; Lanjwani, M.F.; Khuhawar, F.Y. Evaluation of Water Quality of Groundwater of Sanghar District, Sindh, Pakistan: Chemical and Multivariate Analysis. *Water* **2024**, *16*, 856. https://doi.org/10.3390/w16060856.
12. Hasan, S.S.; Salem, Z.E.; Sefelnasr, A. Assessment of Hydrogeochemical Characteristics and Seawater Intrusion in Coastal Aquifers by Integrating Statistical and Graphical Techniques: Quaternary Aquifer, West Nile Delta, Egypt. *Water* **2023**, *15*, 1803. https://doi.org/10.3390/w15101803.

References

1. Gleeson, T.; Befus, K.M.; Jasechko, S.; Luijendijk, E.; Cardenas, M.B. The global volume and distribution of modern groundwater. *Nat. Geosci.* **2016**, *9*, 161–167. [CrossRef]
2. Mukherjee, A.; Scanlon, B.; Aureli, A.; Langan, S.; Guo, H.; McKenzie, A. *Global Groundwater: Source, Scarcity, Sustainability, Security and Solutions*, 1st ed.; Elsevier: Amsterdam, The Netherlands, 2020; p. xxv.
3. Han, D.; Currell, M.J. Review of drivers and threats to coastal groundwater quality in China. *Sci. Total Environ.* **2022**, *806*, 150913. [CrossRef] [PubMed]
4. Small, C.; Nicholls, R.J. A global analysis of human settlement in coastal zones. *J. Coast. Res.* **2003**, *19*, 584–599.

5. Zhang, F.; Huang, G.; Hou, Q.; Liu, C.; Zhang, Y.; Zhang, Q. Groundwater quality in the Pearl River Delta after the rapid expansion of industrialization and urbanization: Distributions, main impact indicators, and driving forces. *J. Hydrol.* **2019**, *577*, 124004. [CrossRef]
6. Gleeson, T.; Wang-Erlandsson, L.; Porkka, M.; Zipper, S.C.; Jaramillo, F.; Gerten, D.; Fetzer, I.; Cornell, S.E.; Piemontese, L.; Gordon, L.J.; et al. Illuminating water cycle modifications and Earth system resilience in the Anthropocene. *Water Resour. Res.* **2020**, *56*, e2019WR024957. [CrossRef]
7. Lapworth, D.J.; Boving, T.B.; Kreamer, D.K.; Kebede, S.; Smedley, P.L. Groundwater quality: Global threats, opportunities and realising the potential of groundwater. *Sci. Total Environ.* **2022**, *811*, 152471. [CrossRef]
8. Huang, G.; Liu, C.; Li, L.; Zhang, F.; Chen, Z. Spatial distribution and origin of shallow groundwater iodide in a rapidly urbanized delta: A case study of the Pearl River Delta. *J. Hydrol.* **2020**, *585*, 124860. [CrossRef]
9. Huang, G.; Song, J.; Han, D.; Liu, R.; Liu, C.; Hou, Q. Assessing natural background levels of geogenic contaminants in groundwater of an urbanized delta through removal of groundwaters impacted by anthropogenic inputs: New insights into driving factors. *Sci. Total Environ.* **2023**, *857*, 159527. [CrossRef] [PubMed]
10. Huang, G.; Zhang, M.; Liu, C.; Li, L.; Chen, Z. Heavy metal(loid)s and organic contaminants in groundwater in the Pearl River Delta that has undergone three decades of urbanization and industrialization: Distributions, sources, and driving forces. *Sci. Total Environ.* **2018**, *635*, 913–925. [CrossRef] [PubMed]
11. Huang, G.; Pei, L.; Li, L.; Liu, C. Natural background levels in groundwater in the Pearl River Delta after the rapid expansion of urbanization: A new pre-selection method. *Sci. Total Environ.* **2022**, *813*, 151890. [CrossRef] [PubMed]
12. Huang, G.; Sun, J.; Zhang, Y.; Chen, Z.; Liu, F. Impact of anthropogenic and natural processes on the evolution of groundwater chemistry in a rapidly urbanized coastal area, South China. *Sci. Total Environ.* **2013**, *463–464*, 209–221. [CrossRef] [PubMed]
13. Huang, G.; Liu, C.; Sun, J.; Zhang, M.; Jing, J.; Li, L. A regional scale investigation on factors controlling the groundwater chemistry of various aquifers in a rapidly urbanized area: A case study of the Pearl River Delta. *Sci. Total Environ.* **2018**, *625*, 510–518. [CrossRef] [PubMed]
14. Liu, R.; Xie, X.; Hou, Q.; Han, D.; Song, J.; Huang, G. Spatial distribution, sources, and human health risk assessment of elevated nitrate levels in groundwater of an agriculture-dominant coastal area in Hainan Island, China. *J. Hydrol.* **2024**, *634*, 131088. [CrossRef]
15. Huang, G.; Han, D.; Song, J.; Li, L.; Pei, L. A sharp contrasting occurrence of iron-rich groundwater in the Pearl River Delta during the past dozen years (2006–2018): The genesis and mitigation effect. *Sci. Total Environ.* **2022**, *829*, 154676. [CrossRef] [PubMed]
16. Huang, G.; Hou, Q.; Han, D.; Liu, R.; Song, J. Large scale occurrence of aluminium-rich shallow groundwater in the Pearl River Delta after the rapid urbanization: Co-effects of anthropogenic and geogenic factors. *J. Contam. Hydrol.* **2023**, *254*, 104130. [CrossRef] [PubMed]
17. Huang, G.; Liu, C.; Zhang, Y.; Chen, Z. Groundwater is important for the geochemical cycling of phosphorus in rapidly urbanized areas: A case study in the Pearl River Delta. *Environ. Pollut.* **2020**, *260*, 114079. [CrossRef] [PubMed]
18. Hou, Q.; Zhang, Q.; Huang, G.; Liu, C.; Zhang, Y. Elevated manganese concentrations in shallow groundwater of various aquifers in a rapidly urbanized delta, south China. *Sci. Total Environ.* **2020**, *701*, 134777. [CrossRef] [PubMed]
19. Zhang, M.; Huang, G.; Liu, C.; Zhang, Y.; Chen, Z.; Wang, J. Distributions and origins of nitrate, nitrite, and ammonium in various aquifers in an urbanized coastal area, south China. *J. Hydrol.* **2020**, *582*, 124528. [CrossRef]

water

MDPI

Article

Occurrence of and Factors Affecting Groundwater Fluoride in the Western Coastal Area of Hainan Island, South China

Ruinan Liu [1,2], Xiwen Li [3,*], Xiujiu Yang [3,*] and Ming Zhang [4]

1 Institute of Hydrogeology and Environmental Geology, Chinese Academy of Geological Sciences, Shijiazhuang 050061, China; liuruinan1996@163.com
2 School of Environmental Studies, China University of Geosciences, Wuhan 430074, China
3 Haikou Marine Geological Survey Center, China Geological Survey, Haikou 571127, China
4 Faculty of Engineering, China University of Geosciences, Wuhan 430074, China; zhangming8157@126.com
* Correspondence: lxw1818168@163.com (X.L.); yangxiuj9@163.com (X.Y.)

Abstract: Hainan, a well-known center of tropical agricultural production in south China, has received little attention regarding groundwater fluoride contamination. This study investigates the occurrence of fluoride in the western coastal area of Hainan Island and discusses factors affecting groundwater fluoride contamination in various aquifers and areas with different land-use types using hydrochemistry and multivariate statistical analysis. A total of 100 groundwater samples were collected from the western coastal area of Hainan Island. The results show that the groundwater fluoride concentration is as high as 4.18 mg/L and that F^--high (>1 mg/L) groundwater accounts for 9% of total groundwater. The proportion of F^--high fissure water is about two times that of F^--high pore water. Among the different land-use types, the proportion of F^--high groundwater from highest to lowest is as follows: bare land > cultivated land > woodland > construction land > grassland. The main factor affecting fluoride in pore water is the leaching of fluorine/aluminum-containing minerals such as phlogopite and calcite in the vadose zone, which is characterized by the co-enrichment of fluoride and aluminum in pore water. The leading cause of fluoride in fissure water is the leaching of fluorine-containing fertilizers, and continuous irrigation promotes the cation exchange of sodium, strontium, and calcium, which is characterized by the co-enrichment of fluoride with sodium and strontium in fissure water. Consequently, it is advised to minimize the excessive use of fluoride fertilizers and increase groundwater quality monitoring in order to decrease the emergence of F^--high groundwater in the western coastal area of Hainan Island.

Keywords: fluoride; groundwater; occurrence; factors; Hainan Island

Citation: Liu, R.; Li, X.; Yang, X.; Zhang, M. Occurrence of and Factors Affecting Groundwater Fluoride in the Western Coastal Area of Hainan Island, South China. *Water* **2023**, *15*, 3678. https://doi.org/10.3390/w15203678

Academic Editors: Juan José Durán, Cesar Andrade and Dimitrios E. Alexakis

Received: 17 July 2023
Revised: 21 September 2023
Accepted: 18 October 2023
Published: 20 October 2023

1. Introduction

Groundwater is a precious resource that provides drinking water for about 50% of the world's population [1–3]. In China, about one-third of water resources are groundwater, 70% of Chinese citizens drink groundwater, and 90% of urban domestic and industrial water depends on groundwater [4]. Therefore, groundwater is essential for sustaining human life, social progress, and the environment [5–7]. However, there has been a significant issue with groundwater quality in recent years. Fluoride is a common groundwater pollutant, and one study estimates that 200 million people worldwide are at risk of suffering groundwater fluorosis [8]. In China, thirty provinces (cities and districts) are affected by groundwater fluoride poisoning; most of these areas are located in the north, including areas such as the Songnen Plain, the North China Plain, and the Hexi Corridor Basin [4]. Long-term exposure to groundwater with a high fluoride level can impact a person's teeth, bones, and neurological system as well as induce symptoms of nausea, diarrhea, and abdominal pain [8,9]. Due to its high toxicity, fluoride poses a serious risk to human health even at very low concentrations [8].

Groundwater quality issues in Hainan Island, China's southernmost island, have previously received little attention. It is important to note that groundwater fluoride pollution in southern China, particularly in coastal areas, is frequently disregarded. Hainan Island is China's primary production base for tropical agriculture. On the one hand, the need for water resources for the socioeconomic growth of Hainan Province as a pilot free trade zone is increasing, and groundwater supplies are becoming a crucial issue. On the other hand, the standards for building a sustainable civilization are improving. To successfully support the ecological growth of civilization and groundwater safety in Hainan Province, it is necessary to identify the issues with groundwater quality. According to the results of the "Groundwater Resources Investigation and Evaluation Report of Hainan Province" implemented in 2021, the groundwater on Hainan Island shows significant nitrate pollution. However, the impact of fluoride on groundwater quality is still unclear. As a result, it is necessary to identify the status of fluoride contamination in groundwater on Hainan Island.

This study intends to (i) describe the concentration of fluoride in groundwater in various aquifers and land-use types in the western coastal area of Hainan Island, China; (ii) delineate the spatial distribution of fluoride; and (iii) discuss the factors affecting fluoride concentrations in various aquifers and land-use types. The findings provide a scientific foundation for groundwater pollution prevention and control on Hainan Island as well as a guide for managing groundwater resources on other sizable tropical islands around the world that are comparable to Hainan Island. The results reveal for the first time the occurrence of and factors affecting fluoride in groundwater in the western coastal area of Hainan Island.

2. Study Area

2.1. Geographical Overview and Land-Use Types

Hainan Island is located in the south of China. The study area covers about 900 km^2 and is located in the NW coastal area of Hainan Island (108°37′–108°59′ E, 19°9–19°30′71 N) and includes the six major towns: Sigeng, Changhua, Haiwei, Shiyuetian, Wulie, and Sanjia (Figure 1). The mean elevations of the western and eastern parts of the study area are about 15 m and 150 m, respectively. Based on remote sensing interpretation, the land-use types in the study area were divided into seven main categories, namely, water bodies, bare land, cultivated land, construction land, grassland, shrubs, and trees. In this case, shrubs and trees were defined as woodland. By area estimates, woodland and grassland account for the largest share of land-use types, followed by construction land and cultivated land accounting for 20% and 10%, respectively; water bodies and bare land account for the least share. Woodland and grassland are mainly distributed in the mountainous areas of Changhua Daling and Sanjialing, construction land is mainly distributed on both sides of the downstream regions of the Changhua River and Sigeng town, and cultivated land is sporadically distributed across the whole district (Figure S1).

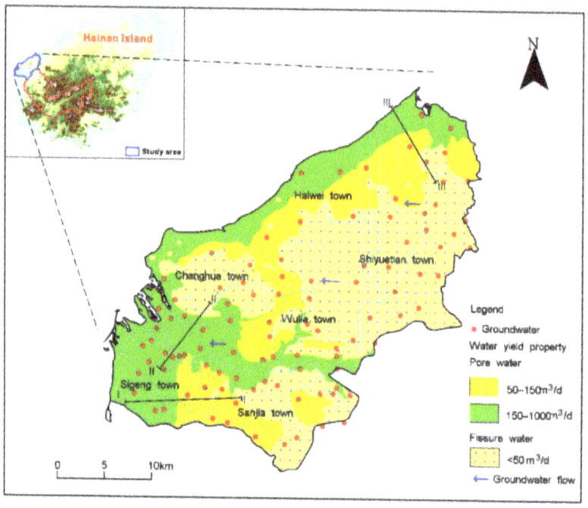

Figure 1. Hydrogeological division and sampling sites in the western coastal area of Hainan Island.

2.2. Geological and Hydrogeological Conditions

The study area is dominated by Middle Proterozoic, Silurian, Carboniferous, Cretaceous, and Quaternary strata in addition to a large distribution of intrusive rocks (Figure S2). Among them, the rock types are mainly biotite plagioclase gneiss, mixed granodiorite, metamorphosed quartz sandstone, sericite slate interbedded with tuff lenses, sericite kyanite, quartz sandstone, silty sandstone, and conglomerate. The acidic-medium acidic intrusive rocks are mainly black mica granite. The lithology of the Quaternary is mainly gravel, sand, clay, and beach rock. The topography of the study area is generally high in the east and low in the west, sloping from east to west. The coastal plains in the west and the mountain plains in the east account for more than 40% of the area, while the rest of the area comprises terraces and hilly areas, mostly distributed in the town of Changhua and the border areas of Haiwei, Shiyuetian, and Wulie. Groundwater in the study area is divided into two categories: pore water of the loose rock type of the Quaternary and bedrock fissure water (Figure 1). A loose rock aquifer is distributed mainly in the coastal accumulation layer, the river alluvial deposit, and the mountain front erosion accumulation layer. The lithology is mainly shell-containing medium-fine sand, gravelly sandy soil, medium-coarse sand, sand, and gravel (Figure S1). The thickness of the aquifer is about 5–15 m and the depth of the water table is 0–2 m. According to the Hainan Island Uniform Survey Report, the specific yield of the medium sand is about 0.2, and the recharge coefficient of precipitation is about 0.471. Furthermore, the area is well watered, and the water yield properties are 50–150 m^3/d and 150–1000 m^3/d. The bedrock fissure water is mainly distributed in the Changhua Daling and Sanjialing mountainous areas, with loose stratum lithology and high permeability. However, the groundwater is poorly enriched, and the water yield property is less than 50 m^3/d. In addition, the annual average precipitation and evaporation in the study area are 1150 mm and 2419 mm, respectively. Moreover, the groundwater in the area is mainly recharged by precipitation and discharged by springs or evaporation. In the Piedmont area, the water table depth of unconfined water is generally higher than that of the confined water, resulting in lateral recharge of confined water by unconfined water.

3. Materials and Methods

3.1. Sampling and Analysis

A total of 100 groundwater samples including 51 pore water (PW) and 49 fissure water (FW) were collected in the study area in 2022 (Figure 1). They were collected at a density of 8–12 per 100 km^2 in the plains areas and 3–6 per 100 km^2 in the hilly areas. The samples were stored at a low temperature (4 °C) and sent to the laboratory for testing. pH, dissolved oxygen (DO), and the oxidation–reduction potential (ORP) were tested using a multi-parameter water quality tester (DZB-712F) at the sampling site after calibration of the instrument. HCO_3^- was measured by the titration method; NH_4^+ by ion chromatography (AQ-1100 ion chromatograph); F^-, NO_3^-, Cl^-, and SO_4^{2-} by ion chromatography (ion chromatography Aqion); NO_2^- by spectrophotometry (visible spectrophotometer 721N); total dissolved solids (TDS) by the weighing method (analytical balance SQP); arsenic (As) and selenium (Se) by the atomic fluorescence method (atomic fluorescence photometer AFS-9800); and K^+, Na^+, Ca^{2+}, Mg^{2+}, aluminum (Al), barium (Ba), manganese (Mn), and strontium (Sr) by inductively coupled plasma emission spectrometry (inductively coupled plasma emission spectrometer ICAP6300). Quality control of the water sample testing was implemented using blank samples, parallel samples, and spiked samples. Among them, the blank and parallel samples accounted for 10–30% of each batch of samples, the blank samples were detected, and the parallel samples were qualified. The recoveries of the spiked samples were 86–113%, which were also within the qualified range of 80–120%.

3.2. Inverse Distance Weighting (IDW) Method

The inverse distance weighting method is a simple and common spatial interpolation method, which is based on the principle that the closer the distance between two sample points is, the more similar the properties. It takes the distance between the point to be estimated and the known sample point as the weight for the weighted average, and the closer the sample point to be estimated is, the greater the weight given to the sample point. The brief steps are as follows: (i) determine the location to be interpolated and the location of the known sample points, and calculate the weight of each sample point based on the distance between the location to be interpolated and the known sample points; and (ii) use the weight of each sample point to perform a weighted average of its function value to obtain the interpolation result.

3.3. Principal Component Analysis (PCA) and Hierarchical Cluster Analysis (HCA)

The principal component analysis (PCA) method extracts the principal factor variables (PCs) by retaining the most important features of the high-dimensional raw data through the concept of dimensionality reduction. The rotation of principal factors is performed by the Varimax method. The principal factors with eigenvalues > 1 are retained for analysis [10]. Hierarchical cluster analysis (HCA) is another multivariate statistical technique for analyzing multidimensional water chemistry datasets [11]. Its main concept is to cluster variables that are close to each other into classes first and those that are farther away into classes later, in order, until each variable is grouped into the appropriate class according to its distance. It can group chemical components of different origins and hydrogeochemical behavior [11]. In this study, principal component analysis and hierarchical cluster analysis were performed using SPSS® version 23.0 software (IBM Corp., Armonk, NY, USA).

4. Results

4.1. Fluoride Concentrations in Different Aquifers and Land-Use Types

As shown in Figure 2 and Table S1, a wide range of groundwater F^- concentrations in the western coastal area of Hainan Island from below the detection limit (<DL) to 4.18 mg/L were found. F^--high groundwater accounted for 9% of the total groundwater in the western coastal area of Hainan Island. The F^- concentrations of PW and FW ranged from <DL to 1.42 mg/L and <DL to 4.18 mg/L, respectively. The maximum concentration of F^- in FW was about three times that in PW. The mean value of F^- in FW was 0.58 mg/L, which was 2.3 times that in PW. Moreover, the F^- coefficients of variation (CVs) in PW and FW were 0.68 and 0.56 (Table S1), respectively, indicating that they have a more uniform spatial distribution of fluoride concentrations. The proportion of F^--high groundwater (PFHG) in FW was 12.2%, which was about 2 times that in PW.

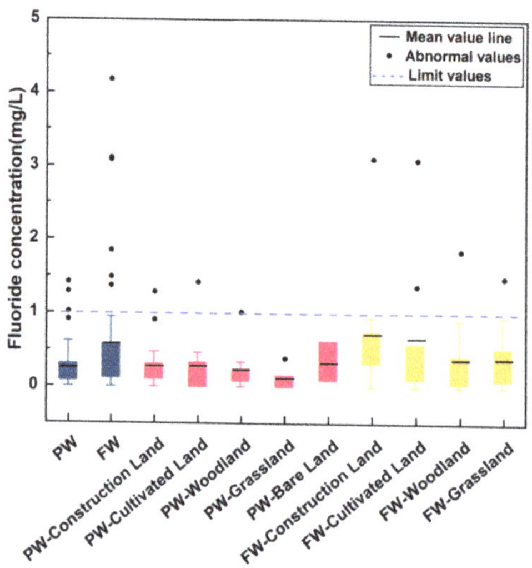

Figure 2. Groundwater fluoride concentrations in various aquifers and land-use types.

The fluoride concentrations in different land-use types are also shown in Figure 2. The F^- concentrations of PW in construction land, cultivated land, woodland, grassland, and bare land ranged from <DL to 1.29, <DL to 1.42, <DL to 1.02, <DL to 0.39, and 0.09 to 0.62 mg/L, respectively. In PW, the highest mean F^- concentration occurred in bare land, followed by cultivated land, construction land, woodland, and grassland. Compared to the CVs for other land-use types, the CVs for construction land and bare land were closer to 1, indicating a relatively high spatial dispersion of fluoride in construction land and bare land (Table S1). The PFHG values of woodland and cultivated land were the highest, with values of 11.1% and 10%, respectively, which were about 2.5 times that of construction land. However, there was no F^--high groundwater in grassland and bare land (Figure 3). In FW, the F^- concentrations in construction land, cultivated land, woodland, and grassland ranged from <DL to 3.11, <DL to 3.09, <DL to 1.85, and <DL to 1.49 mg/L, respectively. The mean F^- concentration in construction land was 0.72 mg/L, while cultivated land, woodland, and grassland had relatively lower mean values of 0.66, 0.38, and 0.39 mg/L, respectively. The CV of grassland in FW was higher than that in construction land, cultivated land, and woodland (Table S1), indicating that the F^- concentration distribution in FW in grassland was relatively more discrete. Except for bare land (not statistically significant, as there was only one data point), the PFHG of cultivated land was the highest, with a value of 20%, which was about two times that of cultivated land and about three times that of woodland and grassland (Figure 3).

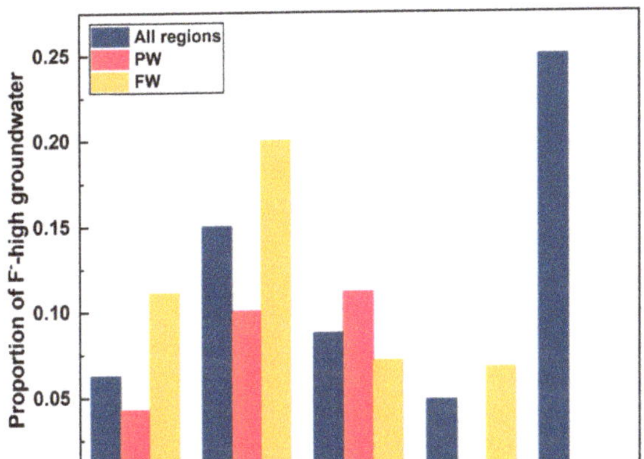

Figure 3. Proportion of F^--high groundwater in different land-use types.

4.2. Hydrochemistry Characteristics of Groundwater with Different Levels of F^-

As shown in Table 1, the pH of both F^--high and F^--poor groundwater in the study area was about 7. In PW, the average concentration order of major cations in F^--high groundwater was $Na^+ > Ca^{2+} > Mg^{2+} > K^+$, while that of major cations in F^--poor groundwater was $Ca^{2+} > Na^+ > K^+ > Mg^{2+}$. Compared to F^--poor groundwater, F^--high groundwater had relatively high sulfate concentrations, with an average value that was about 1.2 times higher than that of F^--poor groundwater. The mean concentration order of major anions in F^--high groundwater was $HCO_3^- > SO_4^{2-} > Cl^-$, while that of major anions in F^--poor groundwater was $HCO_3^- > Cl^- > SO_4^{2-}$. Moreover, the mean value of NH_4^+ and NO_2^- had no significant difference in F^--high and F^--poor groundwater. However, NO_3^- and TDS of F^--high groundwater were relatively low, with mean values of 30.4 mg/L and 320.8 mg/L. In addition, the average concentrations of As, Se, and Ba in F^--high groundwater were larger compared to those in F^--poor groundwater, where the average concentration of Ba was about three times higher than that in F^--poor groundwater. By contrast, the average concentrations of Al, Mn, and Sr in F^--high groundwater were higher than those in F^--poor groundwater.

In FW, the average concentration order of major cations in F^--high groundwater was $Na^+ > Ca^{2+} > Mg^{2+} > K^+$, while that of major cations in F^--poor groundwater was $Ca^{2+} > Na^+ > Mg^{2+} > K^+$. Compared to F^--poor groundwater, the major anion contents of the F^--high groundwater were significantly different, with the average concentration order of $HCO_3^- > Cl^- > SO_4^{2-}$. Moreover, the mean value of NH_4^+ and NO_2^- had no significant difference in F^--high and F^--poor groundwater. However, the mean value of NO_3^- in F^--poor groundwater was 1.5 times that in F^--high groundwater. And the mean value of TDS in F^--high groundwater was relatively higher at about 1.4 times that in F^--poor groundwater. In addition, the As, Mn, and Sr concentrations in F^--high groundwater were higher than those in F^--poor groundwater, and their average values were more than two times those in F^--poor groundwater. By contrast, the average concentrations of Se, Al, and Ba in F^--high groundwater were relatively lower than those in F^--poor groundwater.

Table 1. Mean values of hydrochemical parameters in groundwater with different levels of F^-.

Groundwater	pH	K^+	Na^+	Ca^{2+}	Mg^{2+}	Cl^-	SO_4^{2-}	HCO_3^-	NH_4^+
FHG in PW	7.3	7.4	44.1	44.0	14.0	38.2	55.1	128.0	0.1
FPG in PW	7.1	23.0	40.5	41.0	16.5	63.1	46.6	117.0	0.1
FHG in FW	7.1	2.3	95.6	53.2	18.1	108.1	37.5	242.8	<DL
FPG in FW	7.2	13.1	42.4	43.2	20.3	53.6	31.8	158.4	0.1
Groundwater	**NO_3^-**	**NO_2^-**	**TDS**	**As**	**Se**	**Al**	**Ba**	**Mn**	**Sr**
FHG in PW	30.4	<DL	320.8	0.9	0.3	0.9	0.1	0.1	0.4
FPG in PW	47.5	<DL	324.1	1.3	1.3	0.1	0.2	0.1	0.4
FHG in FW	29.7	<DL	442.9	5.7	0.5	<DL	0.1	0.1	0.8
FPG in FW	44.5	<DL	321.5	0.9	1.0	0.1	0.2	<DL	0.3

Notes: Except for pH, As (ug/L), and Se (ug/L), the units of other parameters are mg/L; FHG: F^--high (>1 mg/L); FPG: F^--poor (≤1 mg/L); DL: detection limit.

4.3. Spatial Distribution of Fluoride

As shown in Figure 4, F^--high (>1 mg/L) groundwater accounted for about 5% of the study area distribution, mainly in the eastern part of Sanjia town, the southwestern part of Shiyuetian town, and the southeastern part of Wulie town, with a small amount in Haiwei town and Changhua town, while there was no F^--high groundwater in Sigeng town. In terms of different land-use types, F^--high groundwater was mainly found in bare land, with the proportion in descending order as follows: bare land (25%)> cultivated land (15%)> woodland (8.7%)> construction land (6.3%)> grassland (4.8%) (Figure 3).

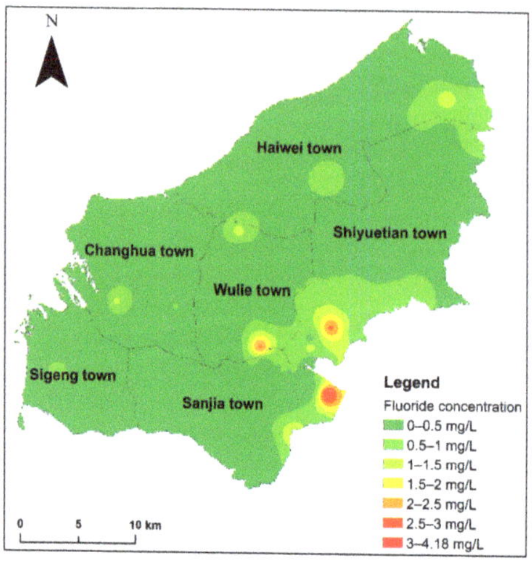

Figure 4. Spatial distribution of fluoride concentrations interpolated using the inverse distance weighting (IDW) method.

4.4. PCA and HCA

As shown in Figure 5, the groundwater chemistry of PA and FA were both controlled by a six-factor model, and the six PCs accounted for 70.8% and 76.1% of the cumulative variance contribution, respectively. In PW, PC4 had moderate positive loading with F^- and Al of 0.70 and 0.5, respectively (Figure 5a). Moreover, PC4 had a moderate negative loading with Ba of −0.67 (Figure 5a). In FW, PC3 had strong positive loading with F^- and Sr of 0.89 and 0.84, respectively (Figure 5b), indicating that they have the same source or geochemical behavior [12]. In addition, PC3 had weak positive loading with Na^+ and weak negative loading with NO_3^-, in which the loading was 0.37 and −0.31, respectively. Hierarchical cluster analysis for pore water and fissure water is shown in Figure 6. As shown in Figure 6a, the twenty groundwater chemical parameters were divided into six clusters at a rescaled distance of 20. Among them, F^- was in a cluster with Al at a rescaled distance of 20. Similarly, as shown in Figure 6b, the twenty groundwater chemical parameters were divided into five clusters at a rescaled distance of 20. And F^- was in a cluster with Sr, at a rescaled distance of 10, indicating that F^- was more closely related to Sr in FW.

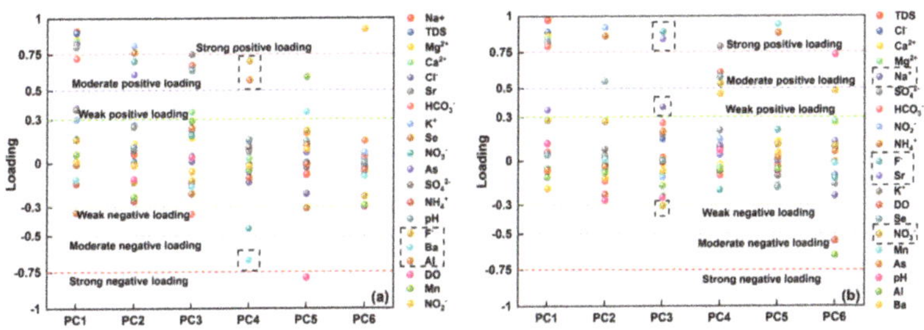

Figure 5. Factor loadings for principal component analysis of (**a**) pore water and (**b**) fissure water.

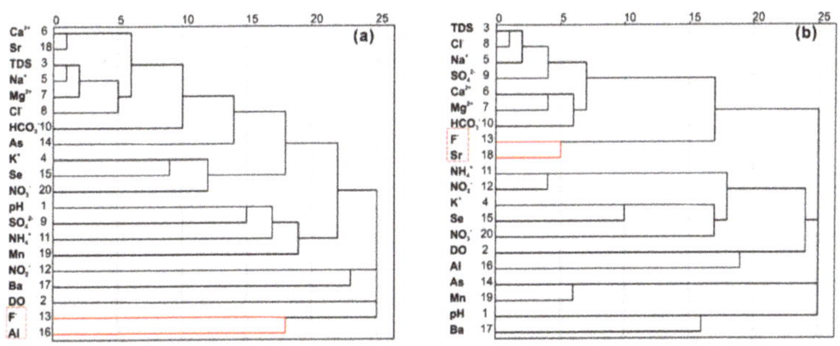

Figure 6. Hierarchical cluster analysis of (**a**) pore water and (**b**) fissure water.

5. Discussion

Both anthropogenic and natural factors frequently contribute to the occurrence of high groundwater fluoride levels. High levels of nitrate and total dissolved solids are frequent signs of different anthropogenic pollution inputs [13]. However, in this study, the F^--high groundwater in the bare land was accompanied by low concentrations of nitrate (<1 mg/L) and total dissolved solids (<200 mg/L), indicating that the geological sources rather than anthropogenic contamination contributed to the F^--high groundwater in bare land because fluoride-containing minerals such as fluorspar and calcite in the ground can dissolve and release fluoride into the groundwater under certain conditions [14]. Additionally, the proportion of F^--high groundwater in cultivated land was two times or higher than that in other land-use types, except for bare land, indicating that long-term agricultural irrigation that causes soil fluoride to leach down and the use of fluoridated pesticides and fertilizers (such as fluoridated phosphate fertilizers) may be the main anthropogenic sources of groundwater fluoride in the study area [15,16]. As shown in Figure 3, the proportion of F^--high groundwater in pore water was higher in woodland than in areas with high-intensity anthropogenic activities such as cultivated land and construction land, indicating that the F^--high pore water in the study area may be mainly from geological sources rather than from anthropogenic contamination. By contrast, the proportion of F^--high groundwater in fissure water was twice or more than that in other land-use types, indicating that the F^--high fissure water in the study area may be more likely to originate from long-term agricultural irrigation resulting in soil fluoride leaching and infiltration and the use of fluoridated pesticides and fertilizers (such as fluoridated phosphate fertilizers).

According to Figure 5a, PC4 in pore water chemistry was characterized as a moderate positive loading of fluoride with aluminum and a moderate negative loading with barium. Fluoride ions can react with aluminum ions in water instead of hydroxide ions under acidic or neutral conditions to generate aluminum fluoride ions such as AlF^{2+}, AlF_2^+, AlF_4^-, and AlF_6^{3-} because, as demonstrated in previous research, fluoride ions are similar in size to hydroxide ions [17]. Correspondingly, the pH of the F^--high pore water in the study area was mostly around 7. Therefore, the fact that fluoride and aluminum are located in the same factor indicated a co-existence relationship/co-source relationship between fluoride and aluminum in pore water, such as the dissolution of concurrent aluminum/fluoride-rich minerals like mica and hornblende [18]. The results of hierarchical cluster analysis, which demonstrated that pore water fluoride was most strongly related to aluminum, further corroborated this conclusion (Figure 6a). At the same time, the study area is mainly distributed with Middle Proterozoic, Silurian, Carboniferous, Cretaceous, and Quaternary strata, and there is also a large area of intrusive rock distribution. The main types of rocks are gneiss, amphibolite, granite, sericite slate, etc., which are rich in fluoride-containing minerals such as mica and hornblende, etc., which can also support the above conclusions. In addition, the fact that F^- and Ba were located in the same factor and had opposite loading relationships may indicate their opposing geochemical behavior since the co-precipitation of barium sulfate made the barium content in groundwater often be controlled by the groundwater sulfate concentration, and high-sulfate-concentration groundwater was often accompanied by low levels of barium [19–21]. Correspondingly, the average sulfate and barium concentrations in F^--high pore water in the study area were 55.10 mg/L and 0.07 mg/L, respectively, which were significantly higher than the average sulfate concentration in F^--poor pore water (46.57 mg/L) and lower than the average concentration of barium (0.19 mg/L).

As shown in Figure 5b, PC3 in the fissure water chemistry was characterized by strong positive loadings of fluoride and strontium, weak positive loadings of sodium, and weak negative loadings of nitrate. The fact that fluoride, strontium, and sodium were located in the same factor and were all positively loaded suggested that this observation may be due to homologous relationships or associated geochemical interactions. On the one hand, considering that minerals rich in fluoride, strontium, and sodium were extremely rare, the possibility of a homologous relationship was presumed to be low; on the other hand, the analysis in the above section considered that soil fluoride leaching by long-term agricultural irrigation may be one of the main sources of F^--high fissure water in the study area. Previous studies have shown that continuous irrigation not only leaches fluoride from the soil but also promotes cation exchange in the soil [16,22], resulting in relatively easy-to-fix cations such as calcium ions being fixed from the irrigation water into the soil medium and replacing relatively difficult-to-fix metal ions such as sodium and strontium ions [23], which ultimately led to the enrichment of fluoride, strontium, and sodium in the groundwater by leaching from the soil layer under the effect of long-term irrigation. Correspondingly, the fluoride and sodium and strontium content in the fissure water of the study area showed highly significant positive correlations (Figure 7).

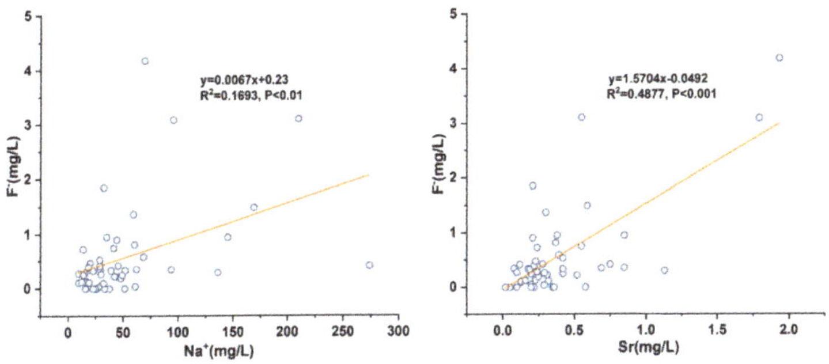

Figure 7. Relationship between the concentrations of F^- and Na^+ and Sr in fissure water.

6. Conclusions

In this paper, the occurrence of F^- in groundwater in the western coastal area of Hainan Island was investigated. F^--high (>1 mg/L) groundwater accounted for about 5% of the total groundwater distribution in the study area, mainly in the eastern part of Sanjia town, the southwestern part of Shiyuetian town, and the southeastern part of Wulie town. The proportion of F^--high groundwater in FW was 12.2%, which was about two times that in PW. The PFHG of PW in decreasing order was woodland > cultivated land > construction land. And the PFHG of FW in decreasing order was cultivated land > construction land > woodland > grassland.

Fluoride in pore water in this study was mainly due to fluoride/aluminum-containing minerals such as phlogopite and calcite in the vadose zone, which are affected by leaching and dissolve in groundwater, thus showing the characteristics of co-enrichment of fluoride and aluminum in pore water. The higher concentration of fluoride in fissure water was mainly influenced by the leaching effect of fluoride-containing fertilizers. Continuous irrigation also promotes the cation exchange of sodium, strontium, and calcium, which ultimately manifests itself in high-fluoride fissure water accompanied by high concentrations of sodium and strontium.

Considering the groundwater fluoride contamination in the western coastal area of Hainan Island revealed in this study, it is recommended that fluoride-containing fertilizers be applied reasonably and that excessive application of fertilizers be limited; additionally, monitoring of groundwater quality in the region should be strengthened to ensure the environmental safety of drinking water sources.

Supplementary Materials: The following supporting information can be downloaded at: https://www.mdpi.com/article/10.3390/w15203678/s1. Table S1: The coefficients of variation of F^--high groundwater; Figure S1: Hydrogeological profile of the study area; Figure S2: Land-use types in the study area.

Author Contributions: Conceptualization, R.L.; methodology, R.L. and X.L.; software, X.Y.; validation, R.L., X.L., X.Y. and M.Z.; formal analysis, R.L.; investigation, M.Z.; resources, X.L.; data curation, R.L.; writing—original draft preparation, R.L.; writing—review and editing, X.L.; visualization, X.Y.; supervision, M.Z. All authors have read and agreed to the published version of the manuscript.

Funding: This research was funded by the China Geological Survey grant number No. ZD20220209, and the China Geological Survey grant number DD20230507.

Data Availability Statement: Not applicable.

Conflicts of Interest: The authors declare that they have no conflict of interest.

References

1. Jha, P.K.; Tripathi, P. Arsenic and fluoride contamination in groundwater: A review of global scenarios with special reference to India. *Groundw. Sustain. Dev.* **2021**, *13*, 100576. [CrossRef]

2. Jadhav, S.V.; Bringas, E.; Yadav, G.D.; Rathod, V.K.; Ortiz, I.; Marathe, K.V. Arsenic and fluoride contaminated groundwaters: A review of current technologies for contaminants removal. *J. Environ. Manag.* **2015**, *162*, 306–325. [CrossRef]
3. Cherry, J. The Groundwater Project: Democratizing Groundwater Knowledge. *Groundwater* **2020**, *58*, 682–683. [CrossRef] [PubMed]
4. Hao, A.B.; Zhang, Y.L.; Zhang, E.Y.; Li, Z.H.; Yu, J.; Wang, H.; Yang, J.F.; Wang, Y. Review: Groundwater resources and related environmental issues in China. *Hydrogeol. J.* **2018**, *26*, 1325–1337. [CrossRef]
5. Gleeson, T.; Cuthbert, M.; Ferguson, G.; Perrone, D. Global Groundwater Sustainability, Resources, and Systems in the Anthropocene. *Annu. Rev. Earth Planet. Sci.* **2020**, *48*, 431–463. [CrossRef]
6. Gleeson, T.; Richter, B. How much groundwater can we pump and protect environmental flows through time? Presumptive standards for conjunctive management of aquifers and rivers. *River Res. Appl.* **2018**, *34*, 83–92. [CrossRef]
7. Medici, G.; Langman, J.B. Pathways and Estimate of Aquifer Recharge in a Flood Basalt Terrain; A Review from the South Fork Palouse River Basin (Columbia River Plateau, USA). *Sustainability* **2022**, *14*, 11349. [CrossRef]
8. Khatri, N.; Tyagi, S. Influences of natural and anthropogenic factors on surface and groundwater quality in rural and urban areas. *Front. Life Sci.* **2015**, *8*, 23–39. [CrossRef]
9. Zhang, L.Q.; Dong, D.L.; Lv, S.T.; Ding, J.; Yan, M.H.; Han, G.L. Spatial evolution analysis of groundwater chemistry, quality, and fluoride health risk in southern Hebei Plain, China. *Environ. Sci. Pollut. Res.* **2023**, *30*, 61032–61051. [CrossRef] [PubMed]
10. Liu, C.Y.; Hou, Q.X.; Chen, Y.T.; Huang, G.X. Hydrogeochemical Characteristics and Groundwater Quality in a Coastal Urbanized Area, South China: Impact of Land Use. *Water* **2022**, *14*, 4131. [CrossRef]
11. Huang, G.X.; Hou, Q.X.; Han, D.Y.; Liu, R.A.; Song, J.M. Large scale occurrence of aluminium-rich shallow groundwater in the Pearl River Delta after the rapid urbanization: Co-effects of anthropogenic and geogenic factors. *J. Contam. Hydrol.* **2023**, *254*, 104130. [CrossRef] [PubMed]
12. Huang, G.X.; Liu, C.Y.; Li, L.P.; Zhang, F.G.; Chen, Z.Y. Spatial distribution and origin of shallow groundwater iodide in a rapidly urbanized delta: A case study of the Pearl River Delta. *J. Hydrol.* **2020**, *585*, 124860. [CrossRef]
13. Huang, G.X.; Pei, L.X.; Li, L.P.; Liu, C.Y. Natural background levels in groundwater in the Pearl River Delta after the rapid expansion of urbanization: A new pre-selection method. *Sci. Total Environ.* **2022**, *813*, 151890. [CrossRef]
14. Kumar, M.; Goswami, R.; Patel, A.K.; Srivastava, M.; Das, N. Scenario, perspectives and mechanism of arsenic and fluoride Co-occurrence in the groundwater: A review. *Chemosphere* **2020**, *249*, 126126. [CrossRef] [PubMed]
15. Kim, Y.; Kim, J.Y.; Kim, K. Geochemical characteristics of fluoride in groundwater of Gimcheon, Korea: Lithogenic and agricultural origins. *Environ. Earth Sci.* **2011**, *63*, 1139–1148. [CrossRef]
16. Young, S.M.; Pitawala, A.; Ishiga, H. Factors controlling fluoride contents of groundwater in north-central and northwestern Sri Lanka. *Environ. Earth Sci.* **2011**, *63*, 1333–1342. [CrossRef]
17. Driscoll, C.T.; Schecher, W.D. The Chemistry of Aluminum in the Environment. *Environ. Geochem. Health* **1990**, *12*, 28–49. [CrossRef]
18. Selinus, O. Fluoride in Natural Waters. In *Essentials of Medical Geology*; Academic Press: Cambridge, MA, USA, 2013; pp. 311–336.
19. Marandi, A.; Karro, E.; Puura, E. Barium anomaly in the Cambrian-Vendian aquifer system in North Estonia. *Environ. Geol.* **2004**, *47*, 132–139. [CrossRef]
20. Underwood, E.C.; Ferguson, G.A.; Betcher, R.; Phipps, G. Elevated Ba concentrations in a sandstone aquifer. *J. Hydrol.* **2009**, *376*, 126–131. [CrossRef]
21. Bondu, R.; Cloutier, V.; Rosa, E.; Roy, M. An exploratory data analysis approach for assessing the sources and distribution of naturally occurring contaminants (F, Ba, Mn, As) in groundwater from southern Quebec (Canada)—ScienceDirect. *Appl. Geochem.* **2020**, *114*, 104500. [CrossRef]
22. Huang, G.X.; Sun, J.C.; Zhang, Y.; Chen, Z.Y.; Liu, F. Impact of anthropogenic and natural processes on the evolution of groundwater chemistry in a rapidly urbanized coastal area, South China. *Sci. Total Environ.* **2013**, *463*, 209–221. [CrossRef] [PubMed]
23. Kaleem, M.; Naseem, S.; Bashir, E.; Shahab, B.; Rafique, T. Discrete geochemical behavior of Sr and Ba in the groundwater of Southern Mor Range, Balochistan, a tracer for igneous and sedimentary rocks weathering and related environmental issues. *Appl. Geochem.* **2021**, *130*, 104996. [CrossRef]

water

MDPI

Article

Hydrogeochemical Characteristics and Groundwater Quality in a Coastal Urbanized Area, South China: Impact of Land Use

Chunyan Liu [1,2], Qinxuan Hou [1,2,*], Yetao Chen [3] and Guanxing Huang [1,*]

1 Institute of Hydrogeology and Environmental Geology, Chinese Academy of Geological Sciences, Shijiazhuang 050061, China
2 Hebei Key Laboratory of Groundwater Remediation, Shijiazhuang 050061, China
3 Hebei Huanxue Environmental Protection Technology Company Limited, Shijiazhuang 050000, China
* Correspondence: houqinxuan@163.com (Q.H.); huangguanxing2004@126.com (G.H.)

Abstract: Land use transformation accompanied with various human activities affects groundwater chemistry and quality globally, especially in coastal urbanized areas because of complex human activities. This study investigated the impact of land use on groundwater chemistry and quality in a coastal alluvial aquifer (CAA) of the Pearl River Delta where urbanization continues. A fuzzy synthetic evaluation method was used to evaluate the groundwater quality. Besides, factors controlling groundwater chemistry and quality in the CAA were discussed by using a principal components analysis (PCA). Nearly 150 groundwater samples were collected. All samples were filtered on-site and stored at 4 °C until the laboratory procedures could be performed. Nineteen chemical parameters including pH, dissolved oxygen, redox potential, total dissolved solids, K^+, Na^+, Ca^{2+}, Mg^{2+}, NH_4^+, HCO_3^-, NO_3^-, SO_4^{2-}, Cl^-, I^-, NO_2^-, Pb, Mn, Fe, and As were analyzed. Results show that groundwater chemistry in the CAA was dominated by Ca-HCO_3 and Ca·Na-HCO_3 facies. In addition, groundwater with NO_3 facies was also present because of more intensive human activities. In the CAA, 61.8% of groundwaters were fit for drinking, and 10.7% of groundwaters were undrinkable but fit for irrigation, whereas 27.5% of groundwaters were unfit for any purpose. Poor-quality groundwaters in urban and agricultural areas were 1.1–1.2 times those in peri-urban areas, but absent in the remaining area. Groundwater chemistry and quality in the CAA was mainly controlled by five factors according to the PCA. Factor 1 is the release of salt and NH_4^+ from marine sediments, and the infiltration of domestic and septic sewage. Factor 2 is agricultural activities related to the irrigation of river water, and the use of chemical fertilizers. Factor 3 is the industrial pollution related to heavy metals and acid deposition. Factor 4 is the input of anthropogenic reducing sewage inducing the reductive dissolution of As-loaded Fe minerals and denitrification. Factor 5 is the I^- contamination from both of geogenic and anthropogenic sources. Therefore, in order to protect groundwater quality in coastal urbanized areas, repairing old sewer systems in urban areas, building sewer systems in peri-urban areas, limiting sewage irrigation and the amount of chemical fertilizers application in agricultural areas, as well as strengthening the supervision of the industrial exhaust gas discharge in urban and peri-urban areas are recommended.

Keywords: groundwater chemistry; groundwater quality; coastal aquifers; land use; factors

Citation: Liu, C.; Hou, Q.; Chen, Y.; Huang, G. Hydrogeochemical Characteristics and Groundwater Quality in a Coastal Urbanized Area, South China: Impact of Land Use. *Water* **2022**, *14*, 4131. https://doi.org/10.3390/w14244131

Academic Editors: Zbigniew Kabala and Dimitrios E. Alexakis

Received: 31 October 2022
Accepted: 15 December 2022
Published: 19 December 2022

1. Introduction

The large scale transformation of agricultural and natural ecosystems to urbanization is one of huge anthropogenic impacts on the groundwater environment [1]. Large-scale urbanization in China has lasted for several decades, especially in coastal areas. For example, the population in urban areas in China shows a pattern, being high in coastal areas and low in inland areas on a national scale [2], because coastal areas connecting the inland and oceans in the earth system are crucial zones, and play an important role for social and economic development. To date, more than two billion people live in coastal areas globally, and the groundwater resource is one of the major sources to supply drinking water for around one billion people in these areas [3]. On the other hand, because of the transformation of agricultural and natural ecosystems to urbanized areas, accompanied by various kinds of human activities [4,5], groundwater chemistry and quality issues are becoming increasingly serious, and threatening drinking water security in coastal areas worldwide [6–9]. For example, Han and Currell reviewed how urban growth and water transfer projects may be responsible for changes observed in coastal groundwater quality in China [10]. Sellamuthu et al. reported that groundwater salinization and nitrate pollution was highly influenced by anthropogenic sources in a rapidly developing urban area in India [11]. As a consequence, it is necessary to investigate the impact of land use transformation from agricultural and natural ecosystems to urbanization on groundwater chemistry and quality in coastal areas [12,13], and to provide suggestions for the protection and management of groundwater resource.

The Pearl River Delta (PRD), adjacent to the South China Sea, is one of the largest coastal urbanized areas in China. The expansion of urbanization in this area has lasted for more than four decades, and land-use change in this area is mainly conversion from agricultural lands to urban areas. For instance, the urban area in the PRD in 2018 was approximately two times and three times of that in 2006 and 1998, respectively [14]. On one hand, urbanization accompanied with a huge influx of population results in the increased importance of groundwater resource in the PRD than before [15]. On the other hand, urbanization accompanied with various human activities has deteriorated not only surface water quality but also shallow groundwater quality in this area [16]. To date, many groundwater environmental issues in this area, such as nitrate and phosphate pollution, and iodine and manganese contamination, had already received attention [17–20]. However, shallow groundwater in the coastal alluvial aquifer is a major source for water supply in this area [21], but knowledge on the influence of large-scale land use conversion on groundwater chemistry and quality in this coastal alluvial aquifer is still limited.

Therefore, the present study aims to analyze the impact of land use on shallow groundwater quality and the hydrogeochemical characteristics of the coastal alluvial aquifer of the PRD, and to discuss factors controlling groundwater chemistry and quality in this coastal alluvial aquifer. Here, a fuzzy synthetic evaluation method (FSEM) combined with the groundwater quality standards of China was used for evaluating groundwater quality in this study [16,22]. The results will contribute to the development and utilization of groundwater resource in the PRD.

2. Study Area

2.1. Geographical and Hydrogeological Settings

The PRD is located in the southern Guangdong Province of China and covers a total area of about 42 thousand km^2 (Figure 1). It is adjacent to the South China Sea in the south and surrounded by hills in the east, west, and north. The climate is typically subtropical marine monsoon and the average annual rainfall and temperature are 1600–2300 mm and 21.4–22.4 °C, respectively [21]. The wet season is from April to September. Three main rivers such as Dongjiang River, Xijiang River, and Beijiang River merge into the Pearl River system and finally discharge into the South China Sea [21]. The PRD can be divided into four groundwater units: coastal alluvial aquifer, alluvial-proluvial aquifer, fissured aquifer, and karst aquifer [23]. As a major aquifer for water supply, the coastal alluvial aquifer is widely distributed in the PRD plain and covers a total area of 8837 km^2 (Figure 1). The Quaternary strata of this area consist of two marine formations and two continental formations. The young marine formation was deposited in the Holocene period and dominated by silt and clay, except the top layer, in which sand is often dominant [24]. The old marine formation and two continental formations were deposited in the Pleistocene. In the former, silt and clay are dominant, whereas the latter two are commonly dominated by sand and gravel [24]. The coastal alluvial aquifer consists of sand/gravel layers in continental formations and the young marine formation. Note that shallow groundwater occurs in sand layers of young marine and continental formations [25]. Shallow groundwater in the coastal alluvial aquifer is mainly recharged by precipitation, agricultural irrigation, and various river waters, and finally discharges into the South China Sea [25]. In addition, shallow groundwater near coastal lines is also often intruded by sea water [26].

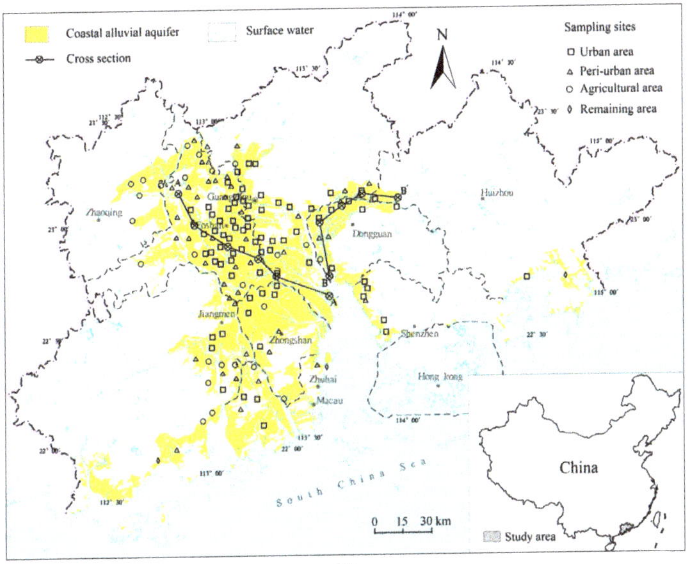

(A)

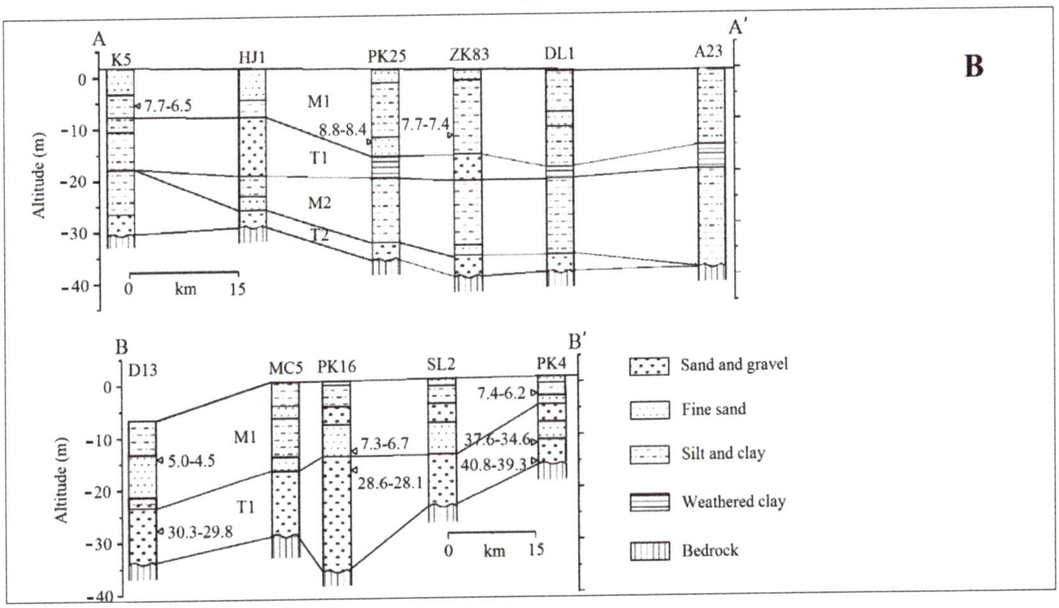

(B)

Figure 1. Hydrogeological setting and sampling sites in the coastal alluvial aquifer of the Pearl River Delta. (**A**) Sampling sites. (**B**) Cross sections.

2.2. Land Use and Human Activity Characteristics

Overlying the coastal alluvial aquifer, the urbanized area had increased to approximately 2300 km^2 in 2006 and accounted for more than one fourth of the total area [27]. The study area can be divided into four areas according to the land use, that is, urban areas (UA), peri-urban areas (PUA), agricultural areas (AA), as well as the remaining area (RA) (Figure 2). UA are large-scale urbanized areas with a high intensity of population and factories [27]. The term PUA refers to regions ~2 km outside of urban areas with a high population and small factories but lacking a sewer system [18]. AA refers to cultivated lands and garden plots, and sometimes sewage irrigation occurs [21]. The RA include surface water bodies, small villages, woodlands, grasslands, and uncultivated lands where human activities are few [14]. The intensity of human activities qualitatively follows the order of UA > PUA > AA > RA [28].

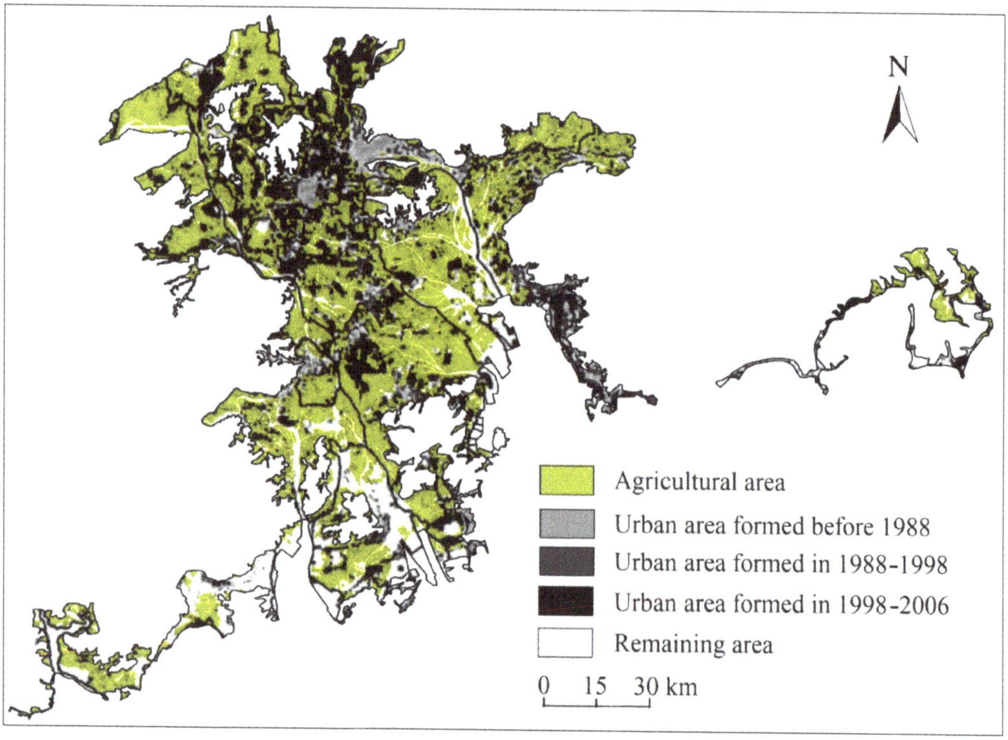

Figure 2. Spatial distribution of land-use types covering the coastal alluvial aquifer of the Pearl River Delta (data related to agricultural land and urbanized areas from [21,27], respectively).

3. Materials and Methods

3.1. Sampling and Analysis

A total of 149 shallow groundwater samples were collected from the coastal alluvial aquifer in the period from August to September of 2006–2007, and the sampling density was 10–20 samples/1000 km^2. Among them, 75 samples, 46 samples, 25 samples, and 3 samples were collected from UA, PUA, AA, and RA, respectively. Three parameters including pH, dissolved oxygen (DO), and redox potential (Eh) were measured on-site using a multi-parameter instrument (WTW Multi 340i/SET, Germany) that was calibrated before measurements were taken. A further 16 parameters, including total dissolved solids (TDS), 5 cations (K^+, Na^+, Ca^{2+}, Mg^{2+}, and NH_4^+), 6 anions (HCO_3^-, NO_3^-, SO_4^{2-}, Cl^-,

I^-, and NO_2^-), and 4 heavy metal(loid)s (Pb, Mn, Fe, and As) in samples were measured in the laboratory. Details for sampling, analysis, and quality control are shown in Section 3.1 of the supplementary material (SM).

3.2. Fuzzy Synthetic Evaluation Method (FSEM)

The FSEM is a common method to overcome the imprecision in the evaluation of groundwater quality [16]. In this study, combining with the groundwater quality standards of China (Table S1) [22], a fuzzy membership function was used to evaluate groundwater quality. Groundwater quality indicators referred to Mn, Fe, NH4+, As, I−, NO3−, TDS, NO2−, Cl−, Na+, Pb, and SO42−. Details are in Section 3.2 of the supplementary material (SM).

3.3. Principal Components Analysis (PCA)

The PCA is a powerful tool for analyzing high-dimensional hydrochemical data sets and reducing a large number of variables to a small number of principal components (PCs) by linearly combining measurements made on the original variables [29]. This multi-step method has been applied successfully to extract PCs and infer the underlying natural and/or anthropogenic processes that control the groundwater chemistry [28,30]. Therefore, in this study, using the software SPSS® Version 23.0 (SPSS Inc., Chicago, IL, USA), the PCA was carried out to reduce hydrochemical datasets of the coastal alluvial aquifer and extract the PCs, and infer the main factors controlling groundwater chemistry and quality in this aquifer. Note that the data below the detection limits were substituted with zero when data of chemical parameters in groundwaters were used for PCA. In the PCA, log-transformed data and a standardized data matrix were used to give each variable equal weight in the multivariate statistical analysis [30]. Rotation of the PCs was carried out using the Varimax method, and PCs with eigenvalues > 1 were retained for analyses. Kaiser–Meyer–Olkin and Bartlett's tests showed significant difference between the correlation coefficient matrix and identity matrix and were suitable for the PCA (Table S2). The absolute PC loadings of >0.75, 0.75–0.5, and 0.5–0.3 were denoted as strong, moderate, and weak, respectively.

4. Results

4.1. Hydrogeochemical Characteristics in the Coastal Alluvial Aquifer

The descriptive statistics for the concentrations of physicochemical parameters in groundwater in the coastal alluvial aquifer are shown in Table 1. Groundwater pH showed acidic to near-neutral values, with the median value of 6.8. DO and Eh values in groundwater showed wide ranges of 0.6–8.2 mg/L and −36–275 mV, respectively. The median concentrations of major cations showed an order of $Ca^{2+} > Na^+ > K^+ > Mg^{2+}$, whereas the median values of $HCO_3^- > Cl^- \approx SO_4^{2-} > NO_3^-$. Groundwater TDS concentrations also showed a wide range of 45–3353 mg/L with the median value of 568 mg/L. Two other nitrogen compounds, NH_4^+ and NO_2^-, were up to 60 mg/L and 33.3 mg/L, respectively, with median concentrations of 0.04 mg/L and 0.03 mg/L, respectively. Groundwater Fe and Mn concentrations were also abnormally high, up to 26.2 mg/L and 7.38 mg/L, respectively, with median concentrations of 0.11 mg/L and 0.12 mg/L, respectively. By contrast, two other heavy metal(loid)s in the groundwater, As and Pb, showed median concentrations of 0.004 mg/L and 0.001 mg/L, respectively. Additionally, groundwater I^- concentrations ranged from below the detection limit to 0.76 mg/L.

Table 1. Descriptive statistics of concentrations of chemical parameters in groundwater of the coastal alluvial aquifer in the Pearl River Delta.

Item	AL	Total Area				Urban Area			Peri-Urban Area			Agricultural Area			Remaining Area		
		Min.	Med.	Max.	PAL (%)	Min.	Med.	Max.	Min.	Med.	Max.	Min.	Med.	Max.	Min.	Med.	Max.
pH		3.5	6.8	7.6		4.4	6.9	7.5	3.5	6.6	7.3	4.7	6.5	7.6	5.5	6.1	6.3
DO (mg/L)		0.6	2.8	8.2		0.6	2.6	8.2	1.2	2.9	5.9	1.4	3.3	6.9	3.1	4.4	5.7
Eh (mV)		−36	17	275		−34	2	221	−26	24	275	−36	67	275	33	51	76
K^+ (mg/L)		<DL	20	88		1	23	88	<DL	14	71	1	22	44	2	12	22
Ca^{2+} (mg/L)		2	79	165		13	95	165	2	75	142	4	48	99	7	19	21
Mg^{2+} (mg/L)		<DL	8	118		1	9	54	<DL	8	94	1	7	118	3	5	9
HCO_3^- (mg/L)		<DL	248	641		3	295	641	<DL	187	616	6	147	459	20	45	55
TDS (mg/L)	1000	45	568	3353	6.7	88	704	1420	45	527	3353	56	406	3152	123	182	208
Cl^- (mg/L)	250	5	50	1631	5.4	8	55	390	5	43	1631	5	37	1620	8	21	35
NO_3^- (mg/L)	88.9	0.3	22.6	184.9	10.1	0.3	28.9	184.9	0.6	19.3	146.8	1.1	6.3	116.4	3.0	26.6	29.3
Na^+ (mg/L)	200	3	37	1009	4.0	4	42	222	4	28	1009	3	24	803	4	13	28
SO_4^{2-} (mg/L)	250	<DL	49	263	1.3	<DL	52	263	<DL	48	255	<DL	35	230	19	29	39
Fe (mg/L)	0.3	<DL	0.11	26.16	33.6	<DL	0.10	16.20	<DL	0.11	26.16	<DL	0.24	11.13	0.02	0.06	0.37
Mn (mg/L)	0.1	<DL	0.12	7.38	53.0	<DL	0.12	2.21	<DL	0.16	7.38	<DL	0.09	2.64	0.01	0.02	0.12
NH_4^+ (mg/L)	0.64	<DL	0.04	60.00	26.8	<DL	0.08	45.00	<DL	0.03	60.00	<DL	0.02	40.00	<DL	0.02	0.02
NO_2^- (mg/L)	3.3	<DL	0.03	33.20	6.0	<DL	0.04	14.72	<DL	0.03	33.20	<DL	0.02	6.90	0.01	0.01	0.04
Pb (mg/L)	0.01	<DL	0.001	0.037	2.7	<DL	0.001	0.017	<DL	0.001	0.037	<DL	0.001	0.009	<DL	0.001	0.006
As (mg/L)	0.01	<DL	0.004	0.303	17.4	<DL	0.004	0.303	<DL	0.003	0.172	<DL	0.001	0.030	<DL	<DL	0.001
I^- (mg/L)	0.08	<DL	0.01	0.76	12.1	<DL	0.01	0.76	<DL	<DL	0.32	<DL	0.03	0.22	<DL	<DL	<DL

Note(s): AL: Allowable limits for drinking purpose in China (GAQSIQPRC, 2017); <DL: below detection limits; PAL: The proportion of samples with the concentration of one chemical above the allowable limit.

In this study, the differences in physicochemical parameter concentrations in groundwater in areas with different types of land use were investigated. As shown in Table 1, the differences of pH value in various areas were insignificant. Median values of DO and Eh in various areas were in the order of UA < PUA < AA/RA, which was opposite to the order of intensity of human activities [28]. This indicates that human activities (e.g., urbanization) result in reducing conditions in groundwater. Similarly, some redox-sensitive parameters such as Mn, NH_4^+, NO_2^-, and As, showed median values in UA and PUA higher than those in AA and RA, which was also likely attributed to the intensity of human activities in different areas. By contrast, another nitrogen compound (NO_3^-) showed median values in the order of UA > RA > PUA > AA. Groundwater TDS and most of the major ions such as Ca^{2+}, Na^+, Mg^{2+}, HCO_3^-, Cl^-, SO_4^{2-} showed median values in the order of UA > PUA > AA > RA, indicating that concentrations of these components in groundwater were positively correlated with the intensity of human activities. By contrast, another major ion (K^+) showed median values in UA and AA nearly two times that in PUA and RA. The median value of groundwater Fe in AA was greater than two times that in UA and PUA and four times that in RA. This is likely ascribed to the common use of Fe-rich river water for irrigation in AA [14]. Similarly, the median value of groundwater I^- in AA was also three times or more that in other areas. Furthermore, median concentrations of groundwater Pb in various areas were the same.

As seen in Figure 3, the number of hydrochemical facies in the coastal alluvial aquifer was up to 43. Ca-HCO_3 facies was the most common hydrochemical facies (38.9%), followed by Ca·Na-HCO_3 facies (10.7%) and Ca·Na-HCO_3·Cl facies (6.7%), while others were <5%. Assuming only one major cation and one major anion remain in hydrochemical facies, the number of hydrochemical facies in the coastal alluvial aquifer was decreased to 10. The main one was also Ca-HCO_3 facies (68.5%), followed by Na-Cl facies (11.4%), Ca-Cl facies (6.7%), and Ca-NO_3 facies (5.4%), and other hydrochemical facies were <4%. Note that groundwaters with NO_3 facies accounted for 6% in the coastal alluvial aquifer; by contrast, this aquifer was free of NO_3 facies groundwater before 1980 [31]. Furthermore, groundwaters with NO_3 facies accounted for 8.7% in PUA and 8% in AA, both were two or more times that in UA, whereas the RA groundwater was free of NO_3 facies (Figure 3). These indicate that human activities resulted in the occurrence of NO_3 facies in the groundwater of this aquifer via wastewater infiltration and sewage irrigation, because sewage irrigation and wastewater infiltration were major driving forces for groundwater NO_3^- contamination in AA and PUA, respectively [11,20].

4.2. Groundwater Quality in the Coastal Alluvial Aquifer

The proportion of groundwaters with a concentration of one chemical above the allowable limit (PAL) was also shown in Table 1. In the study area, Mn showed the highest PAL of 53%, followed by Fe, NH_4^+, As, I^-, NO_3^-, TDS, NO_2^-, Cl^-, Na^+, Pb, and SO_4^{2-}, and PALs of the former six chemicals were higher than 10%. As shown in Figure 4, the groundwater quality for various groundwaters was assessed by the FSEM and classified into five classes. In this coastal alluvial aquifer, classes I, II, III, IV, and V of groundwaters accounted for 30.9%, 11.4%, 19.5%, 10.7%, and 27.5%, respectively (Figure 4). That is, 61.8% groundwaters with good quality (classes I–III) were drinkable and fit for irrigation and other purposes; 10.7% groundwaters (class IV) were undrinkable but fit for irrigation; only 27.5% groundwaters (class V) were undrinkable and unfit for other purposes. Groundwater quality was distinct in various areas. Groundwaters with poor quality (classes IV and V) in UA and AA accounted for 41.3% and 40.0%, respectively, and both were 1.1–1.2 times that in PUA; by contrast, all groundwaters in RA were drinkable (Figure 5). This indicates that human activities such as urbanization and agricultural activities deteriorated groundwater quality in the coastal alluvial aquifer [11].

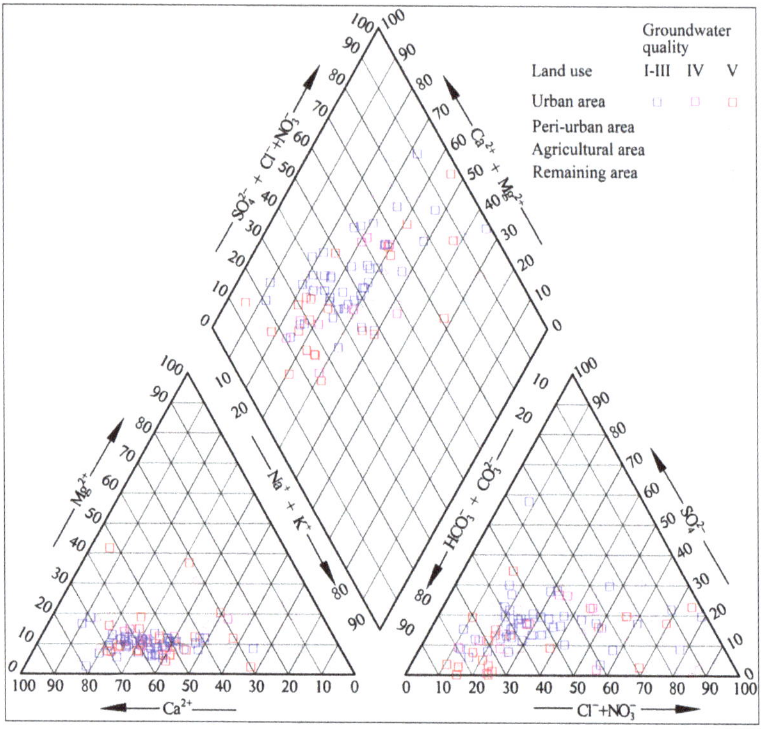

Figure 3. Hydrochemical facies of groundwater in areas with different types of land use of the coastal alluvial aquifer.

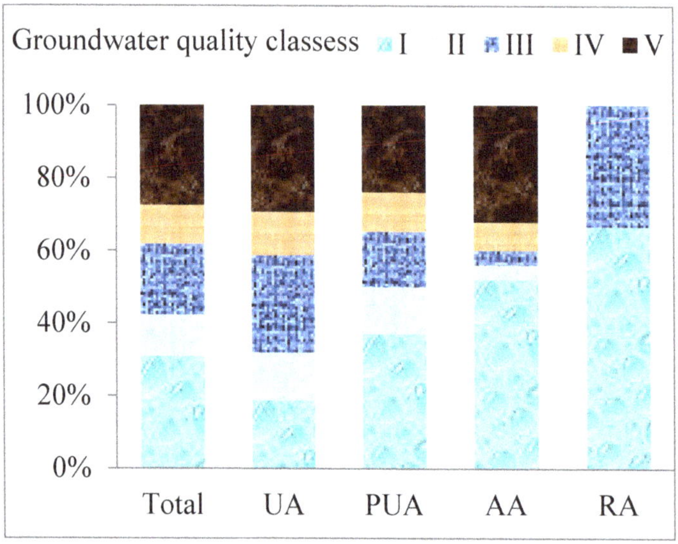

Figure 4. Groundwater quality in areas with different types of land use of the coastal alluvial aquifer.

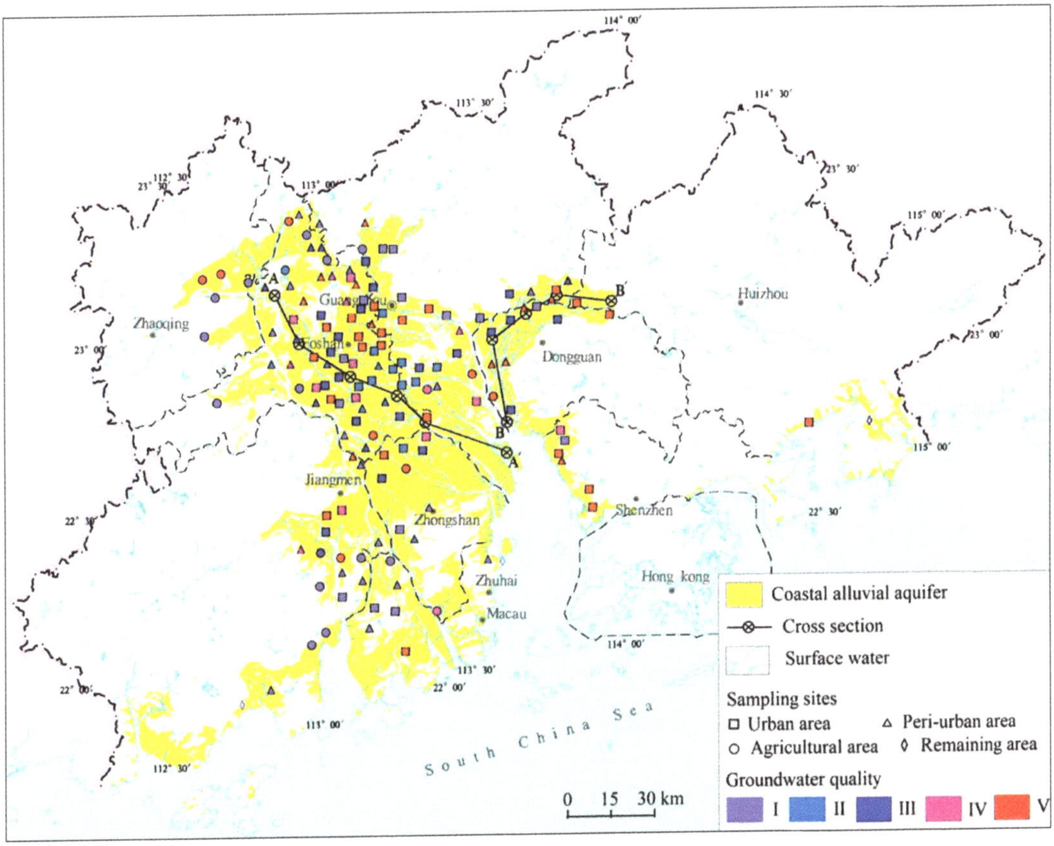

Figure 5. Spatial distribution of groundwater quality in the coastal alluvial aquifer of the Pearl River Delta.

5. Discussion

Generally, in groundwater disturbed by human activities, the PCA technique can distinguish those chemical parameters indicating anthropogenic impact from chemical parameters controlled by natural background [7,32]. In this study, five PCs were extracted by using the PCA technique, and explained 72.5% of the variance in the hydrochemical datasets of the coastal alluvial aquifer (Table 2). Specifically, the PC1, PC2, PC3, PC4, and PC5 explained 25.6%, 16.0%, 12.5%, 10.5%, and 7.8% of the total variance, respectively. Thus, factors controlling groundwater chemistry and quality in the study area are as follows.

Table 2. Principal component (PC) loadings for groundwater chemical parameters in the coastal alluvial aquifer of the Pearl River Delta.

Chemical Parameters	PCs				
	PC1	PC2	PC3	PC4	PC5
Cl^-	**0.973**	−0.005	0.072	0.050	0.017
Na^+	**0.972**	0.055	0.080	0.039	0.043
Mg^{2+}	**0.863**	0.183	0.106	0.123	−0.094

Table 2. *Cont.*

Chemical Parameters	PCs				
	PC1	PC2	PC3	PC4	PC5
TDS	**0.850**	0.488	0.079	0.092	0.082
NH_4^+	**0.622**	0.137	0.205	0.123	0.497
Ca^{2+}	0.020	**0.932**	0.030	0.074	0.021
HCO_3^-	0.304	**0.797**	−0.062	0.416	0.133
SO_4^{2-}	0.074	**0.578**	0.447	−0.404	−0.165
K^+	0.266	**0.570**	−0.085	−0.334	0.007
Pb	0.160	−0.094	**0.894**	−0.099	−0.016
Mn	0.028	0.084	**0.840**	0.181	0.094
NO_3^-	0.003	0.094	−0.078	**−0.796**	0.073
Fe	0.311	0.008	0.453	**0.472**	−0.009
As	0.220	0.177	−0.031	**0.444**	0.121
NO_2^-	0.092	0.191	0.029	−0.235	**0.768**
I^-	−0.049	−0.124	−0.013	0.176	**0.571**
Eigenvalue	4.1	2.6	2.0	1.7	1.3
Explained variance (%)	25.6	16.0	12.5	10.5	7.8
Cumulative % of variance	25.6	41.6	54.1	64.6	72.5

Note(s): Bold numbers = maximum absolute PC loading of one parameter.

5.1. *PC1 (Factor 1)—Release from Marine Sediments and Infiltration of Domestic and Septic Sewage*

Approximately one third of the parameters, including Cl^-, Na^+, Mg^{2+}, TDS, and NH_4^+, are in the PC1. Specifically, the PC1 shows strong positive loadings with Cl^-, Na^+, Mg^{2+}, and TDS and a moderate positive loading with NH_4^+ (Table 2). On one hand, in coastal aquifers of the PRD, co-occurrence of high levels of Cl^-, Na^+, Mg^{2+}, and TDS in groundwater was commonly from three major sources. The first is the seawater intrusion, because seawater is characterized by high concentrations of Cl^-, Na^+, Mg^{2+}, and TDS, and intrusion sometimes occurs in coastal areas of the PRD [33]. The second is the release of trapped seawater in marine sediments entering into groundwater via vertical water flow, because trapped seawater is often retained in marine sediments of Asian deltas [34]. The third is the infiltration of domestic and septic sewage, because domestic and septic sewage is generally enriched with Cl^-, Na^+, Mg^{2+}, and TDS [9], and the leakage of domestic and septic sewage often occurs in UA and PUA of the PRD [21,26]. On the other hand, seawater in the South China Sea is generally at low NH_4^+ concentrations of <0.003 mg/L and less than one tenth of the median concentration of groundwater NH_4^+ in the coastal alluvial aquifer (Table 1) [35]. This indicates that the seawater intrusion is excluded out of the PC1. By contrast, Quaternary marine sediments in the PRD are commonly enriched with organic nitrogen that converted to NH_4^+ under reducing conditions, and finally entering into groundwater via the vertical water flow [36]. This indicates that the PC1 includes the release of seawater and NH_4^+ from marine sediments. In addition, domestic and septic sewage in the PRD was also often enriched with NH_4^+ (>5 mg/L) [21], indicating that the PC1 also includes the infiltration of domestic and septic sewage. Therefore, factor 1 represents the release of salt and NH_4^+ from Quaternary marine sediments, and the infiltration of domestic and septic sewage.

5.2. *PC2 (Factor 2)—Agricultural Activities*

One fourth of the parameters, including Ca^{2+}, HCO_3^-, SO_4^{2-}, and K^+, are in the PC2. Specifically, the PC2 shows strong positive loadings with Ca^{2+} and HCO_3^-, and moderate positive loadings with SO_4^{2-} and K^+ (Table 2). Some studies reported that Ca-HCO_3 and Ca-SO_4 facies were two major hydrochemical facies in river water of the PRD, and irrigation using river water often occurs in agricultural lands of the PRD [21,32]. This indicates that agricultural irrigation may be included in the PC2. However, the median K^+ concentration in river water of the PRD was <10 mg/L and less than half of that in groundwater in the coastal alluvial aquifer [20], indicating that river water in the PRD was often not enriched with K^+. This seems opposite to the result that Ca^{2+}, HCO_3^-, SO_4^{2-}, and K^+ are in the same PC. On the other hand, the wide use of chemical fertilizers such as potassium fertilizers in agricultural activities often results in groundwater in agricultural lands becoming enriched with K^+ [37]. Correspondingly, in this coastal alluvial aquifer, groundwater in AA showed a much higher median K^+ concentration (22 mg/L) in comparison with that in PUA (14 mg/L) and RA (12 mg/L) (Table 1). Therefore, it can be concluded that factor 2 represents agricultural activities related to irrigation of river water and the use of chemical fertilizers.

5.3. PC3 (Factor 3)—Industrial Pollution Related to Heavy Metals and Acid Deposition

The two parameters of Pb and Mn are in the PC3. Specifically, the PC3 has strong positive loadings with Pb and Mn (Table 2). Two studies have already reported that high levels of heavy metals such as Pb and Mn in shallow groundwater of the PRD were mainly attributed to the infiltration of industrial wastewater [16,17], because illegal discharge of industrial wastewater from factories sometimes occurred in UA and PUA and irrigation using river water contaminated by industrial wastewater often occurred in AA of the PRD [32]. Correspondingly, median concentrations of groundwater Mn in UA, PUA, and AA were more than four times that in RA where industrial wastewater is free, and high Pb concentrations (>0.01 mg/L) of groundwaters occurred in UA and PUA, but not in AA and RA (Table 1). On the other hand, two studies showed that high levels of Pb in soils were widely distributed in the PRD owing to the pollution of automobile exhausts and Pb-rich industrial dusts [38,39]. Moreover, acid deposition also widely occurred in the PRD in recent decades [40]. In this case, the release of Pb from soils under acidic conditions entering into groundwater via the water flow is expected. Correspondingly, the PC3 also has a weak positive loading with SO_4^{2-} (a major chemical component in acid rain) (Table 2), and the groundwater Pb concentration had a significantly negative correlation with pH ($p < 0.001$) (Figure 6A). Therefore, it can be concluded that factor 3 represents the industrial pollution related to heavy metals and acid deposition.

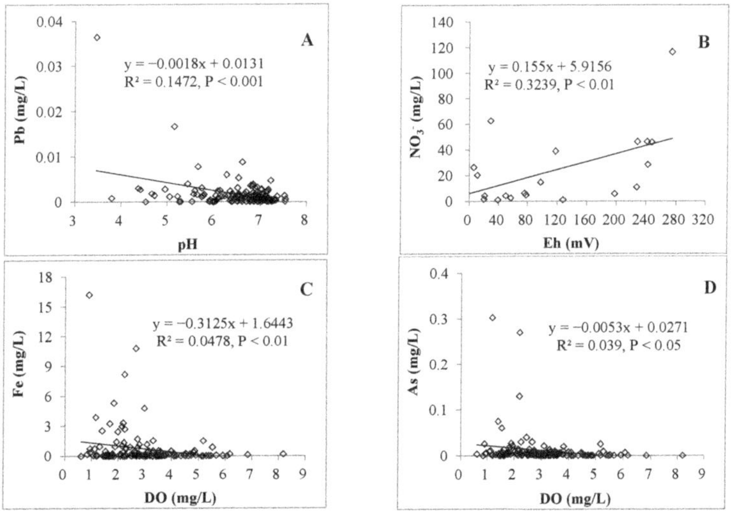

Figure 6. Relationships between concentrations of groundwater chemicals in the coastal alluvial aquifer of the Pearl River Delta. (**A**) Pb and pH; (**B**) NO_3^- and Eh in agricultural areas; (**C**) Fe and DO; (**D**) As and DO.

5.4. PC4 (Factor 4)—Reductive Dissolution of As-Loaded Fe Minerals and Denitrification

The PC4 has a strong negative loading with NO_3^- and weak positive loadings with Fe and As (Table 2). This likely infers the denitrification and the release of Fe and As via reductive dissolution, because groundwater denitrification often occurred in the PRD in 2006 [20], and reductive dissolution of As-loaded Fe minerals in Quaternary sediments induced by reducing sewage was a major driving force for the occurrence of Fe-rich and As-rich groundwaters in the PRD [14,28]. This is also supported by two evidences. One is that groundwater NO_3^- concentrations in AA had a significantly positive correlation with Eh values (Figure 6B). Another is that both Fe and As groundwater concentrations had significantly negative correlations with DO values in this coastal alluvial aquifer (Figure 6C,D). Furthermore, inputs of reducing sewage often occurred in UA, PUA, and AA of the PRD via infiltration of wastewater and irrigation with contaminated river water, respectively [14]. Thus, factor 4 likely represents the input of anthropogenic reducing sewage inducing reductive dissolution of As-loaded Fe minerals and denitrification.

5.5. PC5 (Factor 5)—I^- Contamination

The PC5 shows a strong positive loading with NO_2^- and a moderate positive loading with I^- (Table 2). This probably indicates I^- contamination from both of geogenic and anthropogenic sources, because mineralization of iodine-rich organic matter in marine sediments triggering by nitrogen-rich (e.g., NO_2^-, NH_4^+) sewage was a major geogenic source for high levels of groundwater I^- in the coastal alluvial aquifer, and the leakage of I^--rich sewage and irrigation using I^--rich and nitrogen-rich (e.g., NO_2^-, NH_4^+) river water were main anthropogenic sources for high levels of groundwater I^- in UA and AA of the coastal alluvial plain, respectively [18]. Correspondingly, median values of groundwater I^- in UA and AA were higher than that in PUA and RA (Table 1). Moreover, the PC5 also has a weak positive loading with NH_4^+ (Table 2). Therefore, factor 5 likely represents the I^- contamination from both geogenic and anthropogenic sources.

6. Conclusions

Hydrogeochemical characteristics and groundwater quality in the coastal alluvial aquifer of the PRD were investigated in this study, and were found to be dominated by Ca-HCO_3 and Ca·Na-HCO_3 facies. Groundwater with NO_3 facies occurred in this aquifer with the increase in human activities. Groundwater TDS and most of the major ion (e.g., Ca^{2+}, Na^+, Mg^{2+}, HCO_3^-, Cl^-, and SO_4^{2-}) concentrations were commonly in the order of UA > PUA > AA > RA. Redox-sensitive parameters such as Mn, NH_4^+, NO_2^-, and As in groundwater showed higher median concentrations in UA and PUA than in AA and RA. In this aquifer, 61.8% of groundwaters (classes I–III) were fit for drinking and other purposes, and 10.7% groundwaters (class IV) were undrinkable but fit for irrigation, whereas 27.5% of groundwaters (class V) were unfit for any purpose. Poor-quality groundwater in UA and AA was 1.1–1.2 times that in PUA but absent in RA.

Groundwater chemistry and quality in this aquifer was mainly controlled by five factors according to the PCA method. Factor 1 is the release of salt and NH_4^+ from marine sediments and infiltration of domestic and septic sewage. Factor 2 is agricultural activities related to the irrigation using river water and the use of chemical fertilizers. Factor 3 is the industrial pollution related to heavy metals and acid deposition. Factor 4 is the input of anthropogenic reducing sewage inducing the reductive dissolution of As-loaded Fe minerals and denitrification. Factor 5 is the I^- contamination from both geogenic and anthropogenic sources.

Correspondingly, there are three major suggestions to protect groundwater quality in this coastal aquifer. (1) Repairing old sewer systems in UA and building sewer systems in PUA to reduce illegal discharge of sewage. (2) Limiting sewage irrigation and the use of chemical fertilizers in AA. (3) Strengthening the supervision of the industrial exhaust gas discharge in UA and PUA.

Supplementary Materials: The following supporting information can be downloaded at: https://www.mdpi.com/article/10.3390/w14244131/s1, Table S1: Groundwater quality standards for drinking and irrigation [22]; Table S2: Kaiser-Meyer-Olkin and Bartlett's test for the suitability of principal components analysis (PCA) to groundwater chemical data in coastal alluvial aquifer.

Author Contributions: Conceptualization, C.L.; methodology, C.L. and Q.H.; software, Y.C.; validation, C.L. and G.H.; investigation, C.L. and G.H.; data curation, Q.H.; writing—original draft preparation, C.L.; writing—review and editing, Q.H. and G.H.; visualization, Q.H.; supervision, G.H. All authors have read and agreed to the published version of the manuscript.

Funding: This research was supported by the Natural Science Foundation of Hebei Province, China (D2021504031), the Three-three-three Talent Project of Hebei Province, China (A202105001), and the China Geological Survey Grant (DD20160309).

Institutional Review Board Statement: Not applicable.

Informed Consent Statement: Not applicable.

Data Availability Statement: The datasets generated and/or analyzed during the current study are not publicly available.

Conflicts of Interest: The authors declare no conflict of interest.

References

1. Huang, G.; Liu, C.; Sun, J.; Zhang, M.; Jing, J.; Li, L. A regional scale investigation on factors controlling the groundwater chemistry of various aquifers in a rapidly urbanized area: A case study of the Pearl River Delta. *Sci. Total Environ.* **2018**, *625*, 510–518. [CrossRef] [PubMed]
2. Liu, X.; Xin, L. Assessment of the efficiency of cultivated land occupied by urban and rural construction land in China from 1990 to 2020. *Land* **2022**, *11*, 941. [CrossRef]
3. Small, C.; Nicholls, R.J. A global analysis of human settlement in coastal zones. *J. Coast. Res.* **2003**, *19*, 584–599.
4. Umarani, P.; Ramu, A.; Kumar, V. Hydrochemical and statistical evaluation of groundwater quality in coastal aquifers in Tamil Nadu, India. *Environ. Earth Sci.* **2019**, *78*, 452. [CrossRef]
5. Chucuya, S.; Vera, A.; Pino-Vargas, E.; Steenken, A.; Mahlknecht, J.; Montalván, I. Hydrogeochemical characterization and identification of factors influencing groundwater quality in coastal aquifers, case: La Yarada, Tacna, Peru. *Int. J. Environ. Res. Public Health* **2022**, *19*, 2815. [CrossRef]
6. Armengol, S.; Ayora, C.; Manzano, M.; Bea, S.A.; Martínez, S. The role of loess weathering in the groundwater chemistry of the Chaco-Pampean Plain (Argentina). *J. Hydrol.* **2020**, *587*, 124984. [CrossRef]
7. Kanellopoulos, C.; Argyraki, A. Multivariate statistical assessment of groundwater in cases with ultramafic rocks and anthropogenic activities influence. *Appl. Geochem.* **2022**, *141*, 105292. [CrossRef]
8. Mason, J.P. *Groundwater, Surface-Water, and Water-Chemistry Data, Black Mesa Area, northeastern Arizona—2018–2019. U.S. Geological Survey Open-File Report*; 2022–1086; USGS: Reston, VI, USA, 2022. [CrossRef]
9. Elumalai, V.; Rajmohan, N.; Sithole, B.; Li, P.; Uthandi, S.; Tol, J. Geochemical evolution and the processes controlling groundwater chemistry using ionic ratios, geochemical modelling and chemometric analysis in uMhlathuze catchment, KwaZulu-Natal, South Africa. *Chemosphere* **2023**, *312*, 137179. [CrossRef]
10. Han, D.; Currell, M.J. Review of drivers and threats to coastal groundwater quality in China. *Sci. Total Environ.* **2022**, *806*, 150913. [CrossRef]
11. Sellamuthu, S.; Joseph, S.; Gopalakrishnan, S.; Sekar, S.; Khan, R.; Shukla, S. Appraisal of groundwater quality for drinking and irrigation suitability using multivariate statistical approach in a rapidly developing urban area, Tirunelveli, India. *Environ. Sci. Pollut. Res.* **2022**. [CrossRef]
12. Xing, L.; Guo, H.; Zhan, Y. Groundwater hydrochemical characteristics and processes along flow paths in the North China Plain. *J. Asian Earth Sci.* **2013**, *70–71*, 250–264. [CrossRef]
13. Ri, M.; Guo, H.; Kim, P.; Ri, K.; Ri, G. Influence of seawater intrusion on the hot springs in a coastal area: The case of the Anak-Sinchon Uplift, Korean Peninsula. *J. Hydrol.* **2022**, *607*, 127509. [CrossRef]
14. Huang, G.; Han, D.; Song, J.; Li, L.; Pei, L. A sharp contrasting occurrence of iron-rich groundwater in the Pearl River Delta during the past dozen years (2006–2018): The genesis and mitigation effect. *Sci. Total Environ.* **2022**, *829*, 154676. [CrossRef] [PubMed]
15. Huang, G.; Zhang, M.; Liu, C.; Li, L.; Chen, Z. Heavy metal(loid)s and organic contaminants in groundwater in the Pearl River Delta that has undergone three decades of urbanization and industrialization: Distributions, sources, and driving forces. *Sci. Total Environ.* **2018**, *635*, 913–925. [CrossRef]
16. Zhang, F.; Huang, G.; Hou, Q.; Liu, C.; Zhang, Y.; Zhang, Q. Groundwater quality in the Pearl River Delta after the rapid expansion of industrialization and urbanization: Distributions, main impact indicators, and driving forces. *J. Hydrol.* **2019**, *577*, 124004. [CrossRef]

17. Hou, Q.; Zhang, Q.; Huang, G.; Liu, C.; Zhang, Y. Elevated manganese concentrations in shallow groundwater of various aquifers in a rapidly urbanized delta, south China. *Sci. Total Environ.* **2020**, *701*, 134777. [CrossRef]
18. Huang, G.; Liu, C.; Li, L.; Zhang, F.; Chen, Z. Spatial distribution and origin of shallow groundwater iodide in a rapidly urbanized delta: A case study of the Pearl River Delta. *J. Hydrol.* **2020**, *585*, 124860. [CrossRef]
19. Huang, G.; Liu, C.; Zhang, Y.; Chen, Z. Groundwater is important for the geochemical cycling of phosphorus in rapidly urbanized areas: A case study in the Pearl River Delta. *Environ. Pollut.* **2020**, *260*, 114079. [CrossRef]
20. Zhang, M.; Huang, G.; Liu, C.; Zhang, Y.; Chen, Z.; Wang, J. Distributions and origins of nitrate, nitrite, and ammonium in various aquifers in an urbanized coastal area, south China. *J. Hydrol.* **2020**, *582*, 124528. [CrossRef]
21. Sun, J.; Jing, J.; Huang, G.; Liu, J.; Chen, X.; Zhang, Y. *Report on the Investigation and Assessment of Groundwater Contamination in the Pearl River Delta Area*; Chinese Academy of Geological Sciences; The Institute of Hydrogeology and Environmental Geology: Beijing, China, 2009. (In Chinese)
22. General Administration of Quality Supervision Inspection and Quarantine of the People's Republic of China (GAQSIQPRC). *Standard for Groundwater Quality*; Standards Press of China: Beijing, China, 2017.
23. Bi, P.; Huang, G.; Liu, C.; Li, L. Geochemical factors controlling natural background levels of phosphate in various groundwater units in a large-scale urbanized area. *J. Hydrol.* **2022**, *608*, 127594. [CrossRef]
24. Zong, Y.; Yim, W.W.S.; Yu, F.; Huang, G. Late Quaternary environmental changes in the Pearl River mouth region, China. *Quat. Int.* **2009**, *206*, 35–45. [CrossRef]
25. Lancia, M.; Su, H.; Tian, Y.; Xu, J.; Andrews, C.; Lerner, D.N.; Zheng, C. Hydrogeology of the Pearl River Delta, southern China. *J. Maps* **2020**, *16*, 388–395. [CrossRef]
26. Huang, G.; Pei, L.; Li, L.; Liu, C. Natural background levels in groundwater in the Pearl River Delta after the rapid expansion of urbanization: A new pre-selection method. *Sci. Total Environ.* **2022**, *813*, 151890. [CrossRef]
27. Ye, Y.; Zhang, H.; Liu, K.; Wu, Q. Research on the influence of site factors on the expansion of construction land in the Pearl River Delta, China: By using GIS and remote sensing. *Int. J. Appl. Earth Obs.* **2013**, *21*, 366–373. [CrossRef]
28. Huang, G.; Song, J.; Han, D.; Liu, R.; Liu, C.; Hou, Q. Assessing natural background levels of geogenic contaminants in groundwater of an urbanized delta through removal of groundwaters impacted by anthropogenic inputs: New insights into driving factors. *Sci. Total Environ.* **2023**, *857*, 159527. [CrossRef]
29. Huang, G.; Chen, Z.; Liu, F.; Sun, J.; Wang, J. Impact of human activity and natural processes on groundwater arsenic in an urbanized area (South China) using multivariate statistical techniques. *Environ. Sci. Pollut. R.* **2014**, *21*, 13043–13054. [CrossRef]
30. Güler, C.; Kurt, M.A.; Alpaslan, M.; Akbulut, C. Assessment of the impact of anthropogenic activities on the groundwater hydrology and chemistry in Tarsus coastal plain (Mersin, SE Turkey) using fuzzy clustering, multivariate statistics and GIS techniques. *J. Hydrol.* **2012**, *414–415*, 435–451. [CrossRef]
31. Guangdong Hydrogeological Second Team (GHST). *Regional Hydrogeological Survey Report*; Guangdong Hydrogeological Second Team: Guangzhou, China, 1981. (In Chinese)
32. Huang, G.; Sun, J.; Zhang, Y.; Chen, Z.; Liu, F. Impact of anthropogenic and natural processes on the evolution of groundwater chemistry in a rapidly urbanized coastal area, South China. *Sci. Total Environ.* **2013**, *463–464*, 209–221. [CrossRef]
33. Wang, Y.; Jiao, J.J. Origin of groundwater salinity and hydrogeochemical processes in the confined Quaternary aquifer of the Pearl River Delta, China. *J. Hydrol.* **2012**, *438–439*, 112–124. [CrossRef]
34. Larsen, F.; Tran, L.V.; Van Hoang, H.; Tran, L.T.; Christiansen, A.V.; Pham, N.Q. Groundwater salinity influenced by Holocene seawater trapped in incised valleys in the Red River delta plain. *Nat. Geosci.* **2017**, *10*, 376–381. [CrossRef]
35. Zhu, Y.; Yuan, D.; Huang, Y.; Ma, J.; Feng, S. A sensitive flow-batch system for on board determination of ultra-trace ammonium in seawater: Method development and shipboard application. *Anal. Chim. Acta* **2013**, *794*, 47–54. [CrossRef] [PubMed]
36. Jiao, J.; Wang, Y.; Cherry, J.A.; Wang, X.; Zhi, B.; Du, H.; Wen, D. Abnormally high ammonium of natural origin in a coastal aquifer-aquitard system in the Pearl River Delta, China. *Environ. Sci. Technol.* **2010**, *44*, 7470–7475. [CrossRef] [PubMed]
37. Shamrukh, M.; Corapcioglu, M.Y.; Hassona, F.A.A. Modeling the Effect of Chemical Fertilizers on Ground Water Quality in the Nile Valley Aquifer, Egypt. *Groundwater* **2001**, *39*, 59–67. [CrossRef]
38. Zhu, B.; Chen, Y.; Peng, J. Lead isotope geochemistry of the urban environment in the Pearl River Delta. *Appl. Geochem.* **2001**, *16*, 409–417.
39. Yin, G.; Zhu, H.; Chen, Z.; Su, C.; He, Z.; Chen, X.; Qiu, J.; Wang, T. Spatial distribution and source apportionment of soil heavy metals in Pearl River Delta, China. *Sustainability* **2021**, *13*, 9651. [CrossRef]
40. Liu, F.; Sun, J.; Wang, J.; Zhang, Y. Groundwater acidification in shallow aquifers in Pearl River Delta, China: Distribution, factors, and effects. *Geochem. J.* **2017**, *51*, 373–384. [CrossRef]

water

Article

Factors Controlling Natural Background Levels of Ammonium and Iodide in Shallow Groundwater of Coastal Aquifers, South China

Lixin Pei [1], Xin Lu [2], Xiwen Li [1,*], Ming Zhang [3] and Heqiu Wu [4]

1 Haikou Marine Geological Survey Center, China Geological Survey, Haikou 570100, China
2 Zhejiang Dibo Survey and Design Co., Ltd., Hangzhou 310000, China
3 Faculty of Engineering, China University of Geosciences, Wuhan 430000, China
4 Zhejiang Engineering Geophysical Survey and Design Institute Co., Ltd., Hangzhou 310005, China
* Correspondence: lxw1818168@163.com

Abstract: Assessing natural background levels (NBLs) in groundwater is crucial for evaluating groundwater pollution and the use of groundwater resources in coastal areas. This study assessed NBLs of iodide and ammonium in the shallow groundwater of the Pearl River Delta (PRD) by using a preselection method with Grubbs' test, and discussed factors controlling NBLs in various groundwater units. Here, the preselection method consists of Cl/Br mass ratios versus Cl concentrations and the oxidation capacity, and the PRD is divided into four groundwater units. Results showed that NBL-iodide in groundwater unit A was 0.14 mg/L and >2 times greater than that in other groundwater units. Similarly, NBL-ammonium in groundwater unit A was 0.32 mg/L and also >2 times greater than that in other groundwater units. The release of iodide from both of organic-iodine in the vadose zone and iodine-rich minerals in aquifer sediments were the two main sources for the higher NBL-iodide in groundwater unit A compared to other units. By contrast, the occurrence of ammonium from organic-nitrogen in the vadose zone was the major source for the higher NBL-ammonium in groundwater unit A compared with the other units. Soluble iodide resulted from the mineralization of organic-iodine in Quaternary marine formation, and the release of iodide accompanied with reductive dissolution of iodide-loaded Fe (oxyhydr) oxides in aquifer sediments was the main driving force controlling the higher NBL-iodide in groundwater unit A compared with the other units. By contrast, the release of soluble ammonium from the mineralization of organic-nitrogen in marine formation entering into groundwater was the main driving force controlling the higher NBL-ammonium in groundwater unit A relative to the other units. These results enhance the knowledge on groundwater NBLs in coastal areas and improve groundwater resources management in coastal areas such as the PRD.

Keywords: natural background levels; iodide; ammonium; shallow groundwater; coastal areas

Citation: Pei, L.; Lu, X.; Li, X.; Zhang, M.; Wu, H. Factors Controlling Natural Background Levels of Ammonium and Iodide in Shallow Groundwater of Coastal Aquifers, South China. *Water* **2022**, *14*, 3737. https://doi.org/10.3390/w14223737

Academic Editor: Micòl Mastrocicco

Received: 30 September 2022
Accepted: 28 October 2022
Published: 17 November 2022

1. Introduction

More than 80% of countries in global have coastlines, and areas located at coastlines have been centers of human activity for long time. For example, China has a long coastline with more than 18,000 km, and the population of coastal areas accounted for >20% of total population in China [1]. Groundwater in coastal aquifers is often a very important water resource for human beings in coastal areas because of the ever-increasing water demand resulting from the increase in population and the development of economy [2–4]. For instance, groundwater is being used to supplement the available surface water for water demand in the Pearl River Delta (PRD) adjacent to the South China Sea owing to the large scale urbanization and industrialization, resulting in the growth of the population and shortage of available surface water [5]. However, Zhang et al. reported that a significant portion of groundwater in the PRD is unsuitable for drinking, agricultural, or industrial purposes due to the contamination of several contaminants such as iodide (I^-), ammonium (NH_4^+), nitrate (NO_3^-), etc [6]. This limits the use of groundwater in coastal aquifers such as the PRD. Therefore, it is necessary to understand natural background levels (NBLs) and their geogenic factors of contaminants in groundwater for assessing the status of groundwater contamination and quality in coastal areas, and to thereby support the reference basis for groundwater resource management. Here, groundwater NBL represents the concentration of a certain component in groundwater derived from natural sources [7].

NBLs as part of anthropogenic pollutants in the groundwater of the PRD have already been assessed. For example, Huang et al. used a new preselection method to evaluate NBLs of groundwater Cl^- and NO_3^- in the PRD [8]; moreover, Bi et al. used an oxidation capacity method in combination with Grubbs' test to assess groundwater PO_4^{3-} NBL in various aquifers in the PRD [9]. By contrast, NBLs and the geogenic factors of some of their main contaminants, such as I^- and NH_4^+ in groundwater of the PRD, have received little attention, although a few of studies have reported on the spatial distributions and origins of I^--rich and NH_4^+-rich groundwaters in the PRD [10,11].

Therefore, this study aims to evaluate the NBLs of groundwater I^- and NH_4^+ in various aquifers of the PRD by using a new preselection method established by [8], and thereby to discuss factors controlling these NBLs in various aquifers. Unlike the levels of pollutants such as Cl^- and NO_3^-, high levels of I^- and NH_4^+ in the shallow groundwater of the PRD are not only affected by human activities but are also influenced by geogenic factors [10,11]. The results will enhance the knowledge on groundwater NBLs in coastal areas and improve groundwater resources management in coastal areas such as the PRD.

2. Study Area

2.1. Geographical and Land-Use Conditions

The PRD (111°59′–115°25′ E, 21°17′–23°55′ N) is located in Guangdong Province of China and is adjacent to the South China Sea in the south (Figure 1). It is a compound delta formed by sediments brought by water flows of three major rivers (i.e., Xijiang River, Beijiang River, Dongjiang River) and their tributaries (i.e., Tanjiang River, Suijiang River, and Zengjiang River), with nearly half of the area dotted with hills, platforms, and residual hills. The southern coastline is 1059 km long. The average annual precipitation and temperature are 1800–2200 mm and 21.4–22.4 °C, respectively. The majority of precipitation falls in the wet season between April and September [12,13]. Land use in the PRD can be divided into four types including urbanized area (UA), peri-urban area (PUA), agricultural area (AA), and the remaining area (RA) [14]. The remaining area consists of woodlands, grasslands, small villages, and uncultivated lands.

2.2. Hydrogeological Conditions

Bi et al. reported that the PRD has four types of aquifers and can be divided into four groundwater units (Figure 1) [9]. Briefly, groundwater unit A is mainly distributed in the central-southern part of the PRD and consists of coastal-alluvial plains where continental and marine formations are interbedded with each other [15]. Continental formations are dominated by sandy fluvial deposits or gravel and compose aquifer-I. The vadose zone in unit A consists of marine sediments and is predominantly silt/clay with low permeability. Groundwater unit B is outside of unit A, where valley and interhill plains are distributed, but marine formations are missing. Continental formations in this unit compose aquifer-II. Compared to groundwater unit A, vadose zone in unit B is dominated by coarser media (e.g., sandy clay) with higher permeability. Groundwater unit C is mainly distributed in hilly areas where fractured bedrocks compose aquifer-III. Groundwater unit D is related to karst aquifers (aquifer-IV) where carbonate rocks aged from the Devonian to the Permian are common. Incomplete weathering of bedrocks overlying aquifer-III and aquifer-IV result in vadose zones in groundwater units C and D being composed predominantly of coarse media (e.g., sand, gravel). In general, groundwater in the PRD is mainly recharged by precipitation and agricultural irrigation [16,17]. The direction of groundwater flow in the PRD is from groundwater units B and C to unit A [6].

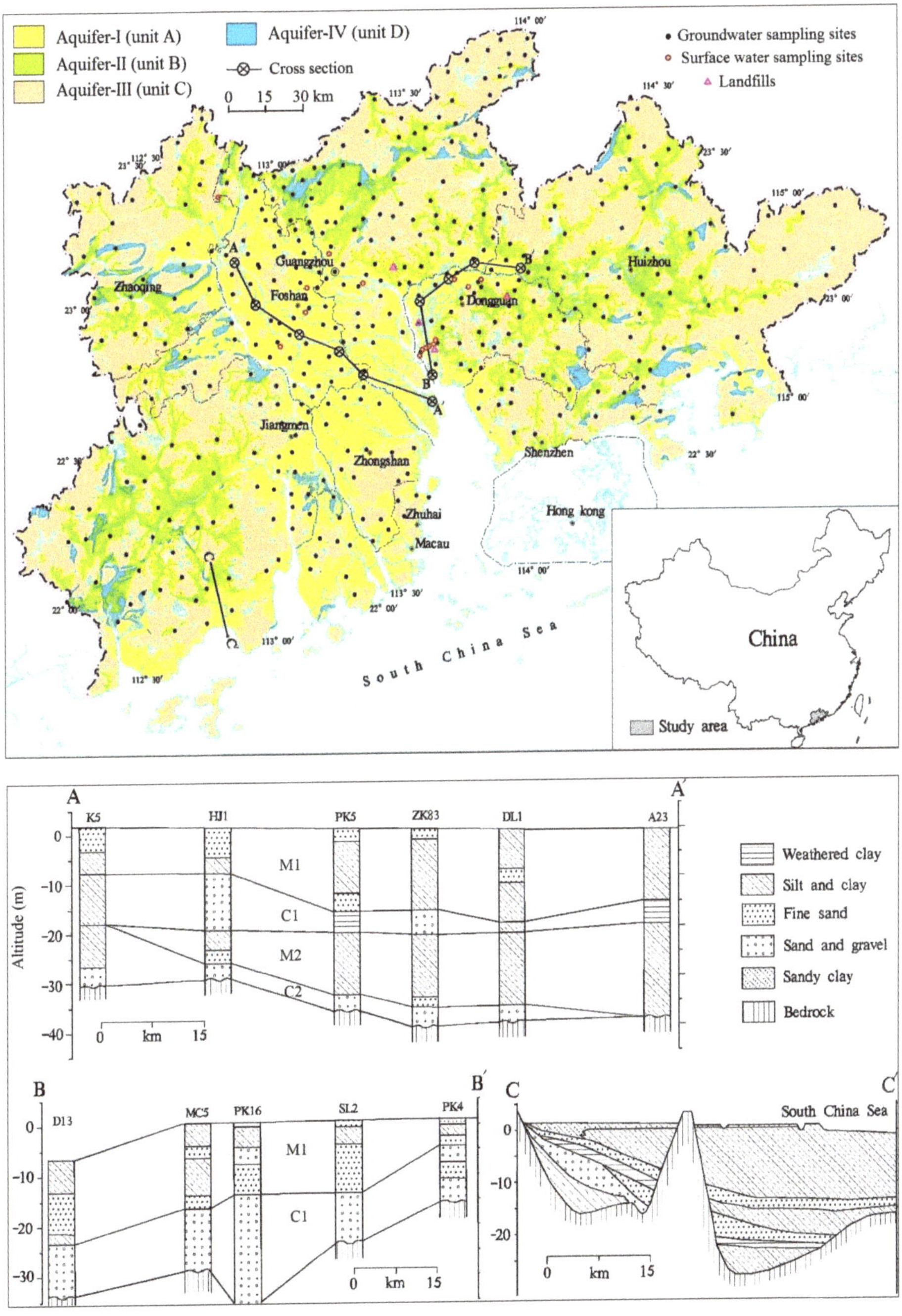

Figure 1. Hydrogeological units and sampling sites in the Pearl River Delta.

3. Materials and Methods

3.1. Sampling and Analysis

In one session, 399 groundwater samples, 15 surface water samples, and 4 leachate samples were collected in the wet season from the PRD. Among them, 124, 134, 132, and 9 sampling sites for groundwater were distributed in groundwater units A to D, respectively. Groundwater samples were collected in situ at the depth of 0.5 m below groundwater level and filtered (0.45 μm membrane) on site. Three physicochemical parameters including pH, redox potential (Eh), and dissolved oxygen (DO) were determined by a multiparameter instrument on site, and multiparametric probes in this instrument were calibrated before measurement. Another 18 parameters including total dissolved solids (TDS), chemical oxygen demand (COD), 5 cations (K^+, Na^+, Ca^{2+}, Mg^{2+}, and NH_4^+), 8 anions (HCO_3^-, NO_3^-, SO_4^{2-}, Cl^-, Br^-, NO_2^-, PO_4^{3-}, and I^-), and 3 heavy metals (Fe, Mn, and Pb) in water samples were measured in the laboratory. Details of the analytical methods have already been described in previous works [8,10].

3.2. NBL Assessment

The method for NBL assessment used in this study was recently established by Huang et al. [8]. It consists of indicators of the oxidation capacity and Cl/Br ratios versus Cl concentrations combined with contamination markers. Briefly, this method included four steps (Figure S1). In step one, eight chemical components including NO_3^-, NO_2^-, PO_4^{3-}, Pb, TDS, Na, NH_4^+, and COD were selected as the candidate contamination markers in groundwater in the PRD owing to the anthropogenic contamination, and contamination markers in various groundwater units were selected by using a hierarchical cluster analysis (HCA) (Figure S2). In step two, contaminated groundwaters were identified from original datasets by using plots of Cl/Br mass ratios versus Cl concentrations combined with contamination markers, and the remaining datasets were denoted as CBM datasets (Figure S3). In step three, groundwaters remaining in the CBM datasets were removed by using the oxidation capacity method, and the remaining datasets after this step were denoted as the PS datasets (Figure S4). In step four, outliers in the PS datasets were deleted by Grubbs' test, and the maximum values in various groundwater units in residual datasets were denoted as NBLs. The details of the above method are shown in Section 3.2 of the Supplementary Materials.

3.3. Principal Components Analysis

Principal component analysis (PCA) is often used to reduce high-dimensional hydrochemical datasets, extract the main factors that control groundwater chemistry, and thereby deduce the geochemical processes indicated by these factors [18–20]. The present study used PCA with SPSS® version 23.0 software to extract PCs that control NBL-I and NBL-NH_4 from residual datasets in various groundwater units. Varimax method was used for rotation of PCs, and eigenvalues > 1 of PCs were retained for analyses. The terms of strong, moderate, and weak refer to the absolute PC loadings of >0.75, 0.75–0.5, and 0.5–0.3, respectively [10].

4. Results

4.1. NBLs of Groundwater I^- and NH_4^+ in the PRD

More than 60% of contaminated groundwaters in the PRD were excluded according to their oxidation capacities and Cl/Br mass ratios versus Cl concentrations (Figures S3 and S4). Then, groundwater NBL-I^- and NBL-NH_4^+ in various units were identified by the Grubbs' test ($\alpha = 0.01$). As shown in Figure 2, the groundwater NBL-I^- in unit A was 0.14 mg/L and was 2.5 times, 3.3 times, and 5.6 times greater than that in units B to D, respectively. Note that the NBL-I^- in groundwater unit A was significantly higher than the allowable limit (0.08 mg/L) of I^- in groundwater recommended by the Chinese government [10], while the NBLs-I^- in other groundwater units were not. This indicates that I^- is likely a more common geogenic contaminant in groundwater unit A rather than it is in other units in the

PRD. Similarly, the groundwater NBL-NH_4^+ in unit A was 0.32 mg/L and was also more than 2 times greater than that in other units. Note that groundwater NBLs-NH_4^+ in all units of the PRD were below the allowable limit (0.5 mg/L as N) of NH_4^+ in groundwater recommended by the Chinese government [11].

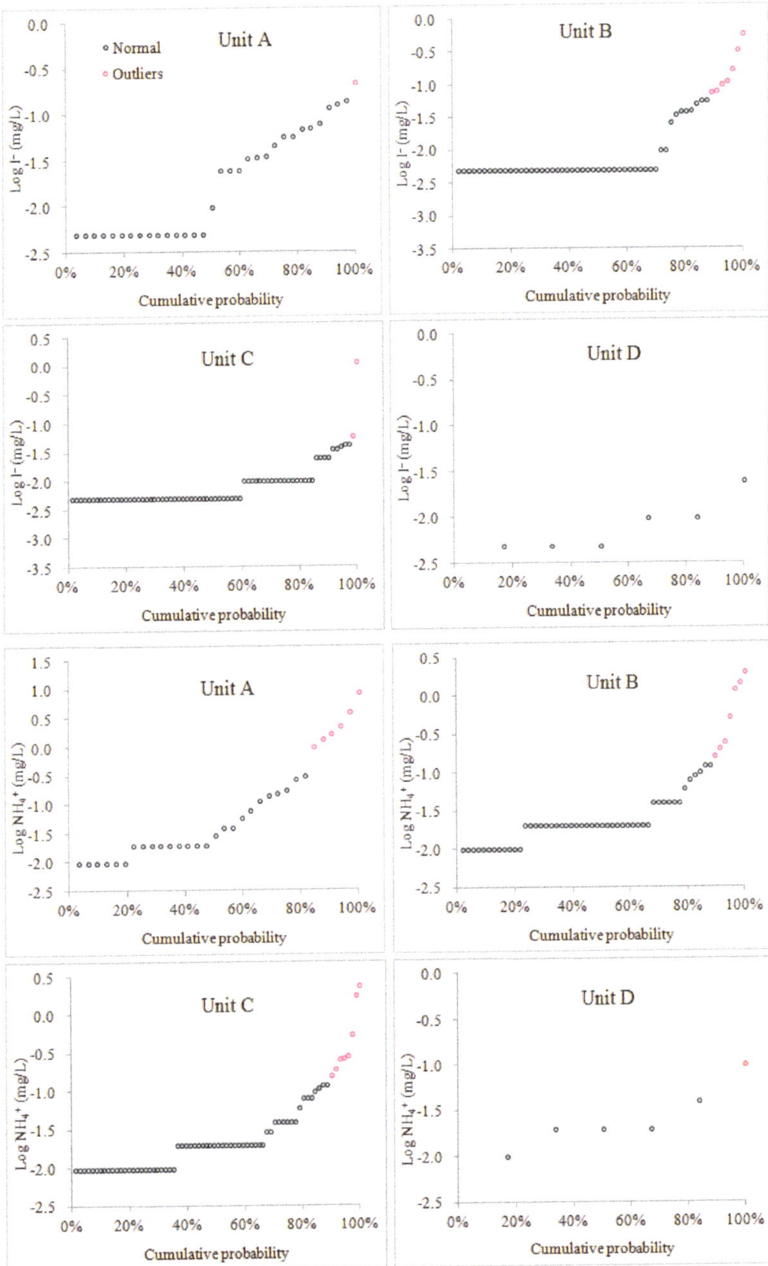

Figure 2. Outliers in the OXC datasets in various groundwater units tested by Grubbs' test.

4.2. Hydrogeochemical Characteristics in Residual Datasets in Various Aquifers

In this study, we focused on the differences of the hydrogeochemical characteristics between unit A and other units because of the much higher NBL-I^- and NBL-NH_4^+ in aquifer-I than in the other aquifers (Figure 2). In the residual datasets for both of NBL-I^- and NBL-NH_4^+, the mean pH values in groundwater unit A were higher than those in other units (Table 1), indicating that the groundwater in unit A had less acidic conditions than did the other units. Mean values of oxidation indicators including Eh, DO, and NO_3^- in groundwater unit A were significantly lower than those in the other units (Table 1). On the contrary, the mean concentrations of reduction indicators including NO_2^-, Fe, and Mn in groundwater unit A were more than 2 times those of other units (Table 1). These results indicate that the groundwater in unit A was a relatively reducing environment in comparison with that in other units. Moreover, the mean concentrations of Na^+, Ca^{2+}, Mg^{2+}, HCO_3^-, SO_4^{2-}, Cl^-, and PO_4^{3-} in groundwater unit A were also about 2 or more times greater than those in the other units (Table 1). This indicates that the groundwater in unit A was enriched with Na^+, Ca^{2+}, Mg^{2+}, HCO_3^-, SO_4^{2-}, Cl^-, and PO_4^{3-} in comparison with the other units.

Table 1. Mean concentrations of physicochemical parameters in residual datasets in various groundwater units.

NBL Datasets		NH_4^+				I^-			
		Unit A	Unit B	Unit C	Unit D	Unit A	Unit B	Unit C	Unit D
pH		6.21	5.89	5.78	5.79	6.37	5.96	5.84	5.87
Eh	mV	70	92	91	103	67	86	83	92
DO	mg/L	3.62	3.92	4.26	3.77	3.74	3.87	4.24	3.45
COD		1.40	1.17	1.17	1.00	1.62	1.25	1.33	1.13
TDS		286	159	103	94	339	164	108	127
K^+		9.0	7.3	4.3	3.2	9.8	7.3	4.1	5.3
Na^+		22.2	10.5	6.2	5.5	22.5	11.0	6.2	7.0
Ca^{2+}		39.5	21.1	12.8	14.1	48.4	21.6	13.8	18.4
Mg^{2+}		6.2	2.9	1.7	1.8	7.6	2.9	1.8	2.6
HCO_3^-		134.2	65.4	41.5	36.0	176.5	69.6	44.5	55.9
Cl^-		38.4	18.1	10.2	10.1	37.4	17.9	9.9	12.0
SO_4^{2-}		16.5	6.8	2.9	6.6	16.1	6.3	3.5	7.4
NO_3^-		8.7	14.3	10.0	10.5	7.1	13.8	10.1	10.3
NO_2^-		0.11	0.01	0.01	0.02	0.38	0.04	0.02	0.02
I^-		0.031	0.022	0.004	0.005	0.031	0.007	0.004	0.004
PO_4^{3-}		0.55	0.07	0.06	0.13	0.54	0.11	0.07	0.11
NH_4^+		0.05	0.02	0.02	0.02	0.66	0.11	0.10	0.03
Fe		0.24	0.10	0.11	0.06	0.60	0.15	0.22	0.07
Mn		0.30	0.08	0.04	0.03	0.50	0.10	0.07	0.06
Pb		0.001	0.002	0.002	0.001	0.001	0.002	0.003	0.006

4.3. Relationships between I^-/NH_4^+ and Other Components in Groundwater Unit A in the Residual Datasets

Correlations between groundwater I^-/NH_4^+ and other components in unit A were investigated because of the higher NBL-I^- and NBL-NH_4^+ in groundwater unit A than those in the other units (Figure 2). As shown in Table 2, in the residual dataset of unit A, the groundwater I^- concentrations had significantly positive correlations with concentrations of Na^+, Ca^{2+}, Mg^{2+}, HCO_3^-, SO_4^{2-}, Cl^-, TDS, NH_4^+, Fe, and Mn, but negative correlations with values of DO and NO_3^-. By contrast, groundwater NH_4^+ concentrations in the residual dataset of unit A had positive correlations with concentrations of NO_2^- and Fe (Table 2). The relationships between I^-/NH_4^+ and other components in groundwater unit A were further investigated by the PCA. As shown in Table 3, in the residual dataset for NBL-I^-, 17 variables were combined to produce 5 significant PCs with eigenvalues > 1, which explained 85.15% of the variance in the hydrochemical dataset. Both PC1 and PC3 were related to I^-. Specifically, the PC1 had strong positive loadings with Cl^-, Na^+, Mg^{2+}, COD, and TDS; a moderate positive loading with HCO_3^-; a weak positive loading with I^-; and a weak negative loading with DO. The PC3 demonstrated strong positive loadings with NH_4^+ and Fe, a moderate positive loading with I^-, and weak negative loadings with NO_3^- and DO. By contrast, in the residual dataset for NBL-NH_4^+, four PCs with eigenvalues > 1 were extracted from the PCA, and the cumulative variance by the four PCs was 80.55% (Table 3). The PC3 was related to NH_4^+. It had strong positive loadings with NH_4^+ and Fe, and a weak negative loading with NO_3^-.

Table 2. Correlation coefficients of concentrations of I^-/NH_4^+ and other chemicals in groundwater unit A.

NBL Datasets	I^-	NH_4^+	NBL Datasets	I^-	NH_4^+
pH	0.283	0.334	NO_3^-	−0.369 *	−0.371
DO	−0.412 *	−0.068	PO_4^{3-}	−0.001	0.043
Eh	−0.236	−0.222	Pb	−0.181	−0.060
K^+	0.173	−0.049	NO_2^-	−0.090	0.610 **
Na^+	0.480 **	0.257	COD	0.309	0.356
Ca^{2+}	0.518 **	0.305	Fe	0.386 *	0.844 **
Mg^{2+}	0.566 **	0.362	TDS	0.588 **	0.314
HCO_3^-	0.561 **	0.288	Mn	0.388 *	0.042
Cl^-	0.471 **	0.251	NH_4^+	0.514 **	1
SO_4^{2-}	0.078	0.367	I^-	1	0.064

*: significance level at $p < 0.05$; **: significance level at $p < 0.01$.

Table 3. Principal component (PC) loadings for groundwater physicochemical parameters in unit A in the residual datasets.

Physicochemical Parameters	PCs—Residual Dataset for I^-					Physicochemical Parameters	PCs—Residual Dataset for NH_4^+			
	PC1	PC2	PC3	PC4	PC5		PC1	PC2	PC3	PC4
Cl^-	***0.974***	0.052	−0.024	0.018	0.014	Cl^-	***0.969***	0.097	0.052	0.082
Na^+	***0.973***	0.129	−0.030	0.012	0.011	Na^+	***0.965***	0.150	0.046	0.029
Mg^{2+}	***0.840***	0.458	0.088	0.182	−0.034	Mg^{2+}	***0.856***	0.424	0.152	0.000
COD	***0.820***	0.214	−0.012	0.102	0.318	COD	***0.851***	0.347	0.181	0.146
TDS	***0.720***	0.616	0.149	0.181	0.133	TDS	***0.721***	0.671	0.072	−0.001
Ca^{2+}	0.417	***0.807***	0.159	0.253	0.101	I^-	***0.567***	−0.035	−0.062	−0.074
SO_4^{2-}	0.035	***0.791***	−0.277	−0.109	0.341	DO	***−0.482***	−0.370	0.036	0.419
Eh	−0.172	***−0.767***	−0.067	0.353	0.343	SO_4^{2-}	−0.042	***0.876***	0.217	−0.139
HCO_3^-	0.504	***0.738***	0.238	0.245	0.110	pH	0.119	***0.871***	0.222	0.086
pH	0.137	***0.714***	0.038	0.424	0.304	Ca^{2+}	0.395	***0.862***	0.093	−0.064
NO_3^-	−0.152	***−0.484***	−0.351	−0.168	−0.065	HCO_3^-	0.507	***0.815***	0.074	−0.051
DO	−0.455	***−0.472***	−0.353	0.348	−0.178	K^+	0.135	***0.691***	−0.204	0.430
NH_4^+	−0.040	0.026	***0.955***	0.130	0.048	NO_3^-	−0.161	***−0.431***	−0.415	0.069
Fe	−0.069	0.020	***0.925***	−0.012	0.207	Fe	−0.101	0.059	***0.939***	0.026
I^-	0.476	0.250	***0.573***	0.211	−0.188	NH_4^+	0.187	0.149	***0.932***	−0.041
Mn	0.189	0.143	0.197	***0.902***	0.010	Mn	0.205	0.283	0.009	***0.829***
K^+	0.173	0.225	0.219	0.021	***0.868***	Eh	−0.186	−0.515	−0.055	***0.732***
Eigenvalue	4.80	4.17	2.63	1.54	1.34		5.00	4.89	2.15	1.66
Explained variance (%)	28.23	24.51	15.46	9.07	7.87		29.40	28.75	12.64	9.76
Cumulative % of variance	28.23	52.75	68.21	77.28	85.15		29.40	58.15	70.79	80.55

Bold and italic numbers—maximum absolute PC loading of one parameter.

4.4. *Effectiveness of the Used Method*

In original datasets, the differences between the maximum values of groundwater I^- and NH_4^+ at different land-use types were significant (Table S1). For example, the maximum value of I^- in unit A at UA was 3 times that at AA, and the maximum values of I^- in unit B at UA and PUA were more than 2 times that at RA (Table S1). Similarly, the maximum values of NH_4^+ in unit B at UA and PUA were more than 2 times that at AA and more than 10 times that at RA (Table S1). These findings indicate that groundwater I^- and NH_4^+ concentrations at various land-use types were influenced by human activities.

By contrast, if the used method has utility, in residual datasets, the differences between the maximum values of groundwater I^- and NH_4^+ at different land-use types within a same unit would be insignificant in comparison with those in the original datasets because groundwaters impacted by anthropogenic inputs are excluded via the method used [8,14].

Correspondingly, in the residual datasets, the differences between maximum values of I^- at various land-use types in both the A and B units were all within 0.4 times. Similarly, the differences between maximum values of NH_4^+ at various land-use types in both the A and B units were all within 0.8 times (except RA in unit B). These findings indicate that the differences of maximum values at various land-use types in residual datasets are more insignificant than those in the original datasets and that human activities have little impact on groundwater I^- and NH_4^+ concentrations at different land-use types in residual datasets in comparison with those in the original datasets. Therefore, the preselection method used in this study is effective in evaluating the NBLs of groundwater I^- and NH_4^+ in urbanized areas.

5. Discussion

In this study, we focused on factors controlling groundwater NBLs of I^-/NH_4^+ in unit A rather than in other units due to the much higher NBLs-I^-/NH_4^+ in groundwater unit A relative to those in the other units (Figure 2).

5.1. Main Sources for NBLs-I^-/NH_4^+ in Groundwater Unit A

Generally, after the removal of anthropogenic sources, chemical components in groundwater are from the precipitation, the natural release of sediments in vadose zones and aquifers, and the interaction with other water bodies [21]. Firstly, the iodine in rainfall is commonly at low levels of <0.01 mg/L and is less than one-tenth of groundwater NBL-I^- in unit A (Figure 2) [22]. This indicates that the precipitation has little contribution to the higher NBL-I^- in groundwater unit A compared to other units. By contrast, the precipitation may be an important source for groundwater NBL-NH_4^+ because nitrogen deposition is one of the major components in the rainfall in the PRD [23]. However, in the residual datasets, groundwaters with high levels of NH_4^+ (>0.2 mg/L) were distributed in areas where acid rain pH is relatively high and acid rain frequency is relatively low (representing low nitrogen deposition) (Figure 3). Thus, the contribution of the precipitation for higher NBL-NH_4^+ in groundwater unit A compared to that in other units is also negligible. Secondly, the natural release from the iodine/nitrogen-rich organic matter in the vadose zone is likely to be the major source for the higher NBLs-I^-/NH_4^+ in groundwater unit A compared to those in other units because the vadose zone overlaying aquifer-I consists of marine sediments where organic-iodine and organic-nitrogen are abundant and the mineralization of organic-iodine/nitrogen results in the release of I^-/NH_4^+, whereas iodine/nitrogen-rich marine sediments are missing in the vadose zones overlaying other aquifers [24,25]. Thirdly, the natural release from iodine-rich minerals (e.g., iodine-loaded Fe/Mn oxides) in aquifer sediments is also probably another major contributor to the higher NBLs-I^-/NH_4^+ in groundwater unit A compared to those in the other units because Fe/Mn (oxyhydr) oxides are often the primary reservoirs for coastal sediment iodine owing to their high adsorption capacities for iodine [26], and I^- concentration was positively correlated with concentrations of Fe and Mn in groundwater unit A (Table 2). Fourthly, the interaction with other water bodies may be another important source for shallow groundwater I^-/NH_4^+ in unit A. This is can be explained by two phenomena. One is that deep groundwater in unit A is often enriched with high concentrations of geogenic I^-/NH_4^+ and salinity [24], and these I/NH_4-rich deep groundwaters in coastal areas are sometimes pumped out for farming marine animals (e.g., shrimp), thereby contaminating the shallow groundwater via infiltration [12]. Another is that groundwaters with high levels of NH_4^+ (>0.2 mg/L) in the residual datasets were distributed in the river network areas where surface water was enriched with NH_4^+ (Figure 3) [11], and the interaction between shallow groundwater and surface water often occurs in river network areas of the PRD [12]. By contrast, groundwaters with high levels of I^- (>0.08 mg/L) in residual the datasets were independent of the river network area (Figure 4), and uncontaminated surface water was commonly at low levels of I^- (<0.01 mg/L) [12]. Furthermore, other natural water bodies (e.g., sea water) that interact with the shallow groundwater of the PRD are generally poor in

I^- (<0.02 mg/L) and NH_4^+ (<0.1 mg/L) [10,12]. This suggests that I-rich deep groundwater is a more important source for the higher NBLs-I^- in groundwater unit A relative to other units, but it is not the main one; meanwhile, NH_4-rich deep groundwater and surface water are two important sources for the higher NBL-NH_4^+ in groundwater unit A compared with other units.

5.2. Main Driving Forces for NBLs-I^-/NH_4^+ in Groundwater Unit A

A few studies reported that the mineralization of organic matter in marine sediments is one of the major factors controlling the occurrence of I/NH_4-rich shallow groundwater in the PRD [10,11]. This is also probably the major geogenic force for the higher NBLs-I^-/NH_4^+ in groundwater unit A compared with the other units, which can be explained by the following. Firstly, as mentioned above, the Quaternary marine formation overlaying aquifer-I is enriched with organic matter such as organic-iodine and organic-nitrogen, whereas the Quaternary marine formation is missing in other units (Figure 1) [24]. Secondly, groundwater unit A was shown to be a more reducing environment in both the vadose zone and aquifer in comparison with other units because the vadose zone in unit A consists of marine sediments dominated by silt/clay with low permeability, whereas the vadose zones in other units are dominated by coarser media (e.g., sandy clay, sand) with higher permeability. Similarly, reduction indicators (e.g., NO_2^-, Fe, and Mn) in aquifer-I were shown to have higher concentrations than in the other aquifers, and oxidation indicators (e.g., Eh, DO, and NO_3^-) in aquifer-I had lower values than those in the other aquifers (Table 1). Thirdly, Cl^-, Na^+, Mg^{2+}, TDS, NH_4^+, Fe, HCO_3^-, and I^- in groundwater unit A had the same source or similar geochemical behaviors because the PC1 had positive loadings of Cl^-, Na^+, Mg^{2+}, TDS, HCO_3^-, and I^-, while PC3 had positive loadings of NH_4^+, Fe, and I^- (Table 3). Similarly, NH_4^+ and Fe in groundwater unit A had the same source or similar geochemical behaviors because the PC1 had positive loadings of NH_4^+ and Fe (Table 3). Fourthly, fine-grained marine sediments in Asian deltas often contain trapped seawater enriched with Cl^-, Na, Mg, and TDS [27]. Therefore, the major driving forces controlling higher NBLs-I^-/NH_4^+ in groundwater unit A compared to other units can be outlined as follows. In a reducing environment, HCO_3^- and soluble I^- result from the mineralization of organic-iodine in the Quaternary marine formation, and then Cl^-, Na^+, and Mg^{2+} in trapped seawater accompanied with HCO_3^- and soluble I^- in the marine formation enter into aquifer-I via water flow. This induces the reductive dissolution of I-loaded Fe (oxyhydr) oxides and is accompanied by the release of I^- in aquifer sediments due to the influx of water with reducing conditions. Finally, groundwater in unit A is enriched with I^- and related components, such as Cl^-, Na^+, Mg^{2+}, Fe, and HCO_3^- (Table 1). This is also supported by the evidence that groundwater I^- concentration was positively correlated with concentrations of Cl^-, Na^+, Mg^{2+}, Fe, and HCO_3^- (Table 2). Similarly, the mineralization of organic-nitrogen in the marine formation also leads to an increase in soluble NH_4^+, which enters into aquifer-I and is accompanied by the infiltrated water with reducing conditions. This triggers the reductive dissolution of Fe (oxyhydr) oxides and the denitrification in aquifer sediments because the PC3 has positive loading with NH_4^+ but negative loading with NO_3^- (Table 3). Finally, groundwater in unit A is enriched with NH_4^+ and Fe but is NO_3^- poor (Table 1).

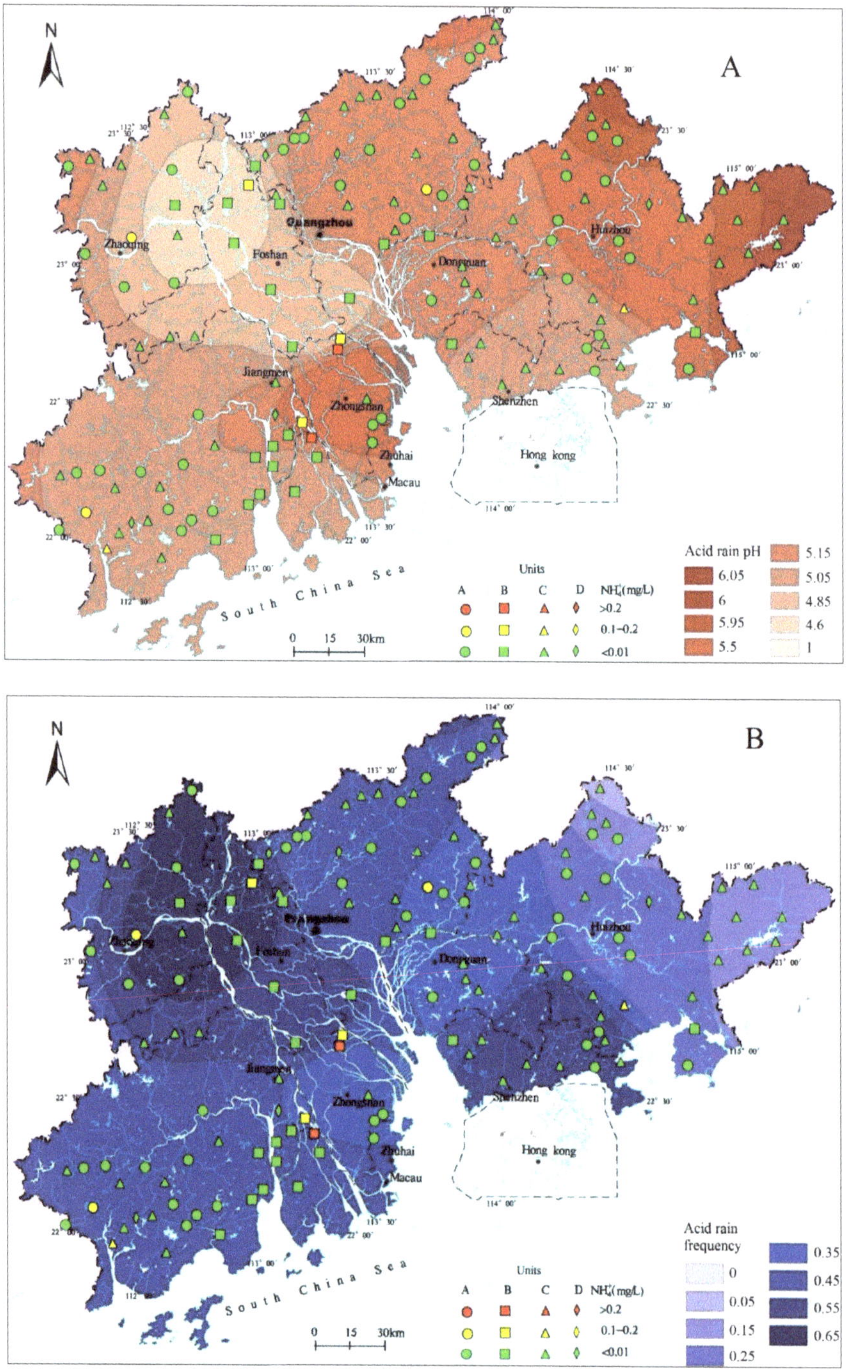

Figure 3. Distribution of NH_4^+ in various groundwater units in residual datasets and acid rain data for the Pearl River Delta (data of acid rain from Liu et al. [23]). (**A**) is acid rain pH, and (**B**) is acid rain frequency.

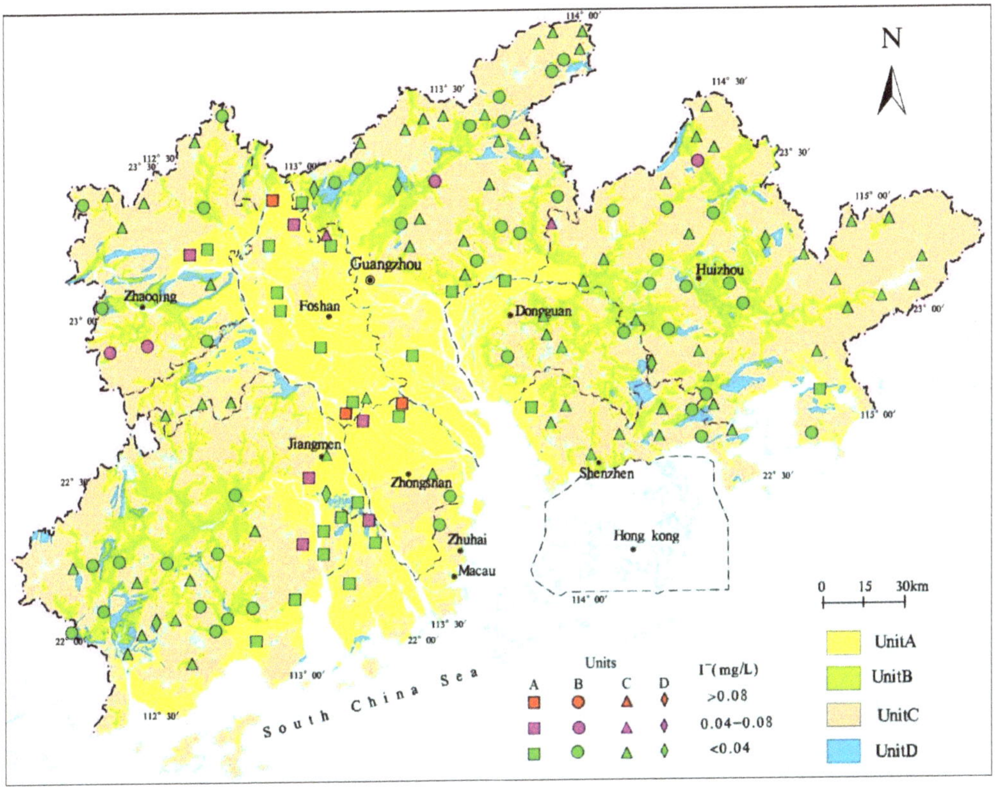

Figure 4. Distribution of I^- in various groundwater units in the residual datasets.

6. Conclusions

The NBLs of I^- and NH_4^+ in the shallow groundwater of the PRD was assessed by a preselection method with Grubbs' test. Here, the preselection method consisted of Cl/Br mass ratios versus Cl concentrations and the oxidation capacity. In the evaluation of four groundwater units, NBL-I^- in groundwater unit A was 0.14 mg/L, which is nearly 2 times greater than the allowable limit of I^- in groundwater recommended by the Chinese government, and was 2.5 times, 3.3 times, and 5.6 times higher than those in groundwater units B to D, respectively. The NBL-NH_4^+ in groundwater unit A was 0.32 mg/L, which was also 2 times greater than that in the other groundwater units.

The natural release of I^- from both of organic matter in the vadose zone and iodine-rich minerals (e.g., iodine-loaded Fe/Mn oxides) in aquifer sediments are two major sources for the higher NBLs-I^- in groundwater unit A compared to other units. Additionally, I-rich deep groundwater is another important source but is not the main one. Rather, the occurrence of NH_4^+ from organic matter in the vadose zone is the major source for the higher NBLs-NH_4^+ in groundwater unit A compared to other units. In addition, NH_4-rich deep groundwater and surface water are two other important sources for the higher NBL-NH_4^+ in groundwater unit A.

Soluble I^- results from the mineralization of organic-iodine in the Quaternary marine formation, and the release of I^- accompanied by reductive dissolution of I^--loaded Fe (oxyhydr) oxides in aquifer sediments is the main driving force controlling the higher NBL-I^- in groundwater unit A relative to other units. By contrast, soluble NH_4^+ resulting from the mineralization of organic-nitrogen in marine formation entering into aquifer-I is the main driving force controlling the higher NBL-NH_4^+ in groundwater unit A relative to other units.

In the future, we will use these NBLs to evaluate the I^- and NH_4^+ pollution status in groundwater of the PRD and to provide suggestions for the prevention of groundwater I^- and NH_4^+ pollution. In addition, we will monitor groundwater quality once again and verify these NBLs by using new hydrochemical data.

Supplementary Materials: The following supporting information can be downloaded at: https://www.mdpi.com/article/10.3390/w14223737/s1, Table S1: Maximum concentrations of groundwater I- and NH4+ in units A and B at different land use in original and residual datasets; Figure S1: Procedure of the NBL assessment from [8]; Figure S2: Cl/Br mass ratios and candidate contaminated-markers in various groundwater units in the Pearl River Delta was grouped by the hierarchical cluster analysis (Ward's method and Euclidean distance); Figure S3: Plots of Cl/Br mass ratios versus Cl concentrations combining with concentrations of contaminated-markers in various groundwater units in the Pearl River Delta [8]; Figure S4: Concentrations of NO_3 and SO_4 as well as oxidation capacity in groundwaters in CBM datasets [8].

Author Contributions: Methodology, L.P.; software, X.L. (Xin Lu); validation, L.P. and X.L. (Xiwen Li); formal analysis, M.Z. and H.W.; investigation, L.P.; data curation, L.P; writing—original draft preparation, L.P.; writing—review and editing, X.L. (Xiwen Li); visualization, X.L. (Xin Lu) and M.Z.; supervision, L.P.; project administration, L.P.; funding acquisition, X.L. (Xiwen Li). All authors have read and agreed to the published version of the manuscript.

Funding: This research was supported by the China Geological Survey (grant no. ZD20220209) and Guangxi Karst Resources and Environment Research Center of Engineering Technology funded project (KFKT2022001).

Institutional Review Board Statement: Not applicable.

Informed Consent Statement: Not applicable.

Data Availability Statement: The datasets generated and/or analyzed during the current study are not publicly available.

Conflicts of Interest: The authors declare no conflict of interest.

References

1. Martinez, M.L.; Intralawan, A.; Vazquez, G.; Pérez-Maqueo, O.; Sutton, P.; Landgrave, R. The coasts of our world: Ecological, economic and social importance. *Ecol. Econ.* **2007**, *63*, 254–272. [CrossRef]
2. Khosravi, K.; Barzegar, R.; Golkarian, A.; Busico, G.; Cuoco, E.; Mastrocicco, M.; Colombani, N.; Tedesco, D.; Ntona, M.M.; Kazakis, N. Predictive modeling of selected trace elements in groundwater using hybrid algorithms of iterative classifier optimizer. *J. Contam. Hydrol.* **2021**, *242*, 103849. [CrossRef] [PubMed]
3. Fu, T.; Qi, C.; Wang, Z.; Li, C.; Liu, W.; Fu, Y.; Chen, G.; Su, Q.; Xu, X.; Yu, H. Hydrochemical characteristics and quality assessment of groundwater under the impact of seawater intrusion and anthropogenic activity in the coastal areas of Zhejiang and Fujian provinces, China. *Lithosphere* **2022**, *2022*, 1394857. [CrossRef]
4. Ntona, M.M.; Busico, G.; Mastrocicco, M.; Kazakis, N. Modeling groundwater and surface water interaction: An overview of current status and future challenges. *Sci. Total Environ.* **2022**, *846*, 157355. [CrossRef] [PubMed]
5. Huang, G.; Han, D.; Song, J.; Li, L.; Pei, L. A sharp contrasting occurrence of iron-rich groundwater in the Pearl River Delta during the past dozen years (2006–2018): The genesis and mitigation effect. *Sci. Total Environ.* **2022**, *829*, 154676. [CrossRef]
6. Zhang, F.; Huang, G.; Hou, Q.; Liu, C.; Zhang, Y.; Zhang, Q. Groundwater quality in the Pearl River Delta after the rapid expansion of industrialization and urbanization: Distributions, main impact indicators, and driving forces. *J. Hydrol.* **2019**, *577*, 124004. [CrossRef]
7. Griffioen, J.; Passier, H.F.; Klein, J. Comparison of selection methods to deduce natural background levels for groundwater units. *Environ. Sci. Technol.* **2008**, *42*, 4863–4869. [CrossRef]
8. Huang, G.; Pei, L.; Li, L.; Liu, C. Natural background levels in groundwater in the Pearl River Delta after the rapid expansion of urbanization: A new pre-selection method. *Sci. Total Environ.* **2022**, *813*, 151890. [CrossRef]
9. Bi, P.; Huang, G.; Liu, C.; Li, L. Geochemical factors controlling natural background levels of phosphate in various groundwater units in a large-scale urbanized area. *J. Hydrol.* **2022**, *608*, 127594. [CrossRef]
10. Huang, G.; Liu, C.; Li, L.; Zhang, F.; Chen, Z. Spatial distribution and origin of shallow groundwater iodide in a rapidly urbanized delta: A case study of the Pearl River Delta. *J. Hydrol.* **2020**, *585*, 124860. [CrossRef]

11. Zhang, M.; Huang, G.; Liu, C.; Zhang, Y.; Chen, Z.; Wang, J. Distributions and origins of nitrate, nitrite, and ammonium in various aquifers in an urbanized coastal area, South China. *J. Hydrol.* **2020**, *582*, 124528. [CrossRef]
12. Sun, J.; Jing, J.; Huang, G.; Liu, J.; Chen, X.; Zhang, Y. Report on the investigation and assessment of groundwater contamination in the Pearl River delta area. In *Chinese Academy of Geological Sciences*; The Institute of Hydrogeology and Environmental Geology: Shijiazhuang, China, 2009.
13. Huang, G.; Liu, C.; Sun, J.; Zhang, M.; Jing, J.; Li, L. A regional scale investigation on factors controlling the groundwater chemistry of various aquifers in a rapidly urbanized area: A case study of the Pearl River Delta. *Sci. Total Environ.* **2018**, *625*, 510–518. [CrossRef]
14. Huang, G.; Song, J.; Han, D.; Liu, R.; Liu, C.; Hou, Q. Assessing natural background levels of geogenic contaminants in groundwater of an urbanized delta through removal of groundwaters impacted by anthropogenic inputs: New insights into driving factors. *Sci. Total Environ.* **2022**, *857*, 159527. [CrossRef]
15. Zong, Y.; Yim, W.W.S.; Yu, F.; Huang, G. Late Quaternary environmental changes in the Pearl River mouth region, China. *Quat. Int.* **2009**, *206*, 35–45. [CrossRef]
16. Huang, G.; Zhang, M.; Liu, C.; Li, L.; Chen, Z. Heavy metal(loid)s and organic contaminants in groundwater in the Pearl River Delta that has undergone three decades of urbanization and industrialization: Distributions, sources, and driving forces. *Sci. Total Environ.* **2018**, *635*, 913–925. [CrossRef]
17. Huang, G.; Liu, C.; Zhang, Y.; Chen, Z. Groundwater is important for the geochemical cycling of phosphorus in rapidly urbanized areas: A case study in the Pearl River Delta. *Environ. Pollut.* **2020**, *260*, 114079. [CrossRef]
18. Huang, G.; Sun, J.; Zhang, Y.; Chen, Z.; Liu, F. Impact of anthropogenic and natural processes on the evolution of groundwater chemistry in a rapidly urbanized coastal area, South China. *Sci. Total Environ.* **2013**, *463*, 209–221. [CrossRef]
19. Huang, G.; Chen, Z.; Liu, F.; Sun, J.; Wang, J. Impact of human activity and natural processes on groundwater arsenic in an urbanized area (South China) using multivariate statistical techniques. *Environ. Sci. Pollut. Res.* **2014**, *21*, 13043–13054. [CrossRef]
20. Hou, Q.; Zhang, Q.; Huang, G.; Liu, C.; Zhang, Y. Elevated manganese concentrations in shallow groundwater of various aquifers in a rapidly urbanized delta, south China. *Sci. Total Environ.* **2020**, *701*, 134777. [CrossRef]
21. Gao, Y.; Qian, H.; Huo, C.; Chen, J.; Wang, H. Assessing natural background levels in shallow groundwater in a large semiarid Drainage Basin. *J. Hydrol.* **2020**, *584*, 124638. [CrossRef]
22. Neal, C.; Neal, M.; Wickham, H.; Hill, L.; Harman, S. Dissolved iodine in rainfall, cloud, stream and groundwater in the Plynlimon area of mid-Wales. *Hydrol. Earth Syst. Sci.* **2007**, *11*, 283–293. [CrossRef]
23. Liu, F.; Sun, J.; Wang, J.; Zhang, Y. Groundwater acidification in shallow aquifers in Pearl River Delta, China: Distribution, factors, and effects. *Geochem. J.* **2017**, *51*, 373–384. [CrossRef]
24. Jiao, J.J.; Wang, Y.; Cherry, J.A.; Wang, X.; Zhi, B.; Du, H.; Wen, D. Abnormally high ammonium of natural origin in a coastal aquifer-aquitard system in the Pearl River Delta, China. *Environ. Sci. Technol.* **2010**, *44*, 7470–7475. [CrossRef] [PubMed]
25. Wang, X.; Jiao, J.J.; Wang, Y.; Cherry, J.A.; Kuang, X.; Liu, K.; Lee, C.; Gong, Z. Accumulation and transport of ammonium in aquitards in the Pearl River Delta (China) in the last 10,000 years: Conceptual and numerical models. *Hydrogeol. J.* **2013**, *21*, 961–976. [CrossRef]
26. Li, J.; Zhou, H.; Qian, K.; Xie, X.; Xue, X.; Yang, Y.; Wang, Y. Fluoride and iodine enrichment in groundwater of North China Plain: Evidences from speciation analysis and geochemical modeling. *Sci. Total Environ.* **2017**, *598*, 239–248. [CrossRef]
27. Larsen, F.; Tran, L.V.; Van Hoang, H.; Tran, L.T.; Christiansen, A.V.; Pham, N.Q. Groundwater salinity influenced by Holocene seawater trapped in incised valleys in the Red River delta plain. *Nat. Geosci.* **2017**, *10*, 376–381. [CrossRef]

water

MDPI

Article

Anthropogenic Influences on the Hydrochemical Characteristics of the Groundwater in Xiamen City, China and Their Evolution

Zhenghong Li [1,2], Jianfeng Li [1,2,*], Yuchen Zhu [1,2,*], Yasong Li [1,2] and Qichen Hao [1,2]

1 Fujian Provincial Key Laboratory of Water Cycling and Eco-Geological Processes, Xiamen 361000, China
2 Institute of Hydrogeology and Environmental Geology, CAGS, Shijiazhuang 050061, China
* Correspondence: lijianfeng@mail.cgs.gov.cn (J.L.); zhuyuchen@mail.cgs.gov.cn (Y.Z.)

Abstract: This study analyzed the anthropogenic influences on the hydrochemical composition characteristics of the groundwater in Xiamen City, Fujian province, China, and their evolution. Based on the hydrochemical data of the groundwater of 1993 and 2019–2021, this study identified the indices of the anthropogenic influences using mathematical and statistical analysis methods, such as contrast coefficient, standard deviation, and Mahalanobis distance. The analytical results are summarized as follows: (1) the number of the indices affecting the groundwater quality in Xiamen increased from nine in 1993 to 15 in 2019, and the six increased indicators included NO_3^-, Pb, NH_4^+, Al^{3+}, NO_2^- and Cu (the contribution rates to poor-quality were 26.0%, 16.3%, 10.6%, 4.1%, 0.8% and 0.8%, respectively) which were related to the input of human activities. During this period, the number of hydrochemical types increased from 19 in 1993 to 28 in 2019, with a decrease in the water of the HCO_3 type and an increase in the water of Cl and SO_4 types; (2) In 2019, NO_3^- had higher content than SO_4^{2-} in the groundwater and became a major anion, forming the water of NO_3 type; (2) as indicated by the analytical results obtained using the Mahalanobis distance method, areas with strong anthropogenic influences include densely populated areas and areas with intensively distributed industrial enterprises, while anthropogenic influences are very weak in the northern forest land area.

Keywords: evolution of hydrochemical characteristics; identification of hydrochemical anomalies; human activities; influence degree; Mahalanobis distance

Citation: Li, Z.; Li, J.; Zhu, Y.; Li, Y.; Hao, Q. Anthropogenic Influences on the Hydrochemical Characteristics of the Groundwater in Xiamen City, China and Their Evolution. *Water* **2022**, *14*, 3377. https://doi.org/10.3390/w14213377

Academic Editors: Guanxing Huang and Liangping Li

Received: 14 September 2022
Accepted: 22 October 2022
Published: 25 October 2022

1. Introduction

Safe drinking water, which is closely related to human health, food security, and the orderly development of the social economy, is essential for the sustainable development of a city [1,2]. With the acceleration of urbanization, the area of urban built-up areas and the number and density of the urban population have increased dramatically, leading to a sharp increase in the demand for water resources [3–5]. Groundwater is the main freshwater source for industrial and agricultural production, domestic drinking, and ecological water consumption and is the only source of water supply in arid and semi-arid regions [6,7]. The groundwater quality directly affects the quality of the human living environment and human health, and the groundwater environment is controlled by various factors such as meteorology, hydrogeology, and human activities [8–12]. Urbanization in China presents a sustainable and rapid growing trend, with the urbanization rate increasing from 18.57% in 1978 to 58.52% in 2017 [13]. Moreover, the ability of human beings to transform nature is continuously enhanced, and large-scale human activities have strong influences on climate [14–16], ecological environment [17,18], soil environment [19], and groundwater environment [20–22]. Many cities around the world, especially those in arid and semi-arid regions, are facing dual water stress exerted by groundwater shortage and the deterioration of water quality [23,24].

In 1984, the Commission on Groundwater Quality under the International Association of Hydrogeologists held a symposium entitled *Impact of Agriculture on Groundwater* in

Ireland [25]. The international hydrological community has paid increasing attention to and studied the influence of human activities on the groundwater environment [24,26–31]. When the groundwater environment changes under anthropogenic influences, the hydrochemical characteristics of groundwater will also change accordingly, i.e., hydrochemical anomalies occur [32]. Since various geochemical indices frequently coexist in complex geological systems through mutual influences and constraints, the synergistic effects of various geochemical indices should be fully considered during the identification of geochemical anomalies, aiming to reduce the uncertainty of their identification performance. The multi-element geochemical anomalies can be presently identified using multiple methods, such as the hydrochemical diagrams (Durov diagram, ion ratio plot, et al.) method [33], coefficient of variation [34], affine equivariant [35], and multivariate statistical method [28]. The Mahalanobis distance is a preferred composite index used to identify multi-element geochemical anomalies. This index has been widely used in geochemical fields including geochemical prospecting and metallogenic prediction [36,37], the evaluation of geochemical anomalies of hydrocarbons [38], and the identification of geochemical anomalies of sediments [39]. However, it is rarely used in the identification of hydrochemical anomalies of groundwater.

Although Xiamen has abundant rainfall, it still suffers from a severe shortage of freshwater resources, which is a key problem restricting the sustainable economic development of the city. Groundwater accounts for only about 6% of Xiamen's urban water supply [40], indicating great potential for groundwater exploitation. Moreover, groundwater is an important supplementary and backup emergency water source of Xiamen and its quality directly affects the emergency water supply ability and water supply safety of the city. This study identified the indices of anthropogenic influences using standard deviations, calculated the contrast coefficients of these indices and the Mahalanobis distances between these contrast coefficients, and determined the intensity of anthropogenic influences on the evolution of hydrochemical characteristics of the groundwater. This study qualitatively analyzed the anthropogenic influence on the changes in the groundwater chemistry of the study area since only 172 samples were collected from the large study area. The results of this study will provide some basis for the protection of groundwater and guide the management of groundwater resources to a certain extent.

2. Overview of the Study Area

2.1. Physical Geography and Social Economy

Xiamen is located in southeastern China, with geographical coordinates of 117°53′–118°26′ E and 24°23′–24°54′ N. This city lies in the central portion of the west coast of the Taiwan Strait and the center of the Golden Triangle of Southern Fujian Province. It covers a total area of 1700.61 km^2, including approximately 390 km^2 of seas, with a total coastline length of approximately 234 km. Its administrative region consists of land, seas, and islands. It generally dips to SE, with high and low hills, terraces, marine plains, and tidal flats distributed from NW to SE. Moreover, the Xiamen Island, the Gulangyu Islet, and seas lie in its south. Xiamen has a mild and humid subtropical marine monsoon climate, with an average multiyear temperature of 20.9 °C and average multiyear precipitation of 1100–2000 mm, which increases from SE to NW. Xiamen has developed water systems, a dense dendritic river network, and a runoff direction from NW to SE. The rivers in the city feature small runoff, short flows, narrow channels, and shallow riverbeds. Their water quantity varies greatly with seasons and has low sediment content. Xiamen enjoys abundant marine biological resources but lacks metal mineral resources and freshwater resources.

2.2. Geological and Hydrogeological Conditions

Xiamen is located in the coastal metamorphic zone of southeastern Fujian and is at the intersection of the large EW-trending Nanjing-Xiamen fault and the NES-trending Changdong-Nan'ao deep-seated fault. The exposed strata in the study area consist of the Upper Triassic Wenbinshan Formation, the Lower Jurassic Lishan and Fankeng formations, the Upper Jurassic Nanyuan Formation, and Quaternary strata including the Middle Pleistocene diluvium and alluvial-proluvial strata, the Upper Pleistocene alluvial-proluvial and marine strata, the Holocene alluvial-proluvial and marine strata, and the Quaternary eluvium.

The groundwater in Xiamen mainly occurs in aquifers of unconsolidated rock pores, weathered eluvial pores and fractures, and bedrock fissures (Figure 1). The pore water of unconsolidated rocks is hosted in alluvial-proluvial and marine deposits. The pore water in alluvial-proluvial deposits is distributed in the terraces on both sides of valleys and in piedmont proluvial fans. Its aquifers consist of sands, gravels, and pebbles and have a thickness of 3–9 m and a specific yield of <200 $m^3/d{\cdot}m$ [41]. The depth of water level is 0.5–5.0 m. The pore water in marine deposits is distributed in the coastal and estuarine areas. Its aquifers consist of silty sands and medium-coarse-grained sands and have a small thickness. The pore-fissure water in weathered eluvial deposits is distributed in eluvial platforms and at piedmont slope toes. Its aquifers consist of detritus and breccias and generally have a thickness of 10–20 m and a specific yield of <10 $m^3/d{\cdot}m$. The bedrock fissure water is distributed in the fissures and faults of bedrock in mountainous areas, with the water yield controlled by structural faults. Its water content is extremely uneven, with a specific yield of <10 $m^3/d{\cdot}m$. The groundwater in Xiamen is primarily recharged by vertical infiltration of atmospheric precipitation and agricultural irrigation as well as the lateral flow of groundwater in bedrock mountainous areas during the rainy season. The annual average natural replenishment resource is 1.95×10^3 m^3/a. The regional groundwater flows from the NW and NE to the southeast coast overall.

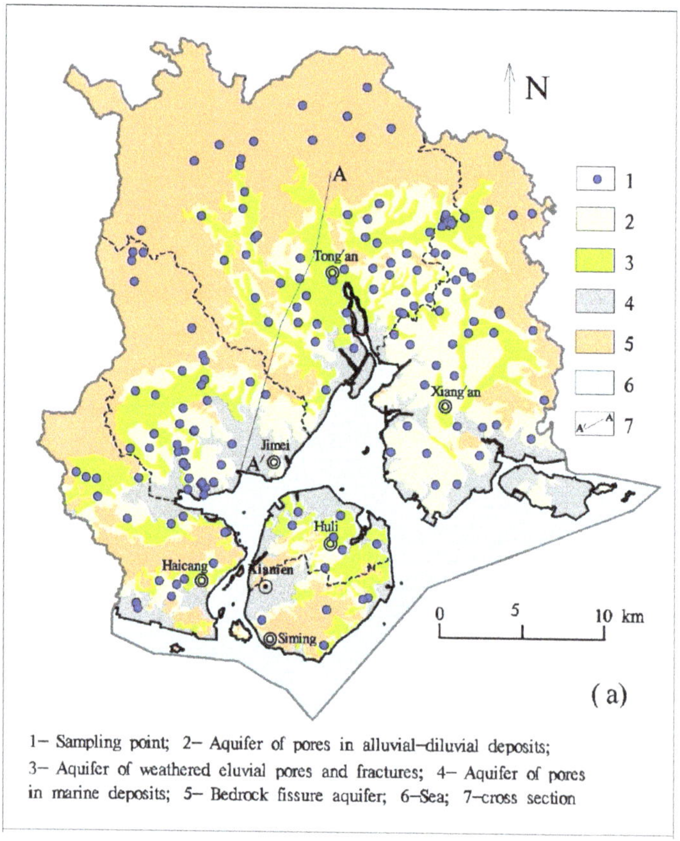

Figure 1. *Cont.*

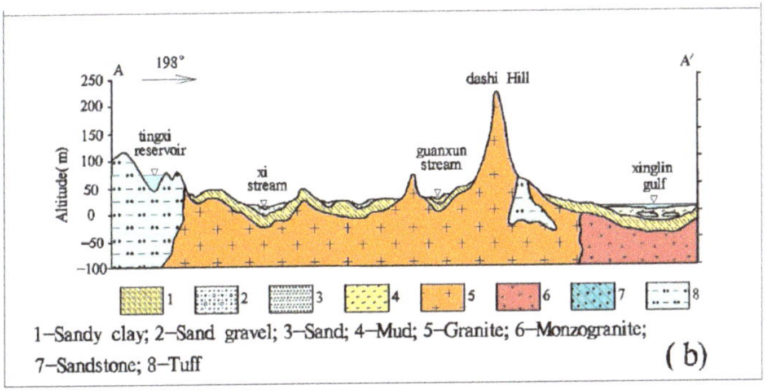

Figure 1. Hydrogeological conditions and sampling points in Xiamen, (**a**) Hydrogeological conditions and sampling points; (**b**) Typical hydrogeological section.

3. Materials and Methods

3.1. Data Sources

Two sets of hydrochemical data were used in this study. One set was the historical hydrochemical data of 1993 obtained from the study results of the study area. It included the data on 84 groundwater samples (19 samples of bedrock fissure water, 27 pore-fissure water samples, and 38 pore water samples), detailing the locations of sampling points, types of water-bearing media, hydrochemical types, pH, COD, the total dissolved solids (TDS) content, and the contents of HCO_3^-, SO_4^{2-}, Cl^-, K^++Na^+, Mg^{2+}, NH_4^+, NO_2^-, NO_3^-, F^-, Fe, Al^{3+}, Cu, Pb, Zn, As, and Cr^{6+}. Unfortunately, the Ca^{2+} content was not recorded.

The other set of original data was 2019–2021 hydrochemical data of groundwater samples surveyed by the project team of this study. These samples were mostly collected from large-diameter wells or small pumping wells in villages and small towns, and 75% of the wells had a depth less than 50 m. The data on 172 groundwater samples were obtained in total, including 58 samples of bedrock fissure water, 72 pore-fissure water samples, and 42 pore water samples. Regarding sampling densities, seven and 8–18 samples were collected per 100 km^2 in bedrock mountainous areas and plains, respectively. This set of data contained all the information that matches the historical data. Two 500 mL polyethylene bottles were used to store filtered (1.2-μm filter membrane) groundwater for the analysis of trace elements and major ions. The bottles were cleaned three times using the water to be sampled before sampling. The water sample in the bottle used to determine trace elements was acidized to pH < 2 using nitric acid, while the water sample in the bottle used to analyze major ions was not acidized. All samples were stored at 4 °C until laboratory procedures could be performed.

3.2. Tests and Analysis

All of the water samples were sent to the Groundwater, Mineral Water and Environmental Monitoring Center of the Institute of Hydrogeology and Environmental Geology, Chinese Academy of Geological Sciences to be tested in strict accordance with GB/T5750-2006 *Standard Examination Methods for Drinking Water* and GB 8538-2016 *National Food Safety Standard–Methods for Examination of Drinking Natural Mineral Water*. Metal ions and trace elements (K^+, Na^+, Mg^{2+}, Fe, Cu, Pb, Zn, As, and Cr^{6+}) were measured using inductively coupled plasma-mass spectrometry (ICP-MS) (Agilent 7500ce ICP-MS, Tokyo, Japan). SO_4^{2-}, Cl^-, NH_4^+, and NO_3^- were measured using ion chromatography (IC) (Shimadzu LC-10ADvp, Japan). NO_2^- and HCO_3^- were determined using the UV spectrophotometry method, and acid-base titration method, respectively. Chemical oxygen demand (COD) and TDS content were measured using potassium dichromate titration and the gravimetric method, respectively. To ensure the accuracy and reliability of the test results, the anion-cation balance was analyzed according to the principle of ionic balance, with analysis errors less than 5%. All chemicals were purchased from Sinopharm Chemical Reagent Co., Ltd., Shanghai, China.

3.3. Multivariate Statistical Method

First, the anomaly thresholds or background values were determined according to the historical hydrochemical data. The indices of both anthropogenic influences and water-rock interactions were then identified using contrast coefficients and standard deviations. Finally, the Mahalanobis distances between the multivariate contrast coefficients of the indices of anthropogenic influences were calculated to analyze the changes in hydrochemical components and determine the intensity of anthropogenic influences on the hydrochemical characteristics of groundwater.

3.3.1. Identification of the Indicative Indices of Both Anthropogenic Influences and Water-Rock Interactions

Standard deviations were used to identify the indices of both anthropogenic influences and water-rock interactions.

A standard deviation (S) is the average of the distances between some data and their average, and a higher standard deviation means higher degrees of fluctuation and dispersion of a variable. The anthropogenic inputs in groundwater have a high degree of spatial dispersion. Therefore, the standard deviation was used to analyze the dispersion degree of groundwater components to determine the indices of the anthropogenic influences on the hydrochemical characteristics of groundwater. The standard deviation is expressed as:

$$S = \sqrt{\frac{1}{N}\sum_{i=1}^{N}(X_i - \overline{X})^2} \quad (1)$$

where S is the standard deviation; N is the number of samples; X_i is the tested value; $\overline{X}$ is the arithmetic average.

The contrast coefficient (V) is a metric that measures the sharpness of anomalies and represents the fluctuation of the anomalies relative to the background values. To compare the anomaly intensity of a certain indicator element in different regions, the contrast coefficient can be used to eliminate the influences caused by the phenomenon that the indicator element has different background values in different regions. First, the ratio of each variable to the anomaly threshold or background value was calculated using Equation (2), and multiple variables with unequal averages were transformed into those with equal averages. The standard deviations of the contrast coefficients of various variables were then calculated and finally compared.

$$V = \frac{X_i}{T} \quad (2)$$

where V is the contrast coefficient; X_i is the tested value; T is the anomaly threshold or background value.

The anomaly threshold or background value (T) is a metric that reflects the geochemical background. The groundwater background values refer to the natural contents of physical and chemical elements in groundwater under uncontaminated conditions. However, a natural environment with no anthropogenic influence barely exists in the strict sense [42]. To analyze the anthropogenic influences on the hydrochemical characteristics of the groundwater in the study area in the past three decades, this study determined the range of background values using the quartile method based on the hydrochemical data of 1993 [43,44].

3.3.2. Intensity of Anthropogenic Influences

The Mahalanobis distance (MD) is a metric that measures multivariate distance. It is used for multivariate statistics of observed objects by characterizing the distance between the objects and their distribution center, during which covariance is considered [45]. This metric is frequently used to calculate the similarity of various indices, identify and quantify anomalies, and sometimes select representative data from a large quantity of measurement data [46]. This study analyzed the changes in hydrochemical components by calculating the Mahalanobis distances between the contrast coefficients of the indices of anthropogenic influences in order to further determine the intensity of anthropogenic influence on the hydrochemical characteristics of groundwater. The calculation formula for Mahalanobis distance is as follows:

$$D^2 = (x_i - \overline{x})S^{-1}(x_i - \overline{x})' \quad (3)$$

where D^2 is the Mahalanobis distance; x_i is an index value of a sample; $\overline{x}$ is the arithmetic average of various indices; S is the standard deviation.

This study used the Kriging interpolation method of MapGis K9 to analyze the intensity grades of anthropogenic influences on the hydrochemical characteristics of the groundwater and then plotted the map of regional anthropogenic influence intensity.

4. Results and Discussion

4.1. Characteristics of Groundwater Quality Indicators

The descriptive statistics for the indicators analyzed in groundwater samples in Xiamen City are summarized in Table 1. Among anions in the groundwater of Xiamen, NO_3^- showed the highest median concentration of 42.23 mg/L, followed by Cl^-, SO_4^{2-} and F^-. High concentrations of NH_4^+ occurred in groundwater in Xiamen, and up to 107.0 mg/L. Na^+ ranged from 2.68 mg/L to 4043.8 mg/L, with the median of 32.55 mg/L. Among trace metal(loid)s in the groundwater of Xiamen, Fe showed the highest median concentration of 0.03 mg/L, followed by Zn, Pb, Cu, As, Cr^{6+} and Al^{3+}. According to the standard for groundwater quality of China (GAQSIQPRC, 2017) [47], the percent of groundwater samples with the concentrations of indicators that exceeded the allowable values (PEV) for drinking purposes are also in Table 1. In Xiamen City, NO_3^- showed the highest PEV of 22.09%, followed by Fe, Pb, NH_4^+, F^-, Cl^-, Al^{3+}, SO_4^{2-}, NO_2^-, Na^+, Cu, and Zn, while the PEVs of the other two indicators were zero.

Table 1. Descriptive statistics of concentrations of groundwater indicators.

Item	Min	Med	Max	SD	Detection Limits	Number of Non-Detects	Allowable Values	PEV (%)
SO_4^{2-} (mg/L)	0.37	24.72	1171.0	107.86	0.2	0	250	1.74
Cl^- (mg/L)	1.29	35.19	7073.0	548.16	0.1	0	250	2.91
NO_3^- (mg/L)	0.39	42.23	2223.14	177.53	0.2	0	88.6	22.09
NO_2^- (mg/L)	ND	ND	5.20	0.55	0.002	110	3.29	1.16
F^- (mg/L)	ND	0.19	11.90	1.04	0.1	49	1.0	7.56
Na^+ (mg/L)	2.68	32.55	4034.0	305.49	0.013	0	200	1.16
NH_4^+ (mg/L)	ND	ND	107.0	11.45	0.04	117	0.64	9.30
Al^{3+} (mg/L)	ND	ND	6.65	0.54	0.02	146	0.2	2.33
Fe (mg/L)	ND	0.03	6.76	0.72	0.02	49	0.3	13.95
Pb (mg/L)	ND	0.001	0.806	0.07	0.001	66	0.01	13.95
Cu (mg/L)	ND	ND	5.51	0.42	0.010	164	1.0	0.58
Zn (mg/L)	ND	0.01	7.74	0.60	0.002	42	1.0	0.58
As (mg/L)	ND	ND	0.009	0.001	0.001	114	0.01	0
Cr^{6+} (mg/L)	ND	ND	0.017	0.002	0.004	161	0.05	0

SD = standard deviation; ND = below the limit of detection; Allowable Values = data from (GAQSIQPRC, 2017); PEV = percent of groundwater samples with the concentration of one index exceeded the allowable value.

4.2. Hydrochemical Changes in Groundwater

4.2.1. Changes in Groundwater Chemical Composition

The evaluation indices used in this study were the water quality indices specified in GB/T14848-2017 *Standard for groundwater quality*. They were contained in the hydrochemical data of 1993 and consisted of 17 indices, namely pH, COD, TDS content, SO_4^{2-}, Cl^-, Na^+, NH_4^+, NO_2^-, NO_3^-, F^-, Fe, Al^{3+}, Cu, Pb, Zn, As, and Cr^{6+}. We next analyzed the changes in the chemical composition of the groundwater in Xiamen in the past three decades by comparing the number and contribution rates of the influencing factors of groundwater quality between 1993 and 2019–2021.

The following evaluation results of the groundwater quality of Xiamen in the past three decades were obtained based on the number of groundwater samples: (1) the proportion of Class I to III potable groundwater decreased from 40% in 1993 to 28.5% during 2019–2021; (2) the total proportion of poor-quality and undrinkable groundwater (Class IV and V) increased from 60% in 1993 to 71.5% in 2019–2021; (3) the proportion of Class IV water that can become portable water after proper treatment increased from 28.2% in 1993 to 51.2% in 2019–2021, and the proportion of non-portable Class V water decreased from 31.8% in 1993 to 20.3% in 2019–2021 (Table 2).

Table 2. Comparison of the evaluation results of groundwater quality between 1993 and 2019–2021.

Year	Number of Samples	Proportion/%					Water Inferior to Class III	
		Class I	Class II	Class III	Class IV	Class V	Number of Samples	Proportion/%
1993	85	12.9	11.8	15.3	28.2	31.8	51	60.0
2019	172	1.2	4.7	22.7	51.2	20.3	123	71.5

The number of the indices affecting the groundwater quality in Xiamen increased from nine in 1993 to 15 in 2019–2021 (Figure 2). The additional six indices during 2019–2021 consisted of NO_3^-, Pb, NH_4^+, Al^{3+}, NO_2^-, and Cu (the contribution rates to poor-quality groundwater were 26.0%, 16.3%, 10.6%, 4.1%, 0.8% and 0.8%, respectively), among which NO_3^- and Pb had contribution rates second only to pH.

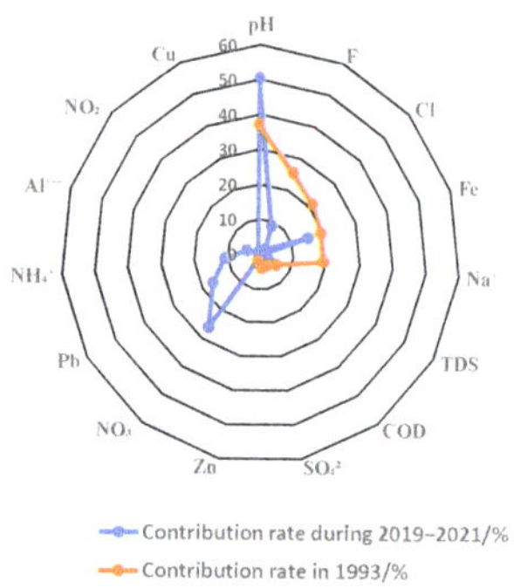

Figure 2. Main indices affecting groundwater quality and their contribution rates.

4.2.2. Changes in the Hydrochemical Types of Groundwater

Human activities have not only changed the groundwater quality but also affected and changed the natural hydrochemical field of groundwater due to a large number of anthropogenic inputs [48].

The groundwater of Xiamen had simple hydrochemical types in 1993. Specifically, 19 types were identified in 67 samples, and seven types were concentrated in 70% of the groundwater samples, namely Na—HCO_3, Ca·Na–HCO_3, Na–HCO_3·Cl, Na·Ca–HCO_3,

Na–Cl, Na·Ca–HCO_3·Cl, and Na–Cl·HCO_3. Regarding anions, the groundwater was dominated by the HCO_3 type, which accounted for 46.3%; followed by the HCO_3·Cl type, which accounted for 25.4%. Moreover, the groundwater of SO_4·Cl and Cl·SO_4 types accounted for a minimum proportion of 1.5% each. In terms of cations, the groundwater in Xiamen in 1993 was dominated by Na and Na·Ca types, which accounted for 41.8% and 25.4%, respectively. They were followed by the Ca·Na type, which accounted for 14.9%. Additionally, the water of Ca, Na·Mg, Ca·Mg, Ca·Na·Mg, and Na·Mg types accounted for 1.5%–6.05% each.

The hydrochemical types of groundwater in Xiamen increased during 2019–2021 compared to 1993. Specifically, 28 hydrochemical types were identified in 172 samples, and seven hydrochemical types were concentrated in 70% of the groundwater samples, namely Ca·Na–HCO_3, Na·Ca–Cl, Na·Ca–HCO_3, Ca·Na–HCO_3·Cl, Na·Ca–HCO_3·Cl, Na·Ca–Cl·HCO_3, Ca–HCO_3, and Na·Ca–Cl·SO_4. Regarding anions, the groundwater was still dominated by the HCO_3 type, whose proportion, however, decreased to 31.4%. By contrast, the Cl^- and SO_4^{2-} contents in the groundwater increased, with the water of Cl type and the hydrochemical types dominated by Cl (Cl·HCO_3, Cl·HCO_3·SO_4 type, and Cl·SO_4 types) accounted for 16.9% and 17.4%, respectively. Moreover, the proportion of the water of SO_4 type or types dominated by SO_4 increased to 5.3%. Regarding cations, the Ca^{2+} content increased, the Mg^{2+} content decreased, and the groundwater was dominated by the Na·Ca type, whose proportion increased to 51.2%. This type was followed by the Ca·Na type, which accounted for 36.0%. Furthermore, the water of Ca and Na types had a low proportion of 6.4% each.

Groundwater in the pore-fissure aquifer showed higher Cl^- and SO_4^{2-} concentrations than that in the bedrock fissure aquifers and pore aquifers (Figure 3). This may have occurred due to the greater number of industrial zones and population located in the area of the pore-fissure aquifer than in the areas of the other two types of aquifers, indicating more serious anthropogenic influences.

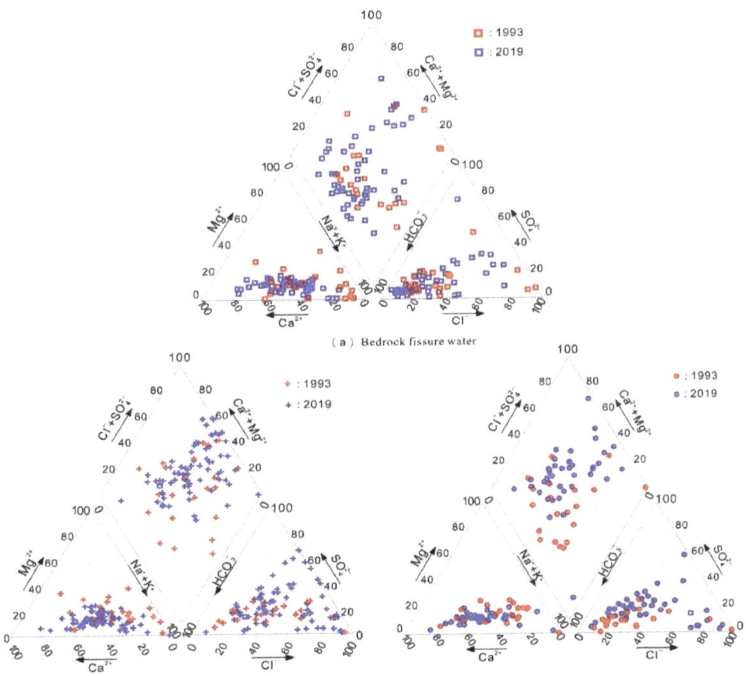

Figure 3. Piper trilinear diagrams.

The increase in the absolute content of nitrates in groundwater affects the groundwater quality, while the increase in their relative content changes the hydrochemical characteristics of groundwater. As indicated by the hydrochemical test results of the groundwater in Xiamen in 2019, NO_3^- was a major anion in Xiamen's groundwater, its relative content exceeded that of SO_4^{2-} (Table 3), and thus more water of NO_3 type exists in the groundwater. It is known that groundwater NO_3^- in Xiamen mainly originated from anthropogenic sources, including infiltration of domestic sewage and landfill leachate [40].

Table 3. Characteristic parameters of the relative contents of major anions in the groundwater of Xiamen.

Item	Year	Sample Number	Range	Med	Average	SD
HCO_3^-	1993	45	0.9–85.8	59.9	52.2	25.6
	2019	172	0.4–89.8	35.0	36.9	24.5
SO_4^{2-}	1993	45	0.8–49.4	9.8	11.7	7.9
	2019	172	0.4–64.3	12.1	13.9	10.8
Cl^-	1993	45	6.5–97.5	21.9	31.8	23.2
	2019	172	2.8–88.2	25.7	27.3	15.9
NO_3^-	1993	45	0–22.5	2.5	4.3	5.1
	2019	172	0.1–84.7	20.2	21.8	16.8

4.3. *Anthropogenic Influences on the Hydrochemical Characteristics of Groundwater*

4.3.1. Indicative Indices of the Anthropogenic Influences

Determination of the Background Values of Groundwater in Different Water-Bearing Media

This study statistically analyzed the data of 1993, obtaining the background values of the groundwater environment in aquifers of bedrock fissures, weathered pores and fissures, and pores in alluvial-proluvial deposits, respectively (Table 4). Furthermore, the background values of indices Cu, Pb, As, and Cr^{6+} were determined to be 0, since there were a few samples of these indices and they were not detected in the water of various water-bearing media.

Identification of the Indicative Indices of Anthropogenic Influences

To conduct a multivariate contrast, various variables were normalized first. In other words, the upper limits of the background values of various indices in Table 3 were used as the anomaly threshold or background value (T) to calculate the contrast coefficients (V) and the standard deviations of various indices of the groundwater of Xiamen in 2019.

Anthropogenic inputs in groundwater generally have high degrees of spatial dispersion and fluctuation [48]. Therefore, standard deviations can be used to distinguish between groundwater components originating from anthropogenic inputs and those from water-rock interactions. The contrast coefficients of pH and HCO_3^- had a standard deviation of 0.09 and 0.53, respectively, indicating very weak anthropogenic influences. By contrast, the contrast coefficients of Cl^-, NO_3^-, and NH_4^+ had a standard deviation of 18.86, 35.83, and 42.76, respectively, indicating very strong anthropogenic influences. The standard deviations of the contrast coefficients of the 15 indices decreased in the order of pH, HCO_3^-, COD, total hardness, Zn, Fe, NO_2^-, SO_4^{2-}, F^-, TDS, Mg^{2+}, K^++Na^+, Cl^-, NO_3^-, and NH_4^+. These 15 indices in the groundwater were divided into four categories according to their sources determined by the standard deviations of their contrast coefficients, as shown in Table 5.

Table 4. Background values of groundwater environment in different water-bearing media in Xiamen.

Item \ Background Value Scope	Bedrock Fissure Water	Weathered Pore Fissure Water	Pore Water in Alluvial Proluvial Deposits
pH	6.57–7.04	6.42–6.93	6.05–7.09
COD	0.41–1.32	0.90–1.22	0.57–1.33
Cl^-	12.0–28.54	27.81–76.09	14.45–76.60
SO_4^{2-}	6.05–23.54	18.97–40.71	5.64–42.81
HCO_3^-	44.39–102.49	38.04–218.15	43.17–166.0
NO_2^-	0.004–0.048	0.021–0.188	0–0.026
NO_3^-	0.50–4.73	6.38–29.38	2.28–13.90
F^-	0.04–0.20	0.04–0.20	0.04–0.20
K^++Na^+	18.05–41.32	30.36–114.59	17.94–70.61
Mg^{2+}	1.32–5.43	4.18–17.63	2.19–6.77
NH_4^+	0.02–0.18	0.03–0.28	0.01–0.10
Fe	0.09–1.20	0.03–0.31	0.03–0.08
Al^{3+}	0–0.026	0–0.037	0
Zn	0.048–0.441	0.019–0.390	0.022–0.350
TDS	96.12–251.79	180.4–379.55	124.9–337.78

Table 5. Standard deviations of the contrast coefficients of indices and their indicative significance.

Index	Standard Deviations	Sources	Anthropogenic Influence Intensity
pH, HCO_3^-	≤1	Water-rock interactions	Very weak
COD, total hardness, Zn, Fe, NO_2^-, SO_4^{2-}, F^-, and TDS content	1–5	Water-rock interactions, supplemented by anthropogenic inputs	Weak
Mg^{2+}, K^++Na^+	5–10	Anthropogenic inputs, supplemented by water-rock interactions	Strong
Cl^-, NO_3^-, NH_4^+	>10	Anthropogenic inputs	Very strong

4.3.2. Anthropogenic Influence Intensity

The Mahalanobis distance is a metric used for multivariate statistics of observed objects. This study combined five indices: Mg^{2+}, K^++Na^+, Cl^-, NO_3^-, and NH_4^+ subject to strong or very strong anthropogenic influences combined into a whole, then calculated the Mahalanobis distances of the contrast coefficients of various groundwater samples collected in 2019, and finally analyzed the anthropogenic influence intensity of various samples. Figure 4 shows the distribution of the Mahalanobis distances of 172 samples. The anthropogenic influence intensity of these samples was divided into four grades according to the Mahalanobis distances 0.25, 0.6, and 6 at inflexion points, which represented the boundaries of anthropogenic influence intensity grades 2, 3, and 4, respectively.

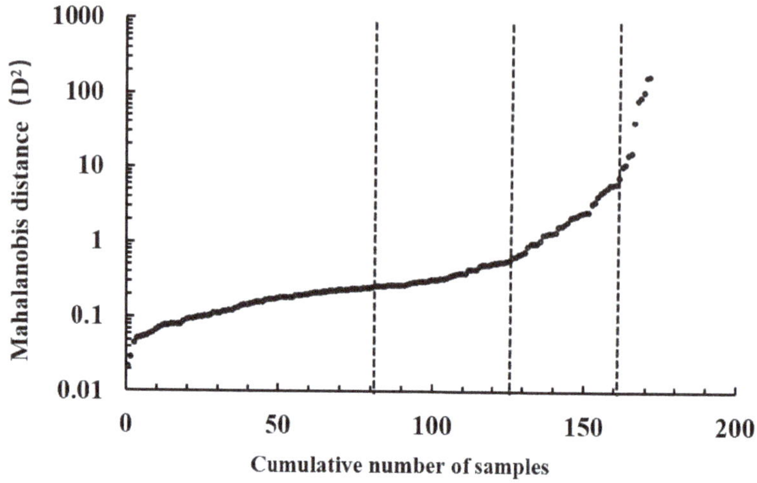

Figure 4. Mahalanobis distances of contrast coefficients of groundwater samples in 2019.

The intensity of anthropogenic influences on the hydrochemical composition of the groundwater in Xiamen was analyzed based on the Mahalanobis distance method. Figure 5 shows the types of land use and the intensity and distribution of anthropogenic influences on the hydrochemical composition of the groundwater in Xiamen. According to this figure, the groundwater subject to strong or relatively strong anthropogenic influences is mostly distributed on artificial surfaces and cultivated land, while the anthropogenic influences on the groundwater sites in forest land areas were mostly rated as weak.

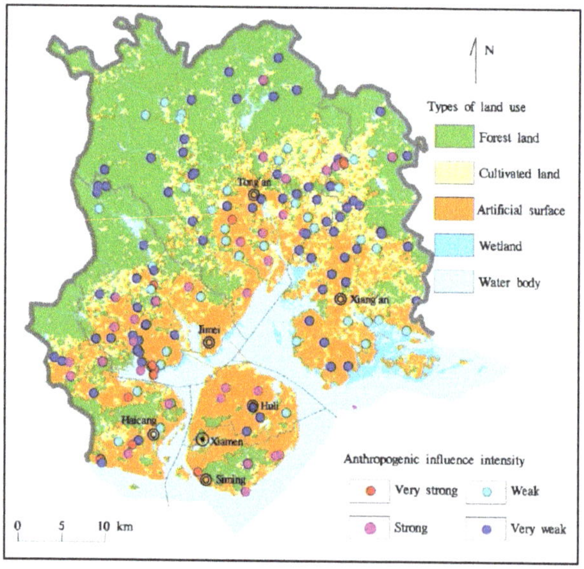

Figure 5. Map showing the distribution of the intensity of anthropogenic influences on groundwater.

Interpolation was conducted for the calculated Mahalanobis distances using the Kriging interpolation method of the MapGis software. The anthropogenic influence intensity was then divided into four grades according to the interpolated Mahalanobis distances, obtaining the map showing the zones of the anthropogenic influence intensity (Figure 6).

According to this map, the zones subject to very strong anthropogenic influences include densely populated areas and areas with intensively distributed industrial enterprises, such as Xinxu Town in the Xiang'an District, Xike Street in the Tong'an District, Houxi Town and Xinglin Street in the Jimei District, Haicang Street in the Haicang District, and Dianqian Street in the Huli District. With low population density and high vegetation coverage, the northern bedrock mountainous area is a zone with very weak anthropogenic influences on groundwater.

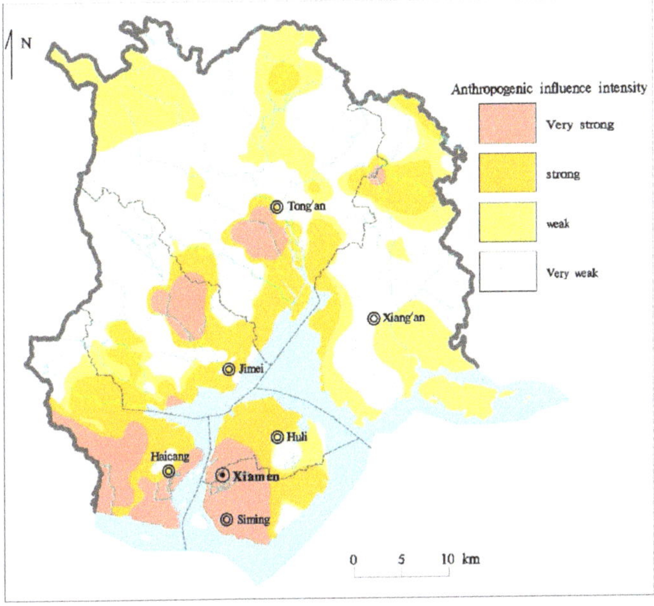

Figure 6. Map showing the zones of the intensity of anthropogenic influences on groundwater.

5. Conclusions

1. The number of indices affecting the groundwater quality in Xiamen increased from nine in 1993 to 15 in 2019, and the six increased indicators included NO_3^-, Pb, NH_4^+, Al^{3+}, NO_2^- and Cu. The number of hydrochemical types increased from 19 in 1993 to 28 in 2019. Moreover, the water of the HCO_3 type has decreased, while the water of the Cl type (including types dominated by Cl) and the SO_4 type (including types dominated by SO_4) have increased, indicating that human activities are the powerful force driving the change in groundwater chemistry in Xiamen City.
2. The increase in the relative contents of nitrates in the groundwater has changed the hydrochemical characteristics of the groundwater. Both the maximum and average relative content of NO_3^- exceed those of SO_4^{2-}. Therefore, NO_3^- has become a major anion affecting the hydrochemical nomenclature of the groundwater in Xiamen, forming the water of NO_3 type (according to Shukarev classification and Designation rules).
3. As indicated by the analysis results of Mahalanobis distances, zones with very strong anthropogenic influences on groundwater include Houxi Town in Jimei District and Haicang Street in Haicang District, while zones with very weak anthropogenic influences on groundwater include the northern bedrock mountainous area. These results are consistent with the land use, industrial layout, and population distribution of Xiamen, indicating that the Mahalanobis distance method can be used to identify the hydrochemical anomalies of groundwater and determine the intensity of anthropogenic influences on groundwater and that the evaluation results obtained using this method is reasonable.

Author Contributions: Conceptualization, Z.L. and J.L.; methodology, Z.L.; software, Y.Z.; validation, Z.L., Y.Z. and Q.H.; formal analysis, Z.L. and J.L.; investigation, Z.L. and Y.Z.; resources, Y.L.; data curation, Z.L. and Q.H.; writing—original draft preparation, Z.L. and J.L.; writing—review and editing, J.L. and Y.L.; visualization, J.L.; supervision, Y.Z.; project administration, Z.L. and Y.Z.; funding acquisition, Z.L. All authors have read and agreed to the published version of the manuscript.

Funding: This study was financially supported by China Geological Survey's project (Grant No. DD20190303 and DD20221773).

Institutional Review Board Statement: Not applicable.

Informed Consent Statement: Not applicable.

Data Availability Statement: The data presented in this study are available on request from the corresponding author.

Conflicts of Interest: The authors declare no conflict of interest.

References

1. Wang, Y.; Zheng, C.; Ma, R. Review: Safe and sustainable groundwater supply in China. *Appl. Hydrogeol.* **2018**, *26*, 1301–1324. [CrossRef]
2. Turhan, A.; Ozmen, N. Influence of Chloride on Growth, Fruit Yield and Quality Parameters of Processing Pepper. *J. Agric. Nat.* **2021**, *24*, 1139–1144.
3. Döll, P.; Müller Schmied, H.; Schuh, C.; Portmann, F.T.; Eicker, A. Global-scale assessment of groundwater depletion and related groundwater abstractions:Combining hydrological modeling with information from well observations and GRACE satellites. *Water Resour. Res.* **2014**, *50*, 5698–5720. [CrossRef]
4. Mirdashtvan, M.; Najafinejad, A.; Malekian, A.; Sadoddin, A. Sustainable water supply and demand management in semi-arid regions: Optimizing water resources allocation based on RCPs scenarios. *Water Resour. Manag.* **2021**, *35*, 5307–5324. [CrossRef]
5. Liu, C.L.; Zheng, J.H.; Li, Z.H.; Li, Y.S.; Hao, Q.C.; Li, J.F. Analysis on the situation and countermeasures of water resources supply and demand in the cities of small and medium-sized river basins along southeast coast of China-taking Xiamen City as an example. *J. Groundw. Sci. Eng.* **2021**, *9*, 350–358.
6. Yang, H.F.; Meng, R.F.; Li, W.P.; Li, Z.Y.; Zhi, C.S.; Bao, X.L.; Li, C.Q.; Liu, F.T.; Wu, H.P.; Ren, Y. Groundwater resources of the Haihe River Basin and its development potential. *Geol. China* **2021**, *48*, 1032–1051.
7. Nasiri, S.; Ansari, H.; Ziaei, A.N. Determination of water balance equation components in irrigated agricultural watersheds using SWAT and MODFLOW models: A case study of Samalqan plain in Iran. *J. Groundw. Sci. Eng.* **2022**, *10*, 44–56.
8. Zhang, Y.L.; Ma, R.; Li, Z.H. Human health risk assessment of groundwater in Hetao Plain (Inner Mongolia Autonomous Region, China). *Environ. Monit. Assess.* **2014**, *186*, 4669–4684. [CrossRef]
9. Tirkey, P.; Bhattacharya, T.; Chakraborty, S.; Baraik, S. Assessment of groundwater quality and associated health risks: A case study of Ranchi city, Jharkhand, India. *Groundw. Sustain. Dev.* **2017**, *5*, 85–100. [CrossRef]
10. Adimalla, N.; Wu, J. Groundwater quality and associated health risks in a semi-arid region of south India: Implication to sustainable groundwater management. *Hum. Ecol. Risk Assess. Int. J.* **2019**, *25*, 191–216. [CrossRef]
11. Adimalla, N. Controlling factors and mechanism of groundwater quality variation in semiarid region of South India: An ap-proach of water quality index (WQI) and health risk assessment (HRA). *Env. Geochem. Health* **2020**, *42*, 1725–1752. [CrossRef] [PubMed]
12. Egbi, C.D.; Anornu, G.K.; Ganyaglo, S.Y.; Appiah-Adjei, E.K.; Li, S.L.; Dampare, S.B. Nitrate contamination of groundwater in the lower volta river basin of Ghana: Sources and related human health risks. *Ecotoxicol. Environ. Saf.* **2020**, *191*, 110227. [CrossRef]
13. Li, X.Y.; Zhang, Y.L.; Li, Z.H.; Wang, R. Response of the groundwater environment to rapid urbanization in Hohhot, the provincial capital of western China. *J. Hydrol.* **2021**, *603*, 127033. [CrossRef]
14. Mote, T.; Lacke, M.; Shepherd, J.M. Radar signatures of the urban effect on precipitation distribution: A case study for Atlanta, Georgia. *Geophys. Res. Lett.* **2007**, *34*, L20710. [CrossRef]
15. Salerno, F.; Gaetano, V.; Gianni, T. Urbanization and climate change impacts on surface water quality: Enhancing the resilience by reducing impervious surfaces. *Water Res.* **2018**, *144*, 491–502. [CrossRef] [PubMed]
16. Zhou, Q.Q.; Leng, G.Y.; Su, J.H.; Ren, Y. Comparison of urbanization and climate change impacts on urban flood volumes: Importance of urban planning and drainage adaptation. *Sci. Total Environ.* **2019**, *658*, 24–33. [CrossRef]
17. Ariken, M.; Zhang, F.; Liu, K.; Fang, C.L.; Kung, H.T. Coupling coordination analysis of urbanization and eco-environment in Yanqi Basin based on multi-source remote sensing data. *Ecol. Indic.* **2020**, *114*, 106331. [CrossRef]
18. Yu, B.B. Ecological effects of new-type urbanization in China. *Renew. Sustain. Energy Rev.* **2021**, *135*, 110239. [CrossRef]
19. Shaharoona, B.; Al-Ismaily, S.; Al-Mayahi, A.; Al-Harrasi, N.; Al-Kindi, R.; Al-Sulaimi, A.; Al-Busaidi, H.; Al-Abri, M. The role of urbanization in soil and groundwater contamination by heavy metals and pathogenic bacteria: A case study from Oman. *Heliyon* **2019**, *5*, e01771. [CrossRef]

20. Li, Z.H.; Zhang, Y.L.; Hu, B.; Wang, L.J.; Zhu, Y.C.; Li, J.F. Driving action of human activities on NO_3—N pollution in confined groundwater of Togtoh County, Inner Mongolia. *Acta Geosci. Sinica* **2018**, *39*, 358–364.
21. Kaur, L.; Rishi, M.S.; Siddiqui, A.U. Deterministic and probabilistic health risk assessment techniques to evaluate non-carcinogenic human health risk (NHHR) due to fluoride and nitrate in groundwater of Panipat, Haryana, India. *Environ. Pollut.* **2020**, *259*, 113711. [CrossRef] [PubMed]
22. Zhu, L.; Yang, M.N.; Zhang, Y.X.; Liu, J.T. Research on distribution and influence factor of nitrate of shallow groundwater in Dongsheng and surrounding area. *J. Water Resour. Water Eng.* **2016**, *27*, 37–41,45.
23. Kangabam, R.D.; Bhoominathan, S.D.; Kanagaraj, S.; Govindaraju, M. Development of a water quality index (WQI) for the Loktak Lake in India. *Appl. Water Sci.* **2017**, *7*, 2907–2918. [CrossRef]
24. Cheng, X.; Chen, L.D.; Sun, R.H.; Jing, Y.C. Identification of regional water resource stress based on water quantity and quality: A case study in a rapid urbanization region of China. *J. Clean. Prod.* **2019**, *209*, 216–223. [CrossRef]
25. Aldwell, C.R. Agriculture and groundwater must co-exist. *Environ. Geol. Water Sci.* **1986**, *9*, 1–2. [CrossRef]
26. Wang, Y.X.; Ma, T.; Guo, Q.H.; Ma, R. Groundwater and environmental change. *Earth Sci. Front.* **2005**, *12*, 14–21.
27. Wang, J.Z.; Zhang, G.H.; Nie, Z.L.; Yan, M.J.; Wang, Y. Disturbing degree of mankind activities to groundwater in the Hutuo River Valley area. *Bull. Soil Water Conserv.* **2010**, *30*, 65–69.
28. Liu, L.; Zhou, X.; Ye, Y.H. Screening of characteristic indexes for shallow groundwater influenced by human activities using multivariate statistics. *Resour. Surv. Environ.* **2014**, *35*, 305–310.
29. Ouyang, Q.C.; Lin, B.Q.; Chen, X.W. Effects of landuse changes on hydrological processes: An analysis based on SWAT model. *J. Subtrop. Resour. Environ.* **2016**, *11*, 65–70.
30. Malki, M.; Bouchaou, L.; Hirich, A.; Brahim, Y.; Choukr-Allah, R. Impact of agricultural practices on groundwater quality in intensive irrigated area of Chtouka-Massa, Morocco. *Sci. Total Environ.* **2017**, *574*, 760–770. [CrossRef]
31. Flem, B.; Reimann, C.; Fabian, K.; Birke, M.; Filzmoser, P.; Banks, D. Graphical statistics to explore the natural and anthropogenic processes influencing the inorganic quality of drinking water, ground water and surface water. *Appl. Geochem.* **2018**, *88*, 133–148. [CrossRef]
32. Zhang, X.W.; He, J.T.; Peng, C.; Zhang, C.Y.; Ni, Z.H. Comparison of identification methods of main component hydrochemical anomalies in groundwater: A case study of Liujiang Basin. *Environ. Sci.* **2017**, *38*, 3225–3234.
33. Peng, C.; He, J.T.; Liao, L.; Zhang, Z.G. Research on the influence degree of human activities on groundwater quality by the method of geochemistry: A case study from Liujiang Basin. *Earth Sci. Front.* **2017**, *24*, 321–331.
34. Zhao, W.; Lin, J.; Wang, S.F.; Liu, J.L.; Chen, Z.R.; Kou, W.J. Influence of human activities on groundwater environment based on coefficient variation method. *Environ. Sci.* **2013**, *34*, 1277–1283.
35. Rocke, D.M.; Woodruff, D.L. Identification of outliers in multivariate data. *J. Am. Stat. Assoc.* **1996**, *91*, 1047–1061. [CrossRef]
36. Ji, H.J. Multivariate method for distinguishing geochemical background from geochemical anomalies. *J. Chang. Univ. Earth Sci.* **1988**, *18*, 311–320.
37. Tian, L.M. *Geochemical Data Processing and Targets Optimization in Eastern Kunlun Orogenic Belt, Qinghai Province*; Wuhan China University of Geosciences: Wuhan, China, 2017.
38. Tang, Y.P.; Liu, Y.L. The Application of Markov distance method to the anomaly Appraisal in oil-gas geochemical prospecting. *Geophys. Geochem. Explor.* **1998**, *22*, 231–233.
39. Chen, X.N.; Wang, L.J.; Wang, H.P. Application of Mahalanobis distance anomaly in geochemical survey: Taking the 1:25000 geochemical survey of somewhere in Dulan County, Qinghai province as an example. *J. Chongqing Univ. Technol. Nat. Sci.* **2018**, *11*, 115–120.
40. Li, Z.H.; Li, Y.S.; Hao, Q.C.; Li, J.F.; Zhu, Y.C. Characteristics, genesis and treatment measures of NO_3-type groundwater in Xiamen, Fujian Province. *Geol. China* **2021**, *48*, 1441–1452.
41. Li, Y.S.; Liu, C.L.; Hao, Q.C. *Report on Comprehensive Geological Survey of Xiamen-Zhangzhou-Quanzhou City*; Chinese Academy of Geological Sciences, The Institute of Hydrogeology and Environmental Geology: Shijiazhuang, China, 2021. (In Chinese)
42. Qiu, H.X.; Huang, Q.Z. The concept of groundwater environment background and its determination. *J. Ocean. Univ. Qingdao* **1994**, 16–19.
43. Helsel, D.R. Less than obvious: Statistical treatment of data below the detection limit. *Environ. Sci. Technol.* **1990**, *24*, 1767–1774 [CrossRef]
44. Lee, L.; Helsel, D. Baseline models of trace elements in major aquifers of the United States. *Appl. Geochem.* **2005**, *20*, 1560–1570. [CrossRef]
45. Song, Y.H.; Li, Z.X.; Sun, L.H.; Jia, D.C.; Bu, X.J. Contrast between Mahalanob is distance and euclidean distance in geochemical exploration processing. *Jinlin Geol.* **2008**, *27*, 117–120.
46. Jouan-Rimbaud, D.; Massart, D.L.; Saby, C.A.; Puel, C. Characterisation of the representativity of selected sets of samples in multivariate calibration and pattern recognition. *Anal. Chim. Acta* **1997**, *350*, 149–161. [CrossRef]
47. *General Administration of Quality Supervision Inspection and Quarantine of the People's Republic of China (GAQSIQPRC)*; Standard for Groundwater Quality. Standards Press of China: Beijing, China, 2017.
48. Cheng, D.H.; Chen, H.H.; He, J.T.; Lin, J.; Yang, D.G.; Yu, H. A study of indicators of anthropogenic influence and water-rock interaction in groundwater system in the urban region of Beijing. *Hydrogeol. Eng. Geol.* **2007**, *34*, 37–42.

water

MDPI

Article

Assessment of Groundwater Quality Using APCS-MLR Model: A Case Study in the Pilot Promoter Region of Yangtze River Delta Integration Demonstration Zone, China

Zi Chen [1,2], Quanping Zhou [1,2], Jinsong Lv [1,2,*], Yuehua Jiang [1,2,3], Hai Yang [1,2], Hui Yang [1,2], Shijia Mei [1,2], Zhengyang Jia [1,2], Hong Zhang [1,2], Yang Jin [1,2,3], Lin Liu [1,2,3] and Rujia Shen [1]

1 Nanjing Center, China Geological Survey, Nanjing 210016, China
2 Key Laboratory of Watershed Eco-Geological Processes, Ministry of Natural Resources, Nanjing 210016, China
3 Chinese Academy of Geological Sciences, Beijing 100037, China
* Correspondence: ljinsong@mail.cgs.gov.cn

Abstract: Groundwater contaminant source identification is an endeavor task in highly developed areas that have been impacted by diverse natural processes and anthropogenic activities. In this study, groundwater samples from 84 wells in the pilot promoter region of the Yangtze River Delta integration demonstration zone in eastern China were collected and then analyzed for 17 groundwater quality parameters. The principal component analysis (PCA) method was utilized to recognize the natural and anthropogenic aspects impacting the groundwater quality; furthermore, the absolute principal component score-multiple linear regression (APCS-MLR) model was employed to quantify the contribution of potential sources to each groundwater quality parameter. The results demonstrated that natural hydro-chemical evolution, agricultural activities, domestic sewage, textile industrial effluent and other industrial activities were responsible for the status of groundwater quality in the study area. Meanwhile, the contribution of these five sources obtained by the APCS-MLR model were ranked as natural hydro-chemical evolution (18.89%) > textile industrial effluent (18.18%) > non-point source pollution from agricultural activities (17.08%) > other industrial activities (15.09%) > domestic sewage (4.19%). It is believed that this contaminant source apportionment result could provide a reliable basis to the local authorities for groundwater pollution management.

Keywords: Yangtze River Delta; groundwater; APCS-MLR; contaminant source apportionment

Citation: Chen, Z.; Zhou, Q.; Lv, J.; Jiang, Y.; Yang, H.; Yang, H.; Mei, S.; Jia, Z.; Zhang, H.; Jin, Y.; et al. Assessment of Groundwater Quality Using APCS-MLR Model: A Case Study in the Pilot Promoter Region of Yangtze River Delta Integration Demonstration Zone, China. *Water* **2023**, *15*, 225. https://doi.org/10.3390/w15020225

Academic Editors: Guanxing Huang and Liangping Li

Received: 2 December 2022
Revised: 31 December 2022
Accepted: 31 December 2022
Published: 4 January 2023

1. Introduction

Groundwater quality degradation by natural processes or anthropogenic activities is widely recognized and has drawn the attention of researchers for decades [1–3]. The natural processes, mainly water–rock interactions, may cause specific ions' accumulations in groundwater, such as arsenic, magnesium and iodine [4–6]. In the meantime, anthropogenic activities can deteriorate groundwater quality and lead to a series of geological environment problems. For instance, agricultural behaviors including fertilization and livestock breeding are implicated in excess of nitrogen, phosphorus and potassium [7–9]. Industry effluent and leakage can lead to increased concentrations of sulfate and some heavy metal ions [10–12], and domestic sewage can lead to high levels of ammonia in groundwater [13]. Under the interaction of natural processes and anthropogenic activities, groundwater contamination management is undoubtedly a vital task. Classifying the contamination into their original sources is the primary step for an authentic assessment of the contaminated aquifer [14,15]. It will not only benefit the status survey, but also provide a consequential basis for future groundwater protection and pollution management.

Various methods have been developed for identifying the groundwater contaminant sources, such as the in situ survey, stable isotope methods, model-based numerical inversion and multivariate statistical approaches [16–20]. The in situ surveys usually attempts to find out the potential sources with labor-intensive work, but often fails to give a quantified analysis of the contamination source apportionment. The stable isotope methods can provide precise apportionment of sources, but only for the isotope-related contaminant. As for the model-based numerical inversion methods, they are believed to be more suitable for local scale research [21–24]. Meanwhile, multivariate statistical approaches have gained popularity among researchers due to their convenience and efficiency for regional scale problems [25–27]. This branch of methods include cluster analysis (CA), principal component analysis (PCA), positive matrix factorization (PMF), absolute principal component score-multiple linear regression (APCS-MLR) and so on. Compared with other approaches, the APCS-MLR method is an especially effective and practical method for identifying pollution sources. It was first developed by Thurston and Spengler [28] and then applied to pollution source apportionment problems on air, surface water and sediments [29–33]. More recently, the APCS-MLR method has started to be employed in groundwater quality research and has proven to be a powerful tool in identifying the impact of natural processes and anthropogenic activities to groundwater quality. For instance, Zhang et al. [34] employed the PCA and APCS-MLR methods to identify groundwater pollution sources and their apportionment in the Hutuo River alluvial-pluvial fan region of northern China. Meng et al. [35] used the APCS-MLR receptor model to assess the potential pollution sources of groundwater from 2006 to 2016 in the Limin Groundwater Source Area in Harbin. Yu et al. [36] applied the APCS-MLR method in nitrate pollution sources apportionment and compared its result with a Bayesian isotope mixing model. Sheng et al. [37] utilized the APCS-MLR receptor model to estimate the source apportionment of heavy metal pollution in an arid oasis region in Northwest China.

Therefore, in this research, we employ the APCS-MLR model to assess the potential sources of groundwater contamination in the pilot promoter region of the Yangtze River Delta integration demonstration zone. The study area is located in the Taihu watershed in eastern China, of which the groundwater has been influenced by intense anthropogenic activities. Previous work has demonstrated the complexity of the groundwater pollution in this area. Various contaminants have been found, including heavy-metal ions, nitrogen and phosphorus compounds [38–41]. Despite the numerous individual investigations for several specific contaminants, the overall groundwater contaminant source apportionment problem is still unsolved.

The objective of this work is to: 1. recognize the natural and anthropogenic aspects that affect the groundwater quality; 2. quantify the contribution of potential sources to each groundwater quality parameter via the APCS-MLR model. Insights from this paper could provide reliable advice for further pollution remediation plans in the pilot promoter region of the Yangtze River Delta integration demonstration zone.

2. Materials and Methods

2.1. Study Area

The pilot promoter region of the Yangtze River Delta integration demonstration zone (30°54′14″–31°09′25″ N and 120°39′47″–121°07′30″ E) is located southeastern of the Taihu Lake with a total area of 653.9 km^2 (Figure 1). It is under a subtropical monsoon climate, with 16.2 °C annual temperature and 1127.1 mm annual rainfall [42]. The groundwater in this region is mainly distributed in the sand and silt layers from Quaternary sediments [43,44]. The thickness of the Quaternary sediments is larger than 150 m according to the borehole materials, and the phreatic aquifer with thickness ranging from 3 to 7 m is the layer closely related to human activities (Figure 2). The aquifer is recharged from precipitation, river infiltration, and irrigation; in the meantime, its discharge mainly occurs through run off to rivers and evaporation since restrictions on groundwater exploitation have been imposed by the government since 1997 [45]. As an interactive area of Jiangsu, Zhejiang Provinces and Shanghai metropolis, this area is one of the most developed regions in China. Based on the a remote sensing survey with field validation, a major part of the land use in this area is cultivated land and residential area. Industries, especially textile factories, are mainly scattered in the south of the study area. What is more, a dense surface water network, including Taipu River and Dianshan Lake, is located in this area, which makes the surface water–groundwater interaction extremely complex. Hence, heavy agricultural activities and industrial sewage pose substantial threats to groundwater security. The application of nitrogen fertilizers and the use of industrial chemicals constitute potential point and non-point sources of groundwater contamination.

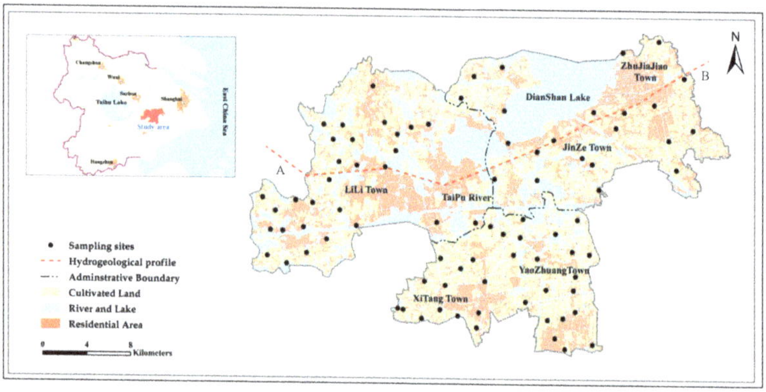

Figure 1. Location of the pilot promoter region of Yangtze River Delta integration demonstration zone and distribution of groundwater sampling sites with the land-use.

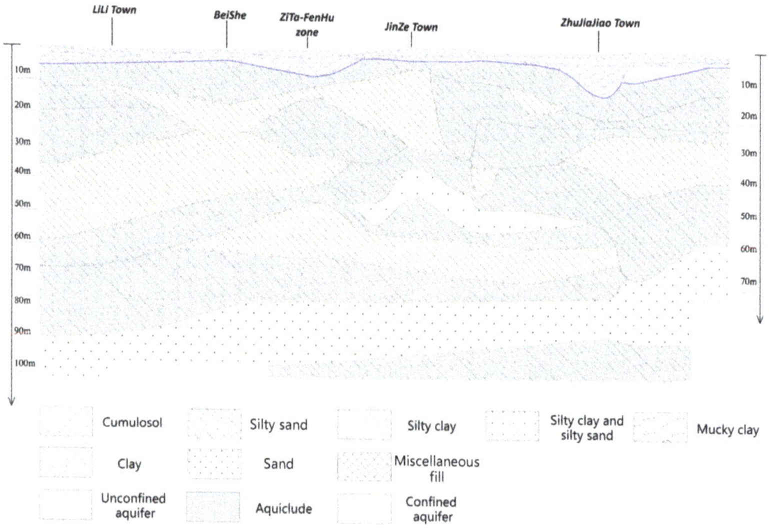

Figure 2. Hydrogeological profile (A-B) of the study area.

2.2. Data Preparation

Groundwater samples were collected from 84 wells in June 2020 and June 2021, and the sampling sites were chosen with the considerations of land-use types (cultivated land, residential land, etc.) and spatial distribution (approximately 1 km^2 per point) (Figure 1). In each well, the water temperature (WT), pH, dissolved oxygen (DO) and total dissolved solids (TDS) were measured by SX-620 pH Testor, SX-630 ORP Testor and Hanna DiST in situ, respectively. Figure 3 shows the landscape surroundings of several wells and the in situ test procedures. Polyethylene containers with a capacity of 1.5 L were used to store groundwater samples, and then brought back to the laboratory for an analysis of total phosphorus (TP), Cl^-, SO_4^{2-}, NH_4-N, NO_3-N, NO_2-N, K^+, Na^+, Ca^{2+}, Mg^{2+}, Mn, I and Sb. Various instruments were used for the analyses (Table 1).

Figure 3. The landscape surroundings of several wells and the in situ test procedures.

Table 1. Water quality parameters, units and analytical methods used for groundwater samples in the pilot promoter region of Yangtze River Delta integration demonstration zone.

Parameters	Abbreviations	Units	Analytical Equipments
Water temperature	WT	°C	SX-620/SX-630
Pondus Hydrogenii	pH	pH unit	SX-620 pH Testor
Dissolved oxygen	DO	mg/L	SX-630 ORP Testor
Total dissolved solids	TDS	mg/L	Hanna DiST
Chloride	Cl^-	mg/L	Dionex-2500
Sulfate	SO_4^{2-}	mg/L	Dionex-2500
Ammonical nitrogen	NH_4-N	mg/L	AutoAnalyzer3
Nitrate nitrogen	NO_3-N	mg/L	Dionex-2500
Nitrite nitrogen	NO_2-N	mg/L	TU-1950
Potassium	K^+	mg/L	ICAP 6300Duo
Sodium	Na^+	mg/L	ICAP 6300Duo
Calcium	Ca^{2+}	mg/L	ICAP 6300Duo
Magnesium	Mg^{2+}	mg/L	ICAP 6300Duo
Manganese	Mn	µg/L	ICAP Q
Total phosphorus	TP	µg/L	ICAP 6300Duo
Iodine	I	µg/L	ICAP Q
Antimony	Sb	µg/L	ICAP Q

2.3. Multivariate Statistical Analysis

For an in-depth analysis of the groundwater chemistry data, APCS-MLR was employed. The ACPS-MLR is a receptor model based on the results of principal component analysis (PCA), together with a multivariate linear regression using the measured contaminant concentrations [46–48]. First, Kaiser-Meyer-Olkin (KMO) criteria and Bartlett's test of sphericity were performed to assess the adequacy of the dataset for PCA [49,50]. We employed the criteria that for the PCA to be considered reliable, the KMO value needs to be larger than 0.5, and the significance level of Bartlett's test of sphericity should be smaller than 0.05. By conducting PCA procedure, the principal components from the related groundwater quality parameters could be obtained, as follows:

$$(A_z)_{ij} = a_{i1}C_{1j} + a_{i2}C_{2j} + \ldots + a_{im}C_{mj}, \quad (1)$$

for $i = 1,2,\ldots,p$, and $j = 1,2,\ldots,n$, where A_z represents the component score; a stands for the component loading; C is the measured concentration of each groundwater quality parameter; p is the number of components; n is the number of samples and m is the number of groundwater quality parameters.

The principal components, with eigenvalues greater than 1.0 in the variance computation, are believed to be able to provide qualitative information about the potential contamination [51,52]. For a clearer interpretation, the original loadings normally need to be rotated until the loadings of PCs are redistributed and polarized. This procedure is normally mentioned as varimax rotation and the obtained new variables are called varifactors (VFs). For each VF, the component loadings reflect relative attribution of the groundwater quality parameters, with absolute loading values >0.75, 0.75–0.5 and 0.5–0.3 defined as strong, medium and weak, respectively [53,54].

Then, the component scores from the PCA are normalized to perform APCS-MLR for groundwater contaminant source apportionment. A detailed description of this method could be found in Thurston and Spengler [28] and Rahman et al. [30]. In brief, the APCS-MLR model assumes that the contaminant sources attribute linearly to the pollutant concentration at each sampling site. Hence, the concentration of each contaminant at each sampling site (C_{kj}) can be calculated by a multiple linear regression of the contribution of contaminant sources through Equation (2):

$$C_{kj} = r_{k0} + \sum_{i=1}^{p} r_{ki} \times APCS_{ij}, \quad (2)$$

where r_{k0} represents the constants term of multiple linear regression for pollutant k; r_{ki} stands for the coefficient of multiple linear regression of the contaminant source i for the pollutant k; $APCS_{ij}$ is the absolute principal component scores and it can be obtained through Equations (3)–(5).

$$(Z_0)_k = \frac{-\bar{C}_k}{\sigma_k}, \quad (3)$$

$$(A_0)_i = \sum_{k=1}^{m} S_{ki} \times (Z_0)_k, \quad (4)$$

$$APCS_{ij} = (A_z)_{ij} - (A_0)_i, \quad (5)$$

where $(Z_0)_k$ stands for the normalized concentration of contaminant k in a non-pollution site; $\bar{C}_k$ represents the mean concentration and σ_k indicates the standard deviation of contaminant k; $(A_0)_i$ is the principal component score in the non-pollution site; S_{ki} represents the score coefficient of component i for pollutant k; $(A_z)_{ij}$ stands for principal component score of sample j in principal component i.

Here, $r_{ki} \times APCS_{ij}$ implies the contribution of contaminant source i to pollutant k in sample j. The average of all samples $r_{ki} \times \overline{APCS_i}$ is established as the contribution of contaminant source i to pollutant k. Notably, negative values may be achieved during the calculation process, which could lead to a total contribution of all pollutants exceeding 100%. Hence, Haji Gholizadeh et al. [46] proposed an absolute value method to calculate the contribution of contaminant sources to water quality parameters, as it is shown in Equations (6) and (7).

$$PC_{ki} = \frac{\left|r_{ki} \times \overline{APCS_i}\right|}{\left|r_{k0}\right| + \sum_{i=1}^{p}\left|r_{ki} \times \overline{APCS_i}\right|} \times 100\%, \quad (6)$$

$$PC_k = \frac{|r_{k0}|}{|r_{k0}| + \sum_{i=1}^{p} |r_{ki} \times \overline{APCS_i}|} \times 100\%, \quad (7)$$

where PC_{ki} stands for the relative contribution rate of contaminant source i to the pollutant k; PC_k indicates the relative contribution rate of unrecognized source to the pollutant k; $\overline{APCS_i}$ is the average value of the absolute principal component scores of all samples.

3. Result and Discussion

3.1. Characteristics of the Groundwater Pollution

The descriptive statistics of physicochemical parameters for all groundwater samples were summarized in Table 2. For several groundwater quality parameters, such as NH_4^+, NO_2^-, NO_3^-, Mn, Ca, I and Sb, the maximum concentration in groundwater samples had exceeded level III of Chinese Groundwater Quality Standard(GB/T 14848-2017). What is more, TP, which had a maximum concentration of 2510 µg/L, also deteriorated the groundwater quality in this region.

Table 2. Summary statistics of physicochemical parameters in groundwater in the pilot promoter region of Yangtze River Delta integration demonstration zone.

Parameters	Min	Max	Mean	Standard Deviation	Coefficients of Variation (%)	National Standards, Class III
WT	18.4	29.7	23.2	2.4	10	/
pH	6.80	8.19	7.20	0.28	4	6.5–8.5
DO	1.4	24.0	3.5	2.8	80	/
TDS	40	792	383	146	38	1000
Cl^-	1.7	170.0	47.9	32.7	68	250
SO_4^{2-}	2.0	146.0	41.4	26.1	63	250
NH_4-N	0.02	4.19	0.36	0.90	250	0.5
NO_3-N	0.01	39.29	4.06	5.90	145	20
NO_2-N	0.002	2.413	0.290	0.538	186	1
K^+	1.3	90.3	20.8	18.6	89	/
Na^+	6.7	174.0	56.3	33.2	59	200
Ca^{2+}	27.6	148.0	70.0	26.0	37	75
Mg^{2+}	3.2	75.2	20.9	14.2	68	150
Mn	0.10	1610.00	78.31	266.27	340	100
TP	6.4	2510.0	484.7	556.7	115	/
I	4.4	446.0	83.3	92.2	111	80
Sb	0.10	6.28	1.63	1.45	89	5

The variation coefficient of groundwater quality parameter is an index presenting the overall variability of the samples, and such variability is believed to be caused by various anthropogenic activities [51,55,56]. Here, the geographic information system technique with inverse distance weighted interpolation method was utilized to generate the spatial distribution of groundwater quality parameters with variation coefficient larger than 80% (Figure 4). The comparison between figures showed that the contaminants could be classified into several groups with similar spatial distribution. For instance, Mn, NH_4^+ and NO_2^- were relatively enriched in the western corner of the study area, while higher concentration of TP and K^+ were detected in the centre and northwestern of the region. As for Sb, it can be found distributed all across the region, but the highest concentrations are found primarily in the southern area, which was consistent with the location of textile industries.

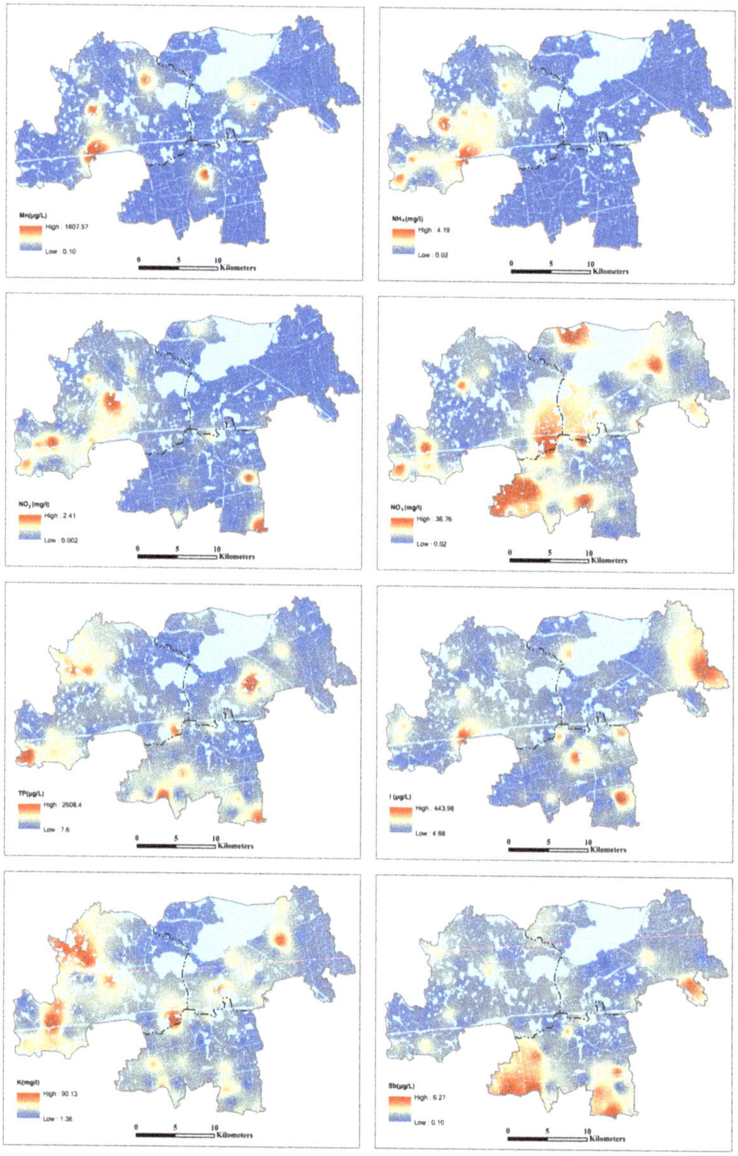

Figure 4. Spatial distribution of Mn, NH_4^+, NO_2^-, NO_3^-, TP, I, $K^+$and Sb in groundwater .

To uncover the linear correlations between physicochemical water quality parameters, the Pearson's correlation coefficient with statistical significance ($p < 0.05$) was performed here. As it was shown in Figure 5, strong correlations (r > 0.5) were observed among Na^+, Ca^{2+}, Mg^{2+} and Cl^-, indicating that they may have the same origin from hydrogeochemical processes. Their high positive correlations with TDS (r > 0.6) implied the chemical composition of groundwater in this region was mainly controlled by these ions. A moderate positive correlation (r = 0.55) was also spotted for NH_4^+ and NO_2^-, illustrating that the nitrification process is occurring in the groundwater. As for K^+, its relations with other physicochemical parameters were mostly uncorrelated, except for TP. With a positive correlation of 0.63, K^+ and TP were likely to be an indication of agricultural fertilization.

Based on the above analysis, we believe the nitrogen and phosphorus compounds could be attributed to the application of fertilizer or waste of livestock breeding. Other contaminants, such as I and Sb, can be ascribed to the natural minerals and industrial effluent, respectively.

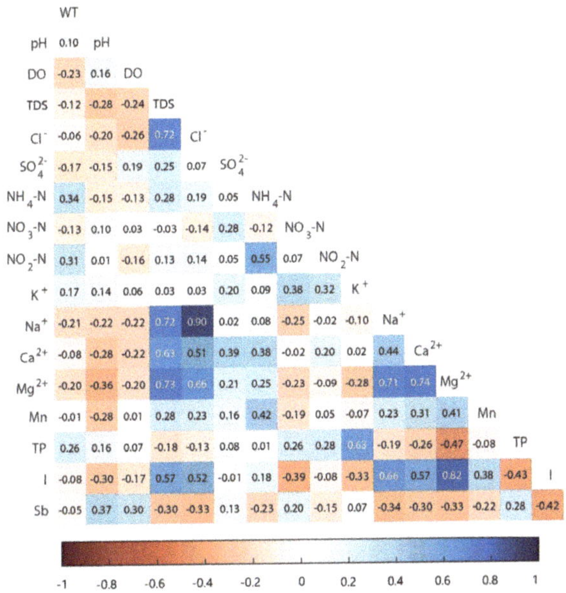

Figure 5. Correlation coefficients between groundwater quality parameters.

3.2. Pollution Sources Identified Based on the PCA Analysis

To provide a more detailed pollution-source analysis, the PCA method was then applied. The Kaiser–Meyer–Olkin(KMO) and Bartlett's test were performed first to check the adequacy of the original dataset. The KMO value was equal to 0.732, while the value of Bartlett's sphericity test was very close to zero ($p < 0.001$), indicating the validity of the PCA application. Based on the criteria, the principal components with eigenvalues exceeding 1.0 were chosen as the potential contaminant sources. As indicated in Table 3, five principal components were selected in this case, accounting for 71.23% of the total variances. The rotated factor loadings for varifactors were shown in Figure 6.

The first varifactor, VF1, accounted for 24.72% of the total variance. It was heavily weighted by TDS (0.83), Cl^- (0.91), Na^+ (0.94) and Mg^{2+} (0.79), while moderately weighted by Ca^{2+} (0.61) and I (0.70). Generally speaking, the contribution of these ions to VF1 was the result of water–rock interactions [47,57]. In addition, the quaternary aquifer marine and lagoon-facies sediments in this area are rich in iodide and serve as a source of dissolved I in groundwater [58,59]. Hence, VF1 can be interpreted as natural hydro-chemical evolution contribution. The second varifactor (VF2) accounted for 15.02% of the total variance, and was mainly characterized by K^+ (0.82) and TP (0.78), with moderate positive loadings of NO_2^- (0.52) and NO_3^- (0.50). These three ions in the groundwater are normally associated to the application of chemical fertilizer [60–62]. During our field survey, lots of aquaculture sites and farmlands could be spotted in the study area, especially near LiLi town. Hence, VF2 is assumed to represent the non-point source pollution from agricultural activities. VF3 (12.09% of the total variance) had a strong positive loading on NH_4^+ (0.85), with moderate positive loadings of NO_2^- (0.54) and Mn (0.63). The most common source of NH_4^+ and NO_2^- in groundwater are domestic sewage, livestock wastes and application of nitrogen fertilizer [48,63]. Dissolved Mn may naturally come from weathering of manganese oxide minerals, but its concentrations can be increased by contamination from industrial effluent and domestic sewage [64,65]. In the pilot promoter region of the Yangtze River Delta integration demonstration zone, as in some rural areas, the sewage collection system is imperfect and municipal sewage may leak and contaminate the groundwater. Especially in the west corner of the study area, several dry toilets and garbage heaps can be spotted during our field surveys. Therefore, VF3 is considered as domestic sewage. VF4 explained 10.02% of the total variance and had the highest loadings on SO_4^{2-} (0.78) and NO_3^- (0.55). Meanwhile, weak negative loadings on WT (−0.45) and pH (−0.41) also could be observed. Since SO_4^{2-} and low pH value (acidic condition) are normally related with industrial effluent [47,66,67], this factor can be identified as the influence of industrial activities. The last component (VF5) accounted for 9.38% of the total variance and it was mainly affected by DO (0.78) and Sb (0.64). The high level of Sb compound in groundwater normally comes from mining or textile industry [68,69]. Since the textile industries are widely distributed in the southern part of our study area, VF5 can be regarded as the impact of textile industrial effluent.

Table 3. The varifactor loadings of 17 parameters after the varimax rotated. Bold values stand for the parameters with medium or strong loadings.

Parameters	VF1	VF2	VF3	VF4	VF5
WT	−0.19	0.31	0.47	−0.45	−0.22
pH	−0.12	0.33	−0.20	−0.41	0.46
DO	−0.19	−0.09	0.04	0.18	**0.78**
TDS	**0.83**	0.03	0.13	0.23	−0.17
Cl^-	**0.91**	0.10	0.03	−0.06	−0.15
SO_4^{2-}	0.13	0.15	0.13	**0.78**	0.22
NH_4-N	0.13	0.12	**0.85**	0.01	−0.15
NO_3-N	−0.19	**0.50**	−0.31	**0.55**	−0.11
NO_2-N	0.02	**0.52**	**0.54**	−0.05	−0.25
K^+	0.01	**0.82**	0.06	0.16	0.05
Na^+	**0.94**	−0.09	−0.06	−0.09	−0.08
Ca^{2+}	**0.61**	−0.07	0.33	0.43	−0.21
Mg^{2+}	**0.79**	−0.39	0.19	0.23	−0.12
Mn	0.26	−0.26	**0.63**	0.20	0.13
TP	−0.16	**0.78**	0.06	−0.06	0.18
I	**0.70**	−0.47	0.22	−0.03	−0.12
Sb	−0.26	0.23	−0.21	0.02	**0.64**
Eigenvalues	4.20	2.55	2.06	1.70	1.60
% of Variance	24.72	15.02	12.09	10.02	9.38
Cumulative %	24.72	39.74	51.83	61.85	71.23

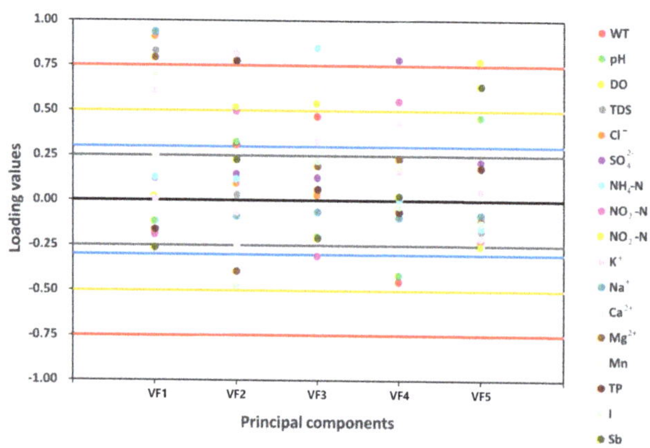

Figure 6. Component loadings for 17 parameters after varimax rotation.

3.3. Source Apportionment Using APCS-MLR

Based on the results of PCA analysis, the APCS-MLR model was then employed to quantify the contribution of each potential contaminant source to all 17 groundwater quality parameters using Equations (6) and (7). The scatter plots of predicted and observed concentrations for main groundwater pollutant parameters by using the APCS-MLR model were shown in Figure 7. Except for *Mn* (0.59) and *Sb* (0.57), the linear regression of other groundwater quality parameters were all well-matched with a R-square values larger than 0.6, indicating the contaminant source apportionment was reliable.

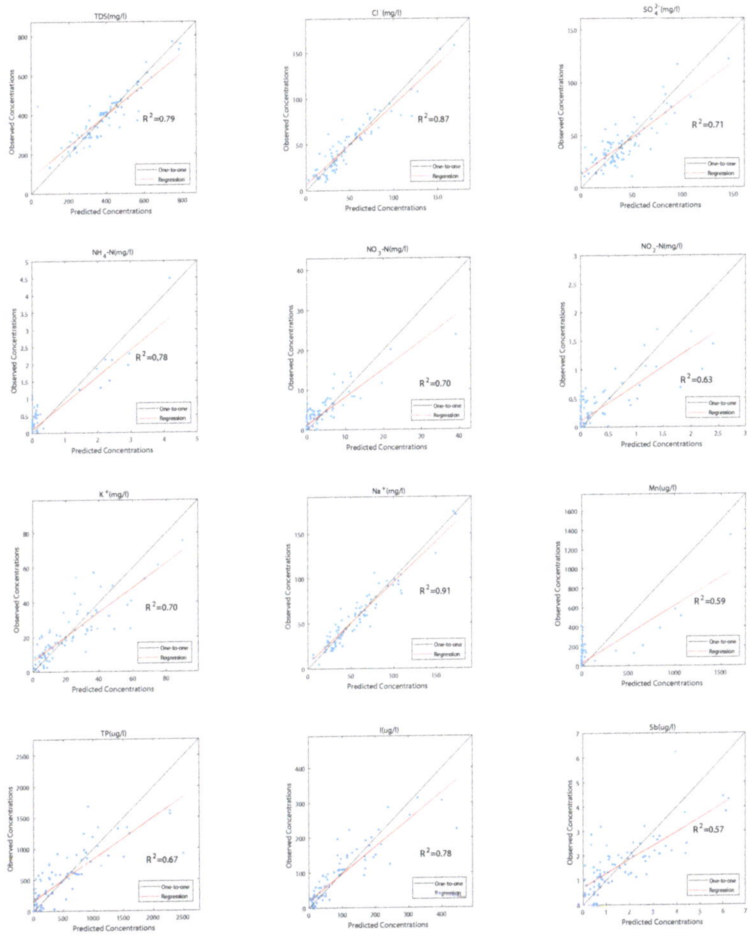

Figure 7. Scatter plots of the predicted and observed concentrations for main groundwater pollutant parameters by using APCS-MLR models.

The outcome of source apportionment via APCS-MLR model were shown in Figure 8. The relative contribution of all pollution sources to each contaminant are calculated according to Equations (6) and (7), and the average relative contribution of pollution sources is also obtained for an overall evaluation. As it is shown in Figure 8B, the hydrochemical characteristics of groundwater in the Yangtze River Delta integration demonstration zone was greatly affected by natural hydro-chemical evolution (VF1), accounting for 18.89% of the total sources and showed high contribution ratios in *TDS* (46%), Cl^- (46%), Na^+ (50%) and *I* (42%) (Figure 8A). The most threatening anthropogenic factor to groundwater quality was textile industrial effluent (VF5), accounting for 18.18% of the total sources and presented high contribution to Sb (44%). Furthermore, non-point source pollution from agricultural activities (VF2) accounted for 17.08% of the entire sources and presented a relatively high contribution to K^+ (49%), TP (36%) and NO_2^- (38%). Other industrial activities (VF4) were also responsible for 15.09% of the total sources, shown as SO_4^{2-} (45%) and NO_3^- (32%), while domestic sewage (VF3, 4.19%) presented relatively high contribution ratio in NH_4^+ (24%). In addition, the contribution of unidentified sources to each groundwater quality parameters, which are mainly due to the complex pollutants evolution processes, ranging from 2% to 59% with the average of 26.58%. Therefore, the contributions of identified sources in the Yangtze River Delta integration demonstration zone were determined as being in the following descending order: natural hydro-chemical evolution > textile industrial effluent > non-point source pollution from agricultural activities > other industrial activities > domestic sewage.

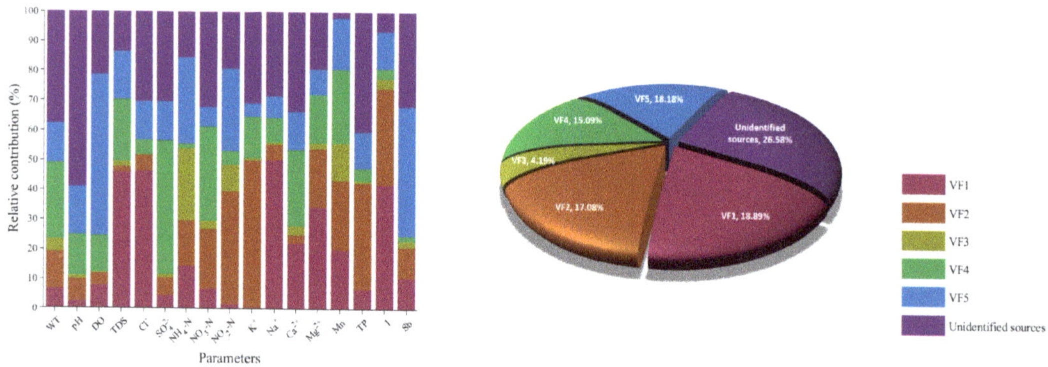

(A) (B)

Figure 8. The contributions on the groundwater quality parameters (**A**) and average contributions (**B**) of pollution sources in the pilot promoter region of Yangtze River Delta integration demonstration zone according to the APCS-MLR model.

4. Conclusions

In the present study, principal component analysis (PCA) was employed to identify the potential natural and anthropogenic aspects impacting the groundwater quality, then the absolute principal component score-multiple linear regression (APCS-MLR) model was used to quantify the contribution of potential sources to 17 groundwater quality parameters in the pilot promoter region of Yangtze River Delta integration demonstration zone, China. Based on the result of PCA analysis and possible sources for each ion, five major sources that affected the groundwater quality were identified, namely natural hydro-chemical evolution, agricultural activities, domestic sewage, textile industrial effluent and other industrial activities. With most linear regression R-square values larger than 0.6, the APCS-MLR model successfully quantified the contribution of potential sources in this study area. The contributions of five potential sources were ranked as natural hydro-chemical evolution (18.89%) > textile industrial effluent (18.18%) > non-point source pollution from agricultural activities (17.08%) > other industrial activities (15.09%) > domestic sewage (4.19%). The results clarified the groundwater in the pilot promoter region of the Yangtze River Delta integration demonstration zone was primarily contaminated by textile industrial effluent and agricultural activities, while domestic sewage only accounted for a small part of the responsibility. These insights could provide reliable advice for groundwater pollution management in highly developed areas. On the other hand, this study still suffered from a shortage of groundwater quality data, especially in temporal scale. Hence, further research with more measurements needs to be conducted and temporal variation could be a possible way to recognize the unidentified sources.

Author Contributions: Conceptualization, Z.C. and J.L.; methodology, Z.C.; validation, H.Y. (Hai Yang), S.M., Z.J. and R.S.; formal analysis, Z.C.; investigation, J.L., H.Y. (Hui Yang), H.Z., Y.J. (Yang Jin) and L.L.; resources, J.L.; data curation, J.L.; writing—original draft preparation, Z.C.; writing—review and editing, J.L. and H.Y. (Hai Yang); supervision, Q.Z. and Y.J. (Yuehua Jiang); project administration, Q.Z. and Y.J. (Yuehua Jiang). All authors have read and agreed to the published version of the manuscript.

Funding: This research was funded by China Geological Survey, grant number DD20190260 and DD20221728.

Institutional Review Board Statement: Not applicable.

Informed Consent Statement: Not applicable.

Data Availability Statement: The data that support the findings of this study are available from the corresponding author, upon reasonable request.

Conflicts of Interest: The authors declare no conflict of interest.

References

1. Han, D.; Currell, M.J.; Cao, G. Deep challenges for China's war on water pollution. *Environ. Pollut.* **2016**, *218*, 1222–1233. [CrossRef] [PubMed]
2. Jia, Y.; Xi, B.; Jiang, Y.; Guo, H.; Yang, Y.; Lian, X.; Han, S. Distribution, formation and human-induced evolution of geogenic contaminated groundwater in China: A review. *Sci. Total. Environ.* **2018**, *643*, 967–993. [CrossRef] [PubMed]
3. Burri, N.M.; Weatherl, R.; Moeck, C.; Schirmer, M. A review of threats to groundwater quality in the anthropocene. *Sci. Total. Environ.* **2019**, *684*, 136–154. [CrossRef] [PubMed]
4. Wen, D.; Zhang, F.; Zhang, E.; Wang, C.; Han, S.; Zheng, Y. Arsenic, fluoride and iodine in groundwater of China. *J. Geochem. Explor.* **2013**, *135*, 1–21. [CrossRef]
5. He, X.; Li, P.; Ji, Y.; Wang, Y.; Su, Z.; Elumalai, V. Groundwater Arsenic and Fluoride and Associated Arsenicosis and Fluorosis in China: Occurrence, Distribution and Management. *Expo. Health* **2020**, *12*, 355–368. [CrossRef]
6. Podgorski, J.; Berg, M. Global threat of arsenic in groundwater. *Science* **2020**, *368*, 845–850. [CrossRef]
7. Datta, P.S.; Deb, D.L.; Tyagi, S.K. Assessment of groundwater contamination from fertilizers in the Delhi area based on 18O, NO3/- and K+ composition. *J. Contam. Hydrol.* **1997**, *27*, 249–262. [CrossRef]
8. Gu, B.; Ge, Y.; Chang, S.X.; Luo, W.; Chang, J. Nitrate in groundwater of China: Sources and driving forces. *Glob. Environ. Chang.* **2013**, *23*, 1112–1121. [CrossRef]
9. Tao, Y.; Deng, Y.; Du, Y.; Xu, Y.; Leng, Z.; Ma, T.; Wang, Y. Sources and enrichment of phosphorus in groundwater of the Central Yangtze River Basin. *Sci. Total. Environ.* **2020**, *737*, 139837. [CrossRef]
10. Bhagure, G.R.; Mirgane, S.R. Heavy metal concentrations in groundwaters and soils of Thane Region of Maharashtra, India. *Environ. Monit. Assess.* **2011**, *173*, 643–652. [CrossRef]
11. Bhutiani, R.; Kulkarni, D.B.; Khanna, D.R.; Gautam, A. Water Quality, Pollution Source Apportionment and Health Risk Assessment of Heavy Metals in Groundwater of an Industrial Area in North India. *Expo. Health* **2016**, *8*, 3–18. [CrossRef]
12. Sharma, M.K.; Kumar, M. Sulphate contamination in groundwater and its remediation: An overview. *Environ. Monit. Assess.* **2020**, *192*, 74. [CrossRef]
13. McCance, W.; Jones, O.A.; Edwards, M.; Surapaneni, A.; Chadalavada, S.; Currell, M. Contaminants of Emerging Concern as novel groundwater tracers for delineating wastewater impacts in urban and peri-urban areas. *Water Res.* **2018**, *146*, 118–133. [CrossRef]
14. Zhou, F.; Huang, G.H.; Guo, H.; Zhang, W.; Hao, Z. Spatio-temporal patterns and source apportionment of coastal water pollution in eastern Hong Kong. *Water Res.* **2007**, *41*, 3429–3439. [CrossRef]
15. Gómez-Hernández, J.J.; Xu, T. Contaminant Source Identification in Aquifers: A Critical View. *Math. Geosci.* **2021**, *54*, 437–458. [CrossRef]
16. Lapworth, D.J.; Baran, N.; Stuart, M.E.; Ward, R.S. Emerging organic contaminants in groundwater: A review of sources, fate and occurrence. *Environ. Pollut.* **2012**, *163*, 287–303. [CrossRef]
17. Zamani, A.A.; Yaftian, M.R.; Parizanganeh, A. Multivariate statistical assessment of heavy metal pollution sources of groundwater around a lead and zinc plant. *Iran. J. Environ. Health Sci. Eng.* **2012**, *9*, 29. [CrossRef]
18. Nijenhuis, I.; Schmidt, M.; Pellegatti, E.; Paramatti, E.; Richnow, H.H.; Gargini, A. A stable isotope approach for source apportionment of chlorinated ethene plumes at a complex multi-contamination events urban site. *J. Contam. Hydrol.* **2013**, *153*, 92–105. [CrossRef]

19. Zhou, H.; Gómez-Hernández, J.J.; Li, L. Inverse methods in hydrogeology: Evolution and recent trends. *Adv. Water Resour.* **2014**, *63*, 22–37. [CrossRef]
20. Cui, J.; Zhou, F.; Gao, M.; Zhang, L.; Zhang, L.; Du, K.; Leng, Q.; Zhang, Y.; He, D.; Yang, F.; et al. A comparison of various approaches used in source apportionments for precipitation nitrogen in a mountain region of southwest China. *Environ. Pollut.* **2018**, *241*, 810–820. [CrossRef]
21. Chen, Z.; Gómez-Hernández, J.J.; Xu, T.; Zanini, A. Joint identification of contaminant source and aquifer geometry in a sandbox experiment with the restart ensemble Kalman filter. *J. Hydrol.* **2018**, *564*, 1074–1084. [CrossRef]
22. Chen, Z.; Xu, T.; Gómez-Hernández, J.J.; Zanini, A. Contaminant Spill in a Sandbox with Non-Gaussian Conductivities: Simultaneous Identification by the Restart Normal-Score Ensemble Kalman Filter. *Math. Geosci.* **2021**, *53*, 1587–1615. [CrossRef]
23. Xu, T.; Gómez-Hernández, J.J.; Chen, Z.; Lu, C. A comparison between ES-MDA and restart EnKF for the purpose of the simultaneous identification of a contaminant source and hydraulic conductivity. *J. Hydrol.* **2021**, *595*, 125681. [CrossRef]
24. Chen, Z.; Xu, T.; Gómez-Hernández, J.J.; Zanini, A.; Zhou, Q. Reconstructing the release history of a contaminant source with different precision via the ensemble smoother with multiple data assimilation. *J. Contam. Hydrol.* **2022**, 104115. [CrossRef] [PubMed]
25. Cloutier, V.; Lefebvre, R.; Therrien, R.; Savard, M.M. Multivariate statistical analysis of geochemical data as indicative of the hydrogeochemical evolution of groundwater in a sedimentary rock aquifer system. *J. Hydrol.* **2008**, *353*, 294–313. [CrossRef]
26. Li, P.; Tian, R.; Liu, R. Solute Geochemistry and Multivariate Analysis of Water Quality in the Guohua Phosphorite Mine, Guizhou Province, China. *Expo. Health* **2019**, *11*, 81–94. [CrossRef]
27. Cheng, G.; Wang, M.; Chen, Y.; Gao, W. Source apportionment of water pollutants in the upstream of Yangtze River using APCS-MLR. *Environ. Geochem. Health* **2020**, *42*, 3795–3810. [CrossRef]
28. Thurston, G.D.; Spengler, J.D. A quantitative assessment of source contributions to inhalable particulate matter pollution in metropolitan Boston. *Atmos. Environ.* **1985**, *19*, 9–25. [CrossRef]
29. Wang, H.; An, J.; Cheng, M.; Shen, L.; Zhu, B.; Li, Y.; Wang, Y.; Duan, Q.; Sullivan, A.; Xia, L. One year online measurements of water-soluble ions at the industrially polluted town of Nanjing , China : Sources , seasonal and diurnal variations. *Chemosphere* **2016**, *148*, 526–536. [CrossRef]
30. Rahman, M.S.; Bhuiyan, S.S.; Ahmed, Z.; Saha, N.; Begum, B.A. Characterization and source apportionment of elemental species in PM2.5 with especial emphasis on seasonal variation in the capital city "Dhaka", Bangladesh. *Urban Clim.* **2021**, *36*, 100804. [CrossRef]
31. Choi, H.; Cho, Y.C.; Kim, S.H.; Yu, S.J.; Kim, Y.S.; Im, J.K. Water Quality Assessment and Potential Source Contribution Using Multivariate Statistical Techniques in Jinwi River Watershed, South Korea. *Water* **2021**, *13*, 2976. [CrossRef]
32. Cho, Y.c.; Choi, H.; Lee, M.G. Sources Using Multivariate Statistical Techniques and APCS-MLR Model to Assess Surface Water Quality. *Water* **2022**, *14*, 793. [CrossRef]
33. Proshad, R.; Kormoker, T.; Al, M.A.; Islam, M.S.; Khadka, S.; Idris, A.M. Receptor model-based source apportionment and ecological risk of metals in sediments of an urban river in Bangladesh. *J. Hazard. Mater.* **2022**, *423*, 127030. [CrossRef]
34. Zhang, Q.; Wang, H.; Wang, Y.; Yang, M.; Zhu, L. Groundwater quality assessment and pollution source apportionment in an intensely exploited region of northern China. *Environ. Sci. Pollut. Res.* **2017**, *24*, 16639–16650. [CrossRef]
35. Meng, L.; Zuo, R.; Sheng Wang, J.; Yang, J.; Guo Teng, Y.; Tao Shi, R.; Zheng Zhai, Y. Apportionment and evolution of pollution sources in a typical riverside groundwater resource area using PCA-APCS-MLR model. *J. Contam. Hydrol.* **2018**, *218*, 70–83. [CrossRef]
36. Yu, L.; Zheng, T.; Yuan, R.; Zheng, X. APCS-MLR model: A convenient and fast method for quantitative identification of nitrate pollution sources in groundwater. *J. Environ. Manag.* **2022**, *314*, 115101. [CrossRef]
37. Sheng, D.; Meng, X.; Wen, X.; Wu, J.; Yu, H.; Wu, M. Contamination characteristics, source identification, and source-specific health risks of heavy metal(loid)s in groundwater of an arid oasis region in Northwest China. *Sci. Total. Environ.* **2022**, *841*, 156733. [CrossRef]
38. Xing, G.; Cao, Y.; Shi, S.; Sun, G.; Du, L.; Zhu, J. N pollution sources and denitrification in waterbodies in Taihu Lake region. *Sci. China Ser. B Chem.* **2001**, *44*, 304–314. [CrossRef]
39. Li, D.; Jiang, X.; Wang, J.; Wang, K.; Zheng, B. Effect of sewage and industrial effluents on bacterial and archaeal communities of creek sediments in the Taihu Basin. *Water* **2017**, *9*, 373. [CrossRef]
40. Yin, X.; Jiang, B.; Feng, Z.; Yao, B.; Shi, X.; Sun, Y.; Wu, J. Comprehensive evaluation of shallow groundwater quality in Central and Southern Jiangsu Province, China. *Environ. Earth Sci.* **2017**, *76*, 400. [CrossRef]
41. Jiang, Y.; Ni, H.; Zhou, Q.; Cheng, Z.; Duan, X.; Zhu, Z.; Wu, J.; Ren, H.; Fan, C.; Yang, J.; et al. Key technology of ecological restoration demonstration in the Yangtze River Economic Zone and its application. *Geol. China* **2021**, *48*, 1305–1333. [CrossRef]
42. Li, L.; Xing, H.X.; Gong, J.S.; Wang, H.S.; Zhou, K.E.; Zhou, Y.E.; Deng, T.T. Hydrochemical characteristics and format mechanism of groundwater in northern Taihu Lake Basin. *East China Geol.* **2022**, *43*, 217–226. [CrossRef]
43. Chen, T.; Wang, Z.H.; Qiang, X.K.; Ma, C.Y.; Zhan, Q. Magnetic properties of minerals recorded by the borehole WJ and Late Quaternary transgressions in the Taihu plain, southern Yangtze Delta. *Chin. J. Geophys. Chin. Ed.* **2013**, *56*, 2748–2759. [CrossRef]
44. Hua, W.; Wang, C.; Chen, G.; Yang, H.; Zhai, Y. Measurement and simulation of soil water contents in an experimental field in delta plain. *Water* **2017**, *9*, 947. [CrossRef]

45. Shi, G.; Ma, P.; Hu, X.; Huang, B.; Lin, H. Surface response and subsurface features during the restriction of groundwater exploitation in Suzhou (China) inferred from decadal SAR interferometry. *Remote Sens. Environ.* **2021**, *256*, 112327. [CrossRef]
46. Haji Gholizadeh, M.; Melesse, A.M.; Reddi, L. Water quality assessment and apportionment of pollution sources using APCS-MLR and PMF receptor modeling techniques in three major rivers of South Florida. *Sci. Total. Environ.* **2016**, *566–567*, 1552–1567. [CrossRef]
47. Li, Q.; Zhang, H.; Guo, S.; Fu, K.; Liao, L.; Xu, Y.; Cheng, S. Groundwater pollution source apportionment using principal component analysis in a multiple land-use area in southwestern China. *Environ. Sci. Pollut. Res.* **2020**, *27*, 9000–9011. [CrossRef]
48. Zhang, H.; Cheng, S.; Li, H.; Fu, K.; Xu, Y. Groundwater pollution source identification and apportionment using PMF and PCA-APCA-MLR receptor models in a typical mixed land-use area in Southwestern China. *Sci. Total. Environ.* **2020**, *741*, 140383. [CrossRef]
49. Kaiser, H.F. An index of factorial simplicity. *Psychometrika* **1974**, *39*, 31–36. [CrossRef]
50. Shrestha, S.; Kazama, F. Assessment of surface water quality using multivariate statistical techniques: A case study of the Fuji river basin, Japan. *Environ. Model. Softw.* **2007**, *22*, 464–475. [CrossRef]
51. Chen, R.; Chen, H.; Song, L.; Yao, Z.; Meng, F.; Teng, Y. Characterization and source apportionment of heavy metals in the sediments of Lake Tai (China) and its surrounding soils. *Sci. Total. Environ.* **2019**, *694*, 133819. [CrossRef]
52. Helena, B.; Pardo, R.; Vega, M.; Barrado, E.; Fernandez, J.M.; Fernandez, L. Temporal evolution of groundwater composition in an alluvial aquifer (Pisuerga River, Spain) by principal component analysis. *Water Res.* **2000**, *34*, 807–816. . [CrossRef]
53. Liu, L.; Tang, Z.; Kong, M.; Chen, X.; Zhou, C.; Huang, K.; Wang, Z. Tracing the potential pollution sources of the coastal water in Hong Kong with statistical models combining APCS-MLR. *J. Environ. Manag.* **2019**, *245*, 143–150. [CrossRef]
54. Li, W.; Wu, J.; Zhou, C.; Nsabimana, A. Groundwater Pollution Source Identification and Apportionment Using PMF and PCA-APCS-MLR Receptor Models in Tongchuan City, China. *Arch. Environ. Contam. Toxicol.* **2021**, *81*, 397–413. [CrossRef]
55. Li, Y.; Wang, S.; Zhang, W.; Yuan, J.; Xu, C. Potential drivers of the level and distribution of nitrogen in the hyporheic zone of Lake Taihu, China. *Water* **2017**, *9*, 544. [CrossRef]
56. Vadde, K.K.; Wang, J.; Cao, L.; Yuan, T.; McCarthy, A.J.; Sekar, R. Assessment of water quality and identification of pollution risk locations in Tiaoxi River (Taihu Watershed), China. *Water* **2018**, *10*, 183. [CrossRef]
57. Li, F.; Zhu, J.; Deng, X.; Zhao, Y.; Li, S. Assessment and uncertainty analysis of groundwater risk. *Environ. Res.* **2018**, *160*, 140–151. [CrossRef]
58. Wei, W.; Nghiem, A.; Ma, R.; Sun, Z.; Gong, X.; Zhou, A.; Prommer, H. Factors controlling iodine enrichment in a coastal plain aquifer in the North Jiangsu Yishusi Plain, China. *J. Contam. Hydrol.* **2021**, *243*, 103894. [CrossRef]
59. Liu, H.; Li, J.; Cao, H.; Xie, X.; Wang, Y. Prediction modeling of geogenic iodine contaminated groundwater throughout China. *J. Environ. Manag.* **2022**, *303*, 114249. [CrossRef]
60. Griffioen, J. Potassium adsorption ratios as an indicator for the fate of agricultural potassium in groundwater. *J. Hydrol.* **2001**, *254*, 244–254. [CrossRef]
61. Kendall, C.; Elliott, E.M.; Wankel, S.D. Tracing Anthropogenic Inputs of Nitrogen to Ecosystems. In *Stable Isotopes in Ecology and Environmental Science*; John Wiley & Sons, Ltd.: Hoboken, NJ, USA, 2007; Chapter 12, pp. 375–449. [CrossRef]
62. Chen, X.; Wo, F.; Chen, C.; Fang, K. Seasonal changes in the concentrations of nitrogen and phosphorus in farmland drainage and groundwater of the Taihu Lake region of China. *Environ. Monit. Assess.* **2010**, *169*, 159–168. [CrossRef] [PubMed]
63. Zanotti, C.; Rotiroti, M.; Fumagalli, L.; Stefania, G.A.; Canonaco, F.; Stefenelli, G.; Prévôt, A.S.; Leoni, B.; Bonomi, T. Groundwater and surface water quality characterization through positive matrix factorization combined with GIS approach. *Water Res.* **2019**, *159*, 122–134. [CrossRef] [PubMed]
64. Abd El-Salam, M.M.M.; El-Ghitany, E.M.A.; Kassem, M.M.M. Quality of bottled water brands in Egypt part I: Physico-chemical analyses. *J. Egypt Public Health Assoc.* **2008**, *83*, 369–388. [PubMed]
65. Huang, H.; Liu, M.; Wang, J.; He, J.; Chen, H. Sources Identification of Nitrogen Using Major Ions and Isotopic Tracers in Shenyang, China. *Geofluids* **2018**, *2018*, 8683904. [CrossRef]
66. Wu, J.; Sun, Z. Evaluation of Shallow Groundwater Contamination and Associated Human Health Risk in an Alluvial Plain Impacted by Agricultural and Industrial Activities, Mid-west China. *Expo. Health* **2016**, *8*, 311–329. [CrossRef]
67. Ghahremanzadeh, H.; Noori, R.; Baghvand, A.; Nasrabadi, T. Evaluating the main sources of groundwater pollution in the southern Tehran aquifer using principal component factor analysis. *Environ. Geochem. Health* **2017**, *40*, 1317–1328. [CrossRef]
68. Wu, F.; Sun, F.; Wu, S.; Yan, Y.; Xing, B. Removal of antimony(III) from aqueous solution by freshwater cyanobacteria Microcystis biomass. *Chem. Eng. J.* **2012**, *183*, 172–179. [CrossRef]
69. Li, Z.; Ma, Z.; van der Kuijp, T.J.; Yuan, Z.; Huang, L. A review of soil heavy metal pollution from mines in China: Pollution and health risk assessment. *Sci. Total Environ.* **2014**, *468–469*, 843–853. [CrossRef]

water

MDPI

Article

Hydrogeochemical Characteristics and Groundwater Quality in Phreatic and Confined Aquifers of the Hebei Plain, China

Yong Qian [1,2], Qinxuan Hou [2,*], Chunxiao Wang [2], Shijun Zhen [1,2], Chen Yue [1,2], Xiangxiang Cui [1,2] and Chunyan Guo [1,2]

1 Institute of Hydrogeology and Environmental Geology, Chinese Academy of Geological Sciences, Shijiazhuang 050061, China; desertqy@163.com (Y.Q.); zhenshijun@mail.cgs.gov.cn (S.Z.); yuechen24@163.com (C.Y.); cuixiangxiang@mail.cgs.gov.cn (X.C.); guochunyan@mail.cgs.gov.cn (C.G.)
2 Hebei Key Laboratory of Groundwater Remediation, Shijiazhuang 050061, China; wangchunxiao126@126.com
* Correspondence: houqinxuan@163.com

Abstract: This study aims to investigate hydrogeochemical characteristics and groundwater quality in the Hebei Plain and to discuss factors controlling the groundwater quality. A total of 54 groundwater samples were collected and analyzed for 31 hydrogeochemical parameters, and a fuzzy synthetic evaluation (FSE) method was used for assessing groundwater quality. Results show groundwater total hardness, total dissolved solids (TDS), and major ions excluding K^+ in phreatic aquifers higher than that in confined aquifers. From the Piedmont plain to the littoral plain, phreatic aquifers towards the reducing environment, and the enhancement of water–rock interaction, ion exchange process, and evaporation probably resulted in the increase in groundwater TDS, major ions (excluding HCO_3^- and SO_4^{2-}), B, and Mn concentrations. Moreover, phreatic groundwater chemistry was mainly controlled by rock weathering changing into evaporite dissolution and seawater intrusion from the Piedmont plain to the littoral plain, according to the Gibbs diagram. The proportion of drinkable groundwater in confined aquifers was 1.6 times that in phreatic aquifers. In phreatic aquifers, the proportion of drinkable groundwater in the Piedmont plain was as high as 68%, but none of the drinkable groundwater occurred in the central and littoral plains. Groundwater quality in phreatic aquifers was mainly controlled by five factors, including the water–rock interaction, the marine geogenic sources, the agricultural pollution, the acidification, and the reductive environment. By contrast, groundwater quality in confined aquifers was mainly controlled by three factors, including the water–rock interaction and redox processes, agricultural pollution, and the input of external water. Therefore, in the Hebei Plain, groundwater in confined aquifers is more suitable for drinking purposes than in phreatic aquifers. Additionally, phreatic groundwater in the Piedmont plain should be protected.

Keywords: hydrogeochemical characteristics; groundwater quality; Hebei Plain; phreatic aquifer; confined aquifer

Citation: Qian, Y.; Hou, Q.; Wang, C.; Zhen, S.; Yue, C.; Cui, X.; Guo, C. Hydrogeochemical Characteristics and Groundwater Quality in Phreatic and Confined Aquifers of the Hebei Plain, China. *Water* **2023**, *15*, 3071. https://doi.org/10.3390/w15173071

Academic Editor: Dimitrios E. Alexakis

Received: 8 July 2023
Revised: 15 August 2023
Accepted: 25 August 2023
Published: 28 August 2023

1. Introduction

Groundwater is one of the major freshwater resources for human beings in coastal plains. It plays an important role in drinking, industrial, and domestic purposes in urbanizing areas, as well as irrigation purposes in agricultural areas, especially in plains (e.g., Hebei Plain) where surface water is short [1]. For instance, Huan et al. reported that groundwater supplied more than 70% of drinking water in the Hebei Plain [2]. However, many intense human activities such as urbanization, industrialization, mining activities, and agricultural intensification have impacted regional groundwater quality in recent decades on a global scale [3–7]. For example, Gan et al. reported that domestic sewage and animal waste were major sources of groundwater nitrate pollution in several alluvial-pluvial fans in the Hebei Plain due to urbanization and agricultural activities [8]. Therefore, it is necessary to understand the current status of hydrogeochemical characteristics and groundwater quality for groundwater management in the Hebei Plain because various human activities likely have already changed hydrogeochemical conditions and water quality in the groundwater of this area [9].

To date, many studies have already investigated hydrogeochemical characteristics and groundwater quality in part areas within the Hebei Plain. For instance, Xing et al. investigated hydrogeochemical characteristics in the Hebei Plain by using two groundwater flow paths [10]. Zhang et al. assessed the natural background levels of chemical components in the groundwater of the Hutuo River catchment area within the Hebei Plain [11]. Zhang et al. revealed the spatial distribution of iodine in groundwater in the Hebei Plain via a groundwater section [12]. Zhang et al. reported the spatial distribution of groundwater chemistry and quality in the southern Hebei Plain [13]. Liu et al. revealed factors controlling groundwater chemical evolution under reduced exploitation in the Heilongjiang region within the Hebei Plain [14]. Hao et al. recently reported the annual variability of fluoride concentrations in deep groundwater of a land subsidence plain in Cangzhou within the Hebei Plain [15]. By contrast, the Hebei Plain is a large-scale geological unit, and few studies focused on the spatial distribution of hydrogeochemical characteristics and groundwater quality in this entire area [16]. Especially in recent decades, the intensity of human activities in the groundwater of this area has increased [9].

Therefore, the present study aims to investigate the current status of hydrogeochemical characteristics and groundwater quality in the whole Hebei Plain, where human activities strengthen and discuss factors controlling groundwater quality in this entire area. In this study, 54 groundwater samples were collected from both phreatic and confined aquifers. The Piper and Gibbs diagrams were used for analyzing hydrogeochemical characteristics in various aquifers and land use types and a fuzzy synthetic evaluation (FSE) method was used for assessing groundwater quality in different aquifers and land use types [17–19]. In addition, a principal components analysis (PCA) was used to extract major factors that control groundwater chemistry and quality in the study area [20]. The conclusions will enhance the groundwater management level for sustainable development in the Hebei Plain.

2. Study Area

2.1. Geographical Conditions

The Hebei Plain includes the entire plains of Beijing, Tianjin, and Hebei Province, with a total area of approximately 9.25×10^4 km^2 (Figure 1A). It is bounded by mountains in the West and North and adjacent to Bohai Bay in the east. The topography generally inclines eastward from an altitude of about 100 m above sea level (ASL) in the west to about 2 m ASL in the east [16]. The climate is typically continental semiarid with a mean annual temperature of 11.3 °C and a mean annual precipitation of approximately 500 mm/year, and the summer monsoon generally contributes more than 70% of the annual precipitation from June to September [21]. It is one of the major granaries of China. Croplands in the Hebei Plain, including 17 large irrigation districts and many small-scale irrigated farms, account for 72% of the total area, and approximately 70% of the total water supply for grain production is supported by groundwater [22].

2.2. Geological and Hydrogeological Conditions

The Hebei Plain is a large Mesozoic and Cenozoic sedimentary basin with a basement of the Sinian bedrock. Alluvial and fluvial sediments originated from the middle and lower reaches of the Yellow River, the Haihe River, the Luanhe River, and their tributaries and formed sedimentary aquifers in this basin. The sediment thickness of Quaternary deposits is about 150–500 m. The Quaternary sediments consist of fluvial deposits in the Piedmont plain, alluvial and lacustrine deposits in the central plain, and alluvial deposits with interbedded marine deposits in the littoral plain [23]. The Piedmont plain has more plentiful groundwater resources than the central and littoral plains due to the mountain-front recharge, waterbody leakage from reservoirs, and groundwater lateral flow [24].

Aquifers in the Hebei Plain can be divided into four groups according to hydrodynamic conditions and the distribution of aquifers and aquitards (Figure 1B). From the top to the bottom, the first aquifer group (aquifer-I) consists of phreatic aquifers in a range of 10–50 m below land surface (BLS) with coarse-grained sand in the Piedmont plain to fine-grained sand in the littoral plain. The second aquifer group (aquifer-II) is composed of multiple semi-confined aquifers with buried depths of 120–210 m BLS, while the third aquifer group (aquifer-III) consists of confined aquifers and has the lower boundary between 170 and 350 m BLS, both of them are dominated by sandy gravel in the Piedmont plain and by medium to fine sand in the central and littoral plains. The fourth aquifer group (aquifer-IV) also consists of confined aquifers with cemented sandy gravel and weathered sand in the Piedmont plain, medium to fine sand in the central plain, and fine sand in the littoral plain, lies below 350 m BLS with a thickness of 50–60 m [25,26]. In recent decades, groundwater recharge from rivers, lakes, and wetlands has reduced significantly because surface water flow to the Hebei Plain is often cut off by reservoirs built upstream, especially during the dry seasons [24]. Groundwater flow velocity ranged from 0.013 to 0.26 m/d in the Piedmont plain and from 0.002 to 0.10 m/d in the central plain [27]. Groundwater flows regionally from the west to the east or the northeast; in other words, it flows from the Piedmont plain via the central plain to the littoral plain [10].

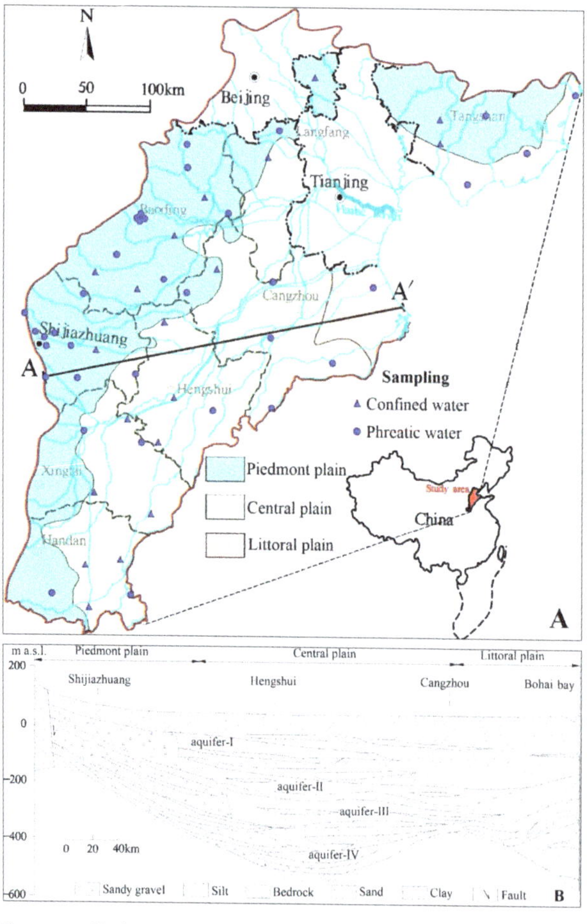

Figure 1. Hydrogeological setting and sampling sites in the Hebei Plain. (**A**) sampling sites, (**B**) cross section.

3. Materials and Methods

3.1. Groundwater Sampling

A total of 54 groundwater samples in the Hebei Plain were collected once in August 2021. Among them, 31 samples, 19 samples, and 4 samples were collected from the Piedmont plain, the central plain, and the littoral plain, respectively. In the Piedmont plain, 22 and 9 samples were collected from phreatic and confined aquifers, respectively. In the central plain, 9 and 10 samples were collected from phreatic and confined aquifers, respectively. In the littoral plain, all samples were collected from phreatic aquifers. In order to ensure samples representing the in-situ conditions were collected after pumping at least 3 well volumes or 30 min. Samples were filtered through 0.45 μm membrane filters to remove suspended solids in the field. Three bottles were used to store groundwater for the analysis of chemical components. Groundwater in a 200-mL brown glass bottle was used to analyze sulfide (S^{2-}), while two 500-mL polyethylene bottles were used to store groundwater for the analysis of trace elements and other inorganic chemicals. One bottle used for trace elements analysis was acidified with nitric acid to a pH of less than 2. All samples were stored at 4 °C until laboratory procedures could be performed.

3.2. Analytical Techniques

All analyses were carried out at the Groundwater Mineral Water and Environmental Monitoring Center of the Institute of Hydrogeology and Environmental Geology, Chinese Academy of Geological Sciences. A multi-parameter portable meter (HANNA, HI 98121, Shanghai, SH, China) was used to measure pH in the field. HCO_3^-, CO_3^{2-}, and the total dissolved solids (TDS) were measured using volumetric and gravimetric methods, respectively. Total hardness (TH) and chemical oxygen demand (COD) were measured by EDTA and potassium dichromate titration methods, respectively. Major cations, Fe, Al, and Mn, were determined by ICP-AES (ICAP6300, Thermo, New York, NY, USA). As, Pb, Hg, Cd, Cr(VI), Se, Ni, Ba, Zn, and B were measured by ICP-MS (Agilent 7500ce ICP-MS, Tokyo, Japan). S^{2-} was measured by the iodometric method. NH_4^+ and other anions (NO_3^-, SO_4^{2-}, Cl^-, NO_2^-, F^-, I^-) were carried out on IC (Shimadzu LC-10ADvp, Kyoto, Japan). To assure data quality for indicators, each groundwater sample was analyzed in triplicate, sample batches were regularly interspersed with standards and blanks, and all data were corrected for instrument drift. The relative errors were less than ±5% for all analyzed indicators.

3.3. Fuzzy Synthetic Evaluation (FSE) Method

In this study, we used a fuzzy membership function with the groundwater quality standards of China (Table 1) to evaluate groundwater quality [28]. Linear membership functions are in the Equation (1).

$$r_{ij} = \begin{cases} 0, \ (C_i \leq S_{ij-1} \text{ or } C_i \geq S_{ij-1}) \\ \frac{C_i - S_{ij-1}}{S_{ij} - S_{ij-1}}, \ (S_{ij-1} < C_i < S_{ij}) \\ \frac{S_{ij+1} - C_i}{S_{ij+1} - S_{ij}}, \ (S_{ij} < C_i < S_{ij+1}) \\ 1, \ (C_i = S_i) \end{cases} \quad (1)$$

where r_{ij} indicates the fuzzy membership of indicator i to class j; every indicator is characterized by five classes (I, II, III, IV, V) according to the groundwater quality standards of China [28]. C_i stands for the analytical value of groundwater quality indicator i, and S_{ij} stands for the allowable value of groundwater quality indicator. The fuzzy membership matrix R consists of groundwater quality indicators and classes.

The weight of groundwater quality indicator is expressed as

$$W_i = \frac{C_i}{S_i} \quad (2)$$

where W_i is the weight of groundwater quality indicator i, C_i is the analytical value of groundwater quality indicator i, and S_i is the arithmetic mean of allowable values of each class.

Table 1. Groundwater quality standards of China for drinking and irrigation.

Items	Class I	Class II	Class III	Class IV	Class V
pH	6.5–8.5	6.5–8.5	6.5–8.5	5.5–6.5 and 8.5–9.0	<5.5 and >9.0
TH (mg/L)	≤150	≤300	≤450	≤650	>650
TDS (mg/L)	≤300	≤500	≤1000	≤2000	>2000
COD (mg/L)	≤1	≤2	≤3	≤10	>10
Cl^- (mg/L)	≤50	≤150	≤250	≤350	>350
NO_3^- (as N, mg/L)	≤2	≤5	≤20	≤30	>30
SO_4^{2-} (mg/L)	≤50	≤150	≤250	≤350	>350
F^- (mg/L)	≤1	≤1	≤1	≤2	>2
NO_2^- (as N, mg/L)	≤0.01	≤0.1	≤1	≤4.8	>4.8
I^- (μg/L)	≤40	≤40	≤80	≤500	>500
S^{2-} (mg/L)	≤0.005	≤0.01	≤0.02	≤0.1	>0.1
NH_4^+ (as N, mg/L)	≤0.02	≤0.1	≤0.5	≤1.5	>1.5
Na^+ (mg/L)	≤100	≤150	≤200	≤400	>400
B (mg/L)	≤0.02	≤0.1	≤0.5	≤2	>2
Al (mg/L)	≤0.01	≤0.05	≤0.2	≤0.5	>0.5
Fe (mg/L)	≤0.1	≤0.2	≤0.3	≤2	>2
Mn (mg/L)	≤0.05	≤0.05	≤0.1	≤1.5	>1.5
Ba (mg/L)	≤0.01	≤0.1	≤0.7	≤4	>4
Zn (mg/L)	≤0.05	≤0.5	≤1	≤5	>5
Pb (μg/L)	≤5	≤5	≤10	≤100	>100
As (μg/L)	≤1	≤1	≤10	≤50	>50
Se (μg/L)	≤10	≤10	≤10	≤100	>100
Ni (μg/L)	≤2	≤2	≤20	≤100	>100
Cd (μg/L)	≤0.1	≤1	≤5	≤10	>10
Cr(VI) (μg/L)	≤5	≤10	≤50	≤100	>100
Hg (μg/L)	≤0.1	≤0.1	≤1	≤2	>2
Suitability	Drinking, Irrigation	Drinking, Irrigation	Drinking, Irrigation	Irrigation	Not suitable

Note: data from reference [28].

The normalized weight of each indicator is calculated by the formula:

$$a_i = \frac{C_i}{S_i} / \sum_{i=1}^{m} \frac{C_i}{S_i} = W_i / \sum_{i=1}^{n} W_i \quad (3)$$

where a_i is the normalized weight of indicator i and W_i is the sum of the weight of all groundwater quality indicators. The fuzzy A consists of the weight of each groundwater quality indicator.

The water quality assessment by fuzzy membership is based on the matrix B,

$$B = A \times R \quad (4)$$

The fuzzy B is the matrix of membership to each groundwater quality class. Groundwater sample is classified into the class with the maximum membership [29].

3.4. *Principal Components Analysis (PCA)*

The PCA is a useful tool for reducing a large number of variables to a small number of principal components (PCs) by linearly combining measurements [30,31]. In this study, the PCA was used to reduce variables and extract the main impact indicators that are responsible for the poor-quality groundwater. In addition, it was also used to extract related variables and infer the underlying natural and/or anthropogenic processes that control the groundwater quality [32]. Rotation of the PCs was conducted using the Varimax method. PCs with eigenvalues > 1 were retained for analyses. The PCA was operated by the SPSS® release 23.0 version.

4. Results and Discussion

4.1. Characteristics of Groundwater Chemistry

As shown in Table 2, groundwater pH in both phreatic and confined aquifers was predominantly near neutral to weak alkaline in the Hebei Plain. Similarly, the median values of groundwater pH in southern Hebei Plain from 2018 to 2020 were 7.6–7.8 [13]. This indicates that groundwater pH in the Hebei Plain in recent years was shown to be stable. Groundwater COD concentrations were a wide range of 0.31–14.11 mg/L in phreatic aquifers but a narrow range of 0.3–1.39 mg/L in confined aquifers. Both groundwater TH and TDS concentrations in phreatic and confined aquifers were shown wide ranges, and their median values in phreatic aquifers were 3.4 times and 1.5 times those in confined aquifers, respectively. Similarly, in the North China Plain containing the Hebei Plain, the median TDS value in shallow groundwater was also approximately 1.5 times that in deep groundwater from 2006 to 2008 [33]. This is probably attributed to anthropogenic inputs and/or the infiltration of minerals dissolution in the vadose zone via the water flow because wastewater with high concentrations of TDS often infiltrates into phreatic aquifers rather than confined aquifers [16,17]. Major ions except NO_3^- and K^+ in phreatic and confined aquifers were shown wide range concentrations. Generally, groundwater major anions concentrations in both phreatic and confined aquifers were orders of $HCO_3^- > SO_4^{2-} > Cl^- > NO_3^-$, and median concentrations of groundwater HCO_3^-, SO_4^{2-}, Cl^-, and NO_3^- in phreatic aquifers were 1.7 times, 2.1 times, 1.3 times, and 8 times of those in confined aquifers, respectively. This is probably due to the anthropogenic inputs of these anions and the infiltration of dissolution of related minerals (e.g., calcite, dolomite, gypsum, and halite) in the vadose zone for phreatic aquifers rather than for confined aquifers [8,16]. Unlike major anions, median values of groundwater major cations in phreatic aquifers were shown in an order of $Ca^{2+} > Na^+ > Mg^{2+} > K^+$, but those in confined aquifers presented an order of $Na^+ > Ca^{2+} > Mg^{2+} > K^+$. Median concentrations of groundwater Ca^{2+}, Mg^{2+}, and K^+ in phreatic aquifers were 4.0 times, 3.1 times, and 1.4 times those in confined aquifers, respectively. By contrast, the median value of groundwater Na^+ in phreatic aquifers was approximately half of that in confined aquifers. These are likely not only due to the anthropogenic input of K^+ and the infiltration of dissolutions of Ca and Mg-containing minerals (e.g., calcite, dolomite) in the vadose zone for phreatic aquifers but also because of the stronger ion exchange process of "Na in sediments replaced by Ca and Mg" in confined aquifers than in phreatic aquifers [9,10]. Median concentrations of trace metal(lion)s including B, Mn, Al, Fe, and Ba in groundwater in both phreatic and confined aquifers were higher than detection limits, while that of others including Zn, Ni, Cd, Pb, As, Se, Hg, and Cr(VI) in groundwater of phreatic and/or confined aquifers were lower than detection limits. Median values of groundwater Al and Ba in phreatic aquifers were approximately double those in confined aquifers. This indicates that groundwater Al and Ba in the Hebei Plain mainly originated from the anthropogenic inputs and/or the infiltration of dissolution of Al- and Ba-containing minerals in the vadose zone [32,34]. By contrast, the median concentration of groundwater F^- in phreatic aquifers was 0.8 times that in confined aquifers. This is consistent with the distribution of groundwater pH in the study area; the median pH value in phreatic aquifers was 0.9 times that in confined aquifers (Table 2). These indicate that geogenic factors such as alkaline conditions likely control the distribution of groundwater F^- in the Hebei Plain because alkalization is in favor of the desorption and dissolution of fluorinated minerals [13,35]. In addition, other chemical components, including I^-, NO_2^-, NH_4^+, and S^{2-} in groundwater of phreatic and/or confined aquifers, were also shown to have median values lower than detection limits.

Table 2. Descriptive statistics of groundwater chemical parameters in phreatic and confined aquifers.

Items	Phreatic Aquifers			Confined Aquifers		
	Min.	Med.	Max.	Min.	Med.	Max.
pH	6.75	7.49	8.22	7.17	8.21	8.82
COD	0.31	0.81	14.11	0.3	0.63	1.39
TH	69.2	469.5	6415	41.5	136.9	3808
TDS	256	686	23,550	185	463	9570
CO_3^{2-}	<DL	<DL	5.9	<DL	5.9	29.7
HCO_3^-	108.7	388.8	1009	73.1	228.9	845.7
Cl^-	9	82.3	13,830	2.5	61.6	2066
NO_3^--N	<DL	3.2	23.9	<DL	0.4	16.9
SO_4^{2-}	7	147.1	2736	6.1	70.3	3964
K^+	0.4	1.9	51	0.4	1.4	3.2
Na^+	5.9	62.5	6465	24	111.4	1940
Ca^{2+}	18.3	98.3	570.8	7.5	24.4	281.8
Mg^{2+}	6.3	41.7	1219	5.4	13.6	759
I^-	<DL	<DL	0.906	<DL	<DL	0.665
NO_2^--N	<DL	<DL	0.223	<DL	<DL	0.081
NH_4^+-N	<DL	<DL	1.02	<DL	<DL	0.37
S^{2-}	<DL	<DL	0.19	<DL	0.005	0.217
F^-	0.16	0.47	2.57	0.15	0.6	2.11
B	<DL	0.06	1.31	0.02	0.06	0.95
Mn	<DL	0.03	3.89	<DL	0.02	0.77
Al	<DL	0.07	0.54	0.01	0.03	0.15
Fe	<DL	0.06	78.31	<DL	0.07	2.94
Ba	0.01	0.07	0.41	0.01	0.04	0.19
Zn	<DL	<DL	36.11	<DL	0.01	2.26
Ni	<DL	0.001	0.007	<DL	<DL	0.006
Cd	<DL	<DL	<DL	<DL	<DL	<DL
Pb	<DL	<DL	0.001	<DL	<DL	0.003
Se	<DL	<DL	0.016	<DL	<DL	0.002
Hg	<DL	<DL	<DL	<DL	<DL	<DL
As	<DL	<DL	0.009	<DL	<DL	0.008
Cr(VI)	<DL	<DL	0.006	<DL	<DL	0.008

Notes: pH without the unit while other parameters with the unit of mg/L; <DL—below detection limits.

Statistics for concentrations of groundwater chemical parameters in phreatic aquifers in Piedmont, central, and littoral plains are present in Table 3. Groundwater pH in phreatic aquifers in all three plains was near neutral to weak alkaline. The median concentration of groundwater COD in phreatic aquifers increased gradually from the Piedmont plain to the littoral plain. This indicates that the groundwater environment from the Piedmont plain to the littoral plain is reducing environment because of the decrease in groundwater flow velocity from the Piedmont plain to the littoral plain [27]. As median and mean concentrations are concerned, TDS and most major ions, including Cl^-, K^+, Na^+, Ca^{2+}, and Mg^{2+} in groundwater in phreatic aquifers, increased gradually from the Piedmont plain to the littoral plain. This is probably attributed to the change in water–rock interaction and evaporation from the Piedmont plain to the littoral plain and the frequent occurrence of seawater intrusion in the littoral plain because ion exchange and evaporation processes for groundwater chemistry from the Piedmont plain to the littoral plain in the Hebei Plain become stronger [10,16,17,36]. For example, Zhan et al. have already pointed out that Ca- and Mg-containing minerals (e.g., calcite, dolomite, and gypsum) in groundwater in the Piedmont plain were shown less negative saturation indices than those in central and littoral plains, and resulting in concentrations of groundwater TDS accompanied by Ca and Mg from the Piedmont plain to the littoral plain increased [16]. In addition, median and mean concentrations of B and Mn in groundwater in phreatic aquifers also increased gradually from the Piedmont plain to the littoral plain. This likely ascribes to the enhancement of the ion exchange process and reductive dissolution from the Piedmont plain to the littoral plain in the Hebei Plain because the enrichment of B in groundwater is commonly accompanied by ion exchange that depletes Ca and enriches Na in groundwater [37], and the reductive dissolution of Mn containing minerals is in favor of the enrichment of Mn in groundwater [38]. By contrast, median and mean concentrations of groundwater HCO_3^-, F^-, Fe, and Zn in phreatic aquifers in the central plain were higher than those in the other two plains. In addition, median and mean concentrations of groundwater NO_3^- in phreatic aquifers in the Piedmont plain were higher than those in other two plains. This indicates that groundwater NO_3^- contamination was more often in the Piedmont plain than in the other two plains because the groundwater environment in the Piedmont plain was more oxidizing than that in the other two plains [9].

Table 3. Descriptive statistics of groundwater chemical parameters in phreatic aquifers in Piedmont, central, and littoral plains.

Items	Piedmont Plain				Central Plain				Littoral Plain			
	Min.	Med.	Mean	Max.	Min.	Med.	Mean	Max.	Min.	Med.	Mean	Max.
pH	7.03	7.53	7.52	8.22	6.75	7.34	7.38	7.88	7.22	7.45	7.53	7.99
COD	0.31	0.58	0.70	1.52	0.69	1.48	1.61	2.82	1.57	2.20	5.02	14.11
TH	69	391	386	662	356	980	1211	2619	297	920	2138	6415
TDS	256	555	575	1118	1394	2356	3109	7110	511	4646	8338	23,550
HCO_3^-	109	311	339	610	296	754	710	1009	293	485	442	507
Cl^-	9	58	61	160	135	464	545	1551	51	2143	4542	13,830
NO_3^--N	<DL	7.1	8.0	19.7	<DL	0.4	2.9	23.9	<DL	0.8	1.2	3.2
SO_4^{2-}	7	66	94	249	302	477	995	2736	21	634	629	1229
K^+	0.4	1.6	1.7	3.0	0.7	1.8	1.8	3.1	6.2	19.5	24.1	51.0
Na^+	6	32	48	200	323	530	648	1565	76	1358	2314	6465
Ca^{2+}	18	91	98	207	41	103	135	326	32	114	208	571
Mg^{2+}	6	34	36	80	57	143	215	523	25	166	394	1219
I^-	<DL	<DL	0.02	0.34	0.04	0.19	0.23	0.45	<DL	0.11	0.28	0.91
NO_2^--N	<DL	<DL	0.01	0.22	<DL	<DL	0.02	0.11	<DL	0.01	0.04	0.15
NH_4^+-N	<DL	<DL	0.01	0.14	<DL	0.06	0.19	1.02	<DL	0.17	0.18	0.37
S^{2-}	<DL	<DL	<DL	0.010	<DL	<DL	0.042	0.190	<DL	0.008	0.014	0.040
F^-	0.16	0.35	0.45	1.92	0.39	0.87	1.13	2.57	0.47	0.83	0.86	1.30
B	<DL	0.04	0.05	0.25	0.16	0.48	0.59	1.31	0.06	0.76	0.60	0.82
Mn	<DL	<DL	0.07	1.11	0.12	0.27	0.49	1.43	0.34	0.42	1.27	3.89
Al	<DL	0.07	0.07	0.16	0.03	0.09	0.15	0.54	0.03	0.09	0.09	0.15
Fe	<DL	0.03	0.10	0.61	0.04	0.76	10.11	78.31	<DL	0.64	0.65	1.33
Ba	0.01	0.09	0.12	0.41	0.01	0.02	0.03	0.07	0.02	0.08	0.11	0.25
Zn	<DL	<DL	0.01	0.09	<DL	0.01	4.02	36.11	<DL	<DL	0.01	0.03
Ni	<DL	<DL	0.001	0.004	<DL	0.003	0.003	0.004	<DL	0.002	0.003	0.007
Cd	<DL	<DL	<DL	<DL	<DL	<DL	<DL	<DL	<DL	<DL	<DL	<DL
Pb	<DL	<DL	<DL	<DL	<DL	<DL	<DL	0.001	<DL	<DL	<DL	<DL
Se	<DL	<DL	0.001	0.005	<DL	<DL	0.002	0.016	<DL	<DL	<DL	<DL
Hg	<DL	<DL	<DL	<DL	<DL	<DL	<DL	<DL	<DL	<DL	<DL	<DL
As	<DL	<DL	<DL	0.005	<DL	<DL	0.002	0.009	0.002	0.003	0.003	0.004
Cr(VI)	<DL	<DL	0.001	0.006	<DL	<DL	<DL	<DL	<DL	<DL	<DL	<DL

Notes: pH without the unit while other parameters with the unit of mg/L; <DL—below detection limits.

As seen in Figure 2, groundwater anions were generally dominated by HCO_3^- in both phreatic and confined aquifers in the Hebei Plain. By contrast, groundwater cations in phreatic aquifers were dominated by Ca^{2+} and Na^+, but confined aquifers, they were almost dominated by Na^+. The numbers of hydrogeochemical facies for groundwater in phreatic and confined aquifers were 18 and 13, respectively. Hydrogeochemical facies of groundwater in phreatic and confined aquifers were dominated by Ca·Mg-HCO_3 facies (25.7%) and Na-HCO_3 facies (15.8%), respectively. In phreatic aquifers, groundwater cations in residential areas were dominated by Ca·Mg facies, which in uncultivated lands were dominated by Na·Mg facies and Na facies, but in agricultural areas were dominated by Ca·Mg facies and Na·Mg facies. On the other hand, groundwater anions in both residential and agricultural areas were dominated by HCO_3 facies and $HCO_3{\cdot}SO_4$ facies, while in uncultivated lands they were dominated by Cl facies and HCO_3 facies. These likely hint that hydrogeochemical facies in residential areas were mainly controlled by geogenic factors, while in uncultivated and agricultural lands they were controlled by both geogenic and anthropogenic factors. Because shallow groundwater in the study area is dominated by Na facies and Cl facies, often ascribed to anthropogenic inputs, while that dominated by Ca facies, Ca·Mg facies, HCO_3 facies, and $HCO_3{\cdot}SO_4$ facies is generally attributed to geogenic sources [11,16]. Unlike in other coastal areas, such as the Pearl River Delta [39], no NO_3 facies groundwater occurred in the Hebei Plain. It is worth mentioning that all groundwater samples with the concentration of NO_3^- accounted for >10% of total major anions concentration (meq) were located in agricultural and residential areas. This indicates that groundwater NO_3^- contamination in the Hebei Plain likely originated from agricultural activities and human wastes [8].

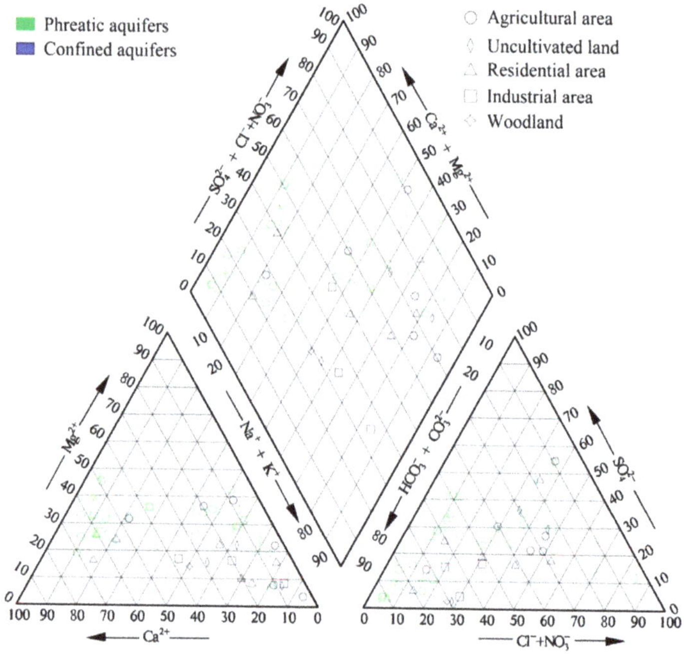

Figure 2. Hydrogeochemical facies of groundwater in phreatic and confined aquifers in the Hebei Plain.

In this study, the Gibbs Diagram was applied to analyze groundwater chemistry in the Hebei Plain. As shown in Figure 3, Na/(Na + Ca) ratios in 86.4% of phreatic groundwaters in the Piedmont plain were less than 0.5 and accompanied by low levels of TDS (<1000 mg/L). This indicates that rock weathering was the dominant mechanism for phreatic groundwater chemistry in the Piedmont plain [18]. By contrast, the confined groundwater in the Piedmont plain was also accompanied by low levels of TDS (<1000 mg/L) but commonly showed higher Na/(Na + Ca) ratios than that in phreatic groundwaters, approximately 67% of confined groundwaters in the Piedmont plain were characterized by Na/(Na + Ca) ratios > 0.5. This infers both rock weathering and ion exchange processes controlling groundwater chemistry in confined aquifers in the Piedmont plain because longer residence times for groundwater in confined aquifers than in phreatic aquifers is in favor of the ion exchange process of sediments adsorbed Na replaced by groundwater Ca [23,26,40]. Compared to the Piedmont plain, Na/(Na + Ca) ratios and

TDS concentrations in both phreatic and confined groundwaters in the central plain were higher. This implies that the evaporation process and/or the dissolution of evaporites with cation exchange for groundwater chemistry in the central plain were likely more important than that in the Piedmont plain because the depth of groundwater level in the central plain was shallower than that in the Piedmont plain and the residence time for groundwater in the central plain was longer than that in the Piedmont plain [26,41]. On the other hand, the depth of groundwater level in the central plain in both phreatic and confined aquifers was generally deeper than 3 m [9,21], indicating that the impact of the evaporation process on groundwater chemistry in the central plain was negligible because evaporation from groundwater is often little when the groundwater level deeper than a few meters [42]. Therefore, evaporite dissolution with cation exchange mainly controlled groundwater chemistry in the central plain. Additionally, in the central plain, the median concentration of TDS in phreatic groundwater was more than three times that in confined groundwater. This indicates that evaporite dissolution for groundwater chemistry in phreatic aquifers in the central plain was more important than confined aquifers [18]. In phreatic aquifers, groundwater Na/(Na + Ca) ratios and TDS concentrations in the littoral plain were commonly higher than in Piedmont and central plains. This is likely attributed to strong evaporite dissolution and the occurrence of seawater intrusion in the littoral plain [9,43].

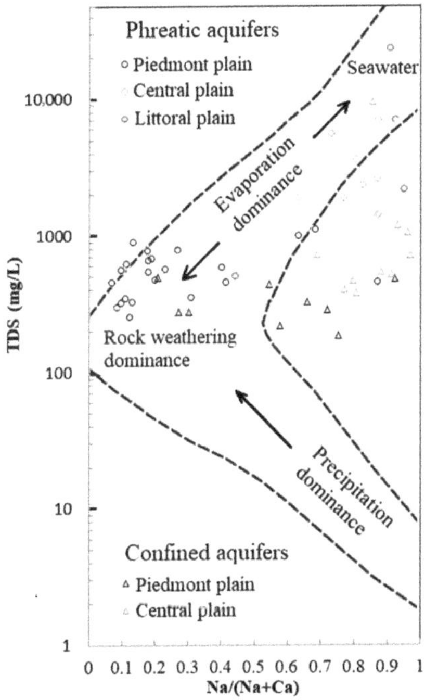

Figure 3. Gibbs diagram of groundwater samples in the Hebei Plain.

4.2. Distribution of Groundwater Quality

Compared to standards for groundwater quality in China [28], proportions of groundwater samples with concentrations of chemical components exceeded allowable limits (PEAL) for drinking purposes are shown in Table 4. In the Hebei Plain, TH was shown the highest PEAL of 54.3% in phreatic aquifers, followed by ten components (Mn, TDS, Na^+, Fe, Cl^-, SO_4^{2-}, I^-, B, F^-, and S^{2-}) with their PEALs > 10% and six components (Al, COD, NO_3^-, NH_4^+, Zn, and Se) with their PEALs of 2–10%, while the PEALs of other nine components in phreatic aquifers were zero. By contrast, in confined aquifers, S^{2-} was shown the highest PEAL of 31.6%, followed by seven components (F^-, Na^+, Mn, TDS, Fe, Cl^-, and I^-) with their PEALs > 10% and five components (TH, SO_4^{2-}, B, Zn, and pH) with their PEALs of 5.3%, while other 13 components with their PEALs of zero. Among them, the PEALs of TH, SO_4^{2-}, and B in phreatic aquifers were more than 10 times, 5 times, and 3 times those in confined aquifers. The PEALs of Mn, TDS, Na^+, Fe, Cl^-, and I^- in phreatic aquifers were 2–3 times those in confined aquifers. In addition, the PEALs of Al, COD, NO_3^-, NH_4^+, and Se in phreatic aquifers were also higher than those in confined aquifers. In contrast, the PEALs of S^{2-}, F^-, pH, and Zn in phreatic aquifers were lower than those in confined aquifers. Within phreatic aquifers, seven components including TH, Mn, Fe, TDS, Na^+, F^-, and I^- in the Piedmont plain showed their PEALs > 0; by contrast, 17 components including Mn, TDS, Na^+, Cl^-, SO_4^{2-}, TH, I^-, Fe, B, F^-, S^{2-}, Al, Zn, COD, Se, NO_3^-, and NH_4^+ in central and littoral plains showed their PEALs > 0. Moreover, PEALs of all these 17 components in central and littoral plains in phreatic aquifers were much higher than those in the Piedmont plain. For example, PEALs of Mn, TDS, Na^+, Fe, I^-, and F^- in central and littoral plains in phreatic aquifers were much more than 7 times, 10 times, 20 times, 5 times, 17 times, and 8 times those in the Piedmont plain, respectively. These indicate that groundwater quality in the Piedmont plain within phreatic aquifers was better than in central and littoral plains.

Table 4. Statistics for proportions of groundwaters with chemical concentrations exceeded allowable limits.

Items	Allowable Limits	Phreatic Aquifers	Confined Aquifers	Piedmont Plain	Central and Littoral Plains
TH	<450 mg/L	54.3%	5.3%	40.9%	76.9%
Mn	<0.1 mg/L	45.7%	15.8%	13.6%	100%
TDS	<1000 mg/L	40.0%	15.8%	9.1%	92.3%
Na^+	<200 mg/L	37.1%	21.1%	4.5%	92.3%
Fe	<0.3 mg/L	34.3%	15.8%	13.6%	69.2%
Cl^-	<250 mg/L	31.4%	15.8%	0	84.6%
I^-	<0.08 mg/L	31.4%	15.8%	4.5%	76.9%
SO_4^{2-}	<250 mg/L	31.4%	5.3%	0	84.6%
B	<0.5 mg/L	20.0%	5.3%	0	53.8%
F^-	<1 mg/L	17.1%	26.3%	4.5%	38.5%
S^{2-}	<0.02 mg/L	14.3%	31.6%	0	38.5%
Al	<0.2 mg/L	5.7%	0	0	15.4%
Zn	<1 mg/L	2.9%	5.3%	0	7.7%
COD	<3 mg/L	2.9%	0	0	7.7%
NO_3^--N	<20 mg/L	2.9%	0	0	7.7%
NH_4^+-N	<0.5 mg/L	2.9%	0	0	7.7%
Se	<0.01 mg/L	2.9%	0	0	7.7%
pH	6.5–8.5	0	5.3%	0	0
NO_2^--N	<1 mg/L	0	0	0	0
Ba	<0.7 mg/L	0	0	0	0
Ni	<0.02 mg/L	0	0	0	0
Cd	<0.005 mg/L	0	0	0	0
Pb	<0.01 mg/L	0	0	0	0
Hg	<0.001 mg/L	0	0	0	0
As	<0.01 mg/L	0	0	0	0
Cr(VI)	<0.05 mg/L	0	0	0	0

Notes: PEAL = proportion of groundwater samples with the concentration of one component exceeded the allowable limit; data of allowable limits are from reference [28].

The groundwater quality (5 classes) of the Hebei Plain was assessed by the FSE method and shown in Figure 4. In phreatic aquifers, the percentages of classes II, III, IV, and V of groundwater samples were 17.2%, 25.7%, 25.7%, and 31.4%, respectively. In other words, 43% and 57% of groundwater samples in phreatic aquifers were good-quality (drinkable, classes I to III) and poor-quality (undrinkable, classes IV and V), respectively. By contrast, the percentages of classes I, II, III, IV, and V groundwater samples in confined aquifers were 15.8%, 21.0%, 31.6%, 15.8%, and 15.8%, respectively. The proportion of drinkable groundwater in confined aquifers was 1.6 times that in phreatic aquifers. In phreatic aquifers, the proportion of drinkable groundwater in the Piedmont plain was as high as 68%, but none of drinkable groundwater occurred in the central and littoral plains. In addition, the proportions of drinkable groundwater in phreatic aquifers in different land use types were in the order of residential area (55.5%) > agricultural area (53.9%) > industrial area (50%) > woodland (33.3%) > uncultivated land (12.5%). This indicates that groundwater quality in phreatic aquifers in the Hebei Plain was probably controlled by natural factors rather than anthropogenic factors because groundwater pollution was generally more frequent in three former land use types than in two latter ones [44].

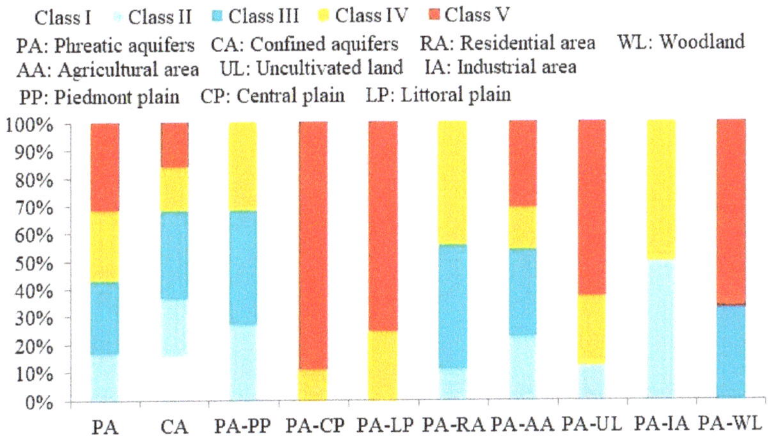

Figure 4. Groundwater quality in various aquifers and different land use types in the Hebei Plain.

4.3. Factors Controlling Groundwater Quality

In this study, the PCA was used to analyze the factors controlling groundwater quality in phreatic and confined aquifers in the Hebei Plain. Here, parameters in the PCA include pH, major ions, and exceeding indicators that concentrations in one or more groundwater samples are higher than their allowable limits. As shown in Table 5, groundwater quality in phreatic aquifers was mainly controlled by five factors. PC1 (factor 1) had strong positive loadings with Cl^-, COD, Na^+, TDS, Mn, TH, Mg^{2+}, K^+, and Ca^{2+}, indicating that the PC1 was indicative of the water–rock interaction (e.g., minerals dissolution, ion exchange) because the dissolution of (evaporate) minerals and the ion exchange process were mainly responsible for enrichments of Cl^-, Na^+, TDS, Mn, TH, Mg^{2+}, K^+, and Ca^{2+} in phreatic groundwater (Figure 3) [16]. PC2 (factor 2) showed high positive loadings of I^-, HCO_3^-, and B and moderate positive loadings of SO_4^{2-} and F^-. This was likely representative of the marine geogenic sources because groundwater SO_4^{2-} and B often originated from the oxidation of pyrite and the dissolution of glauconite in carbonate rocks in marine sediments [45], and the enrichment of groundwater I^- and HCO_3^- accompanied with alkalization (in favor of groundwater F^- enrichment) is often resulted from the mineralization of organic iodine in marine sediments in coastal areas [33,46–48]. PC3 (factor 3) explained 11.6% of the total variance with strong to moderate positive loadings of Se, Zn, and NO_3^-. This probably represented the agricultural pollution because groundwater with high levels of NO_3^- and Zn in phreatic aquifers in the Hebei Plain was generally distributed in agricultural areas (Figure 2) [8], and NO_3^- inputs were in favor of groundwater Se enrichment via the oxidation of reduced Se in sediments [34]. PC4 (factor 4) had a strong negative loading with pH and a moderate positive loading with Fe, indicating that the PC4 likely represented the acidification because the acidification elevates the mobilization of Fe in sediments [49]. PC5 (factor 5) showed a high positive loading of NH_4^+ and weak positive loadings of S^{2-} and Al. This was likely indicative of the reductive environment because the reductive environment is in favor of the enrichment of NH_4^+ and S^{2-} in groundwater [3,50].

Table 5. Principal component (PC) loadings for groundwater chemical parameters in phreatic and confined aquifers in the Hebei Plain.

Items	Phreatic Aquifers				
	PC1	PC2	PC3	PC4	PC5
Cl^-	**0.976**	0.063	−0.045	−0.064	−0.079
COD	**0.976**	0.065	0.000	0.025	0.024
Na^+	**0.969**	0.215	−0.018	−0.004	−0.016
TDS	**0.964**	0.242	0.001	0.065	0.009
Mn	**0.932**	0.183	−0.111	0.089	0.069
TH	**0.931**	0.262	0.012	0.217	0.053
Mg^{2+}	**0.924**	0.312	−0.006	0.161	0.047
K^+	**0.889**	−0.124	−0.058	−0.198	0.052
Ca^{2+}	**0.857**	0.048	0.090	0.416	0.071
I^-	0.161	**0.826**	−0.221	−0.020	0.029
HCO_3^-	0.100	**0.806**	0.123	0.224	0.073
B	0.479	**0.759**	0.184	−0.063	0.215
SO_4^{2-}	0.424	**0.680**	0.180	0.297	0.199
F^-	0.041	**0.574**	−0.082	−0.557	0.144
Se	−0.034	0.066	**0.955**	0.031	−0.008
Zn	0.022	0.184	**0.935**	−0.075	0.109
NO_3^-	−0.098	−0.323	**0.706**	0.173	−0.276
pH	−0.191	−0.281	−0.052	**−0.800**	−0.066
Fe	0.078	0.106	−0.101	**0.617**	0.499
NH_4^+	0.032	0.025	−0.056	0.072	**0.853**
S^{2-}	−0.009	0.349	0.256	0.221	**0.487**
Al	0.000	0.260	−0.155	−0.354	**0.435**
Eigenvalue	8.4	3.5	2.6	2.0	1.6
Explained variance (%)	38.1	15.7	11.6	9.1	7.4
Cumulative % of variance	38.1	53.8	65.4	74.5	81.9

Items	Confined Aquifers		
	PC1	PC2	PC3
TDS	**0.991**	0.015	0.083
Cl^-	**0.991**	0.058	0.033
Na^+	**0.986**	0.072	0.008
Mg^{2+}	**0.985**	−0.018	0.139
TH	**0.982**	−0.054	0.158
SO_4^{2-}	**0.981**	−0.010	0.131
Mn	**0.969**	−0.056	0.100
B	**0.950**	0.223	−0.131
HCO_3^-	**0.945**	−0.105	−0.055
Fe	**0.943**	0.225	0.142
I^-	**0.915**	0.249	−0.086
Ca^{2+}	**0.914**	−0.210	0.239
COD	**0.686**	0.298	−0.328
Zn	0.024	**0.925**	−0.033
S^{2-}	0.611	**0.771**	−0.091
pH	−0.501	**0.648**	−0.253
K^+	0.220	0.074	**0.832**
F^-	0.089	0.289	**−0.805**
Eigenvalue	12.3	2.3	1.7
Explained variance (%)	68.2	12.7	9.5
Cumulative % of variance	68.2	80.9	90.4

Note: Bold numbers = maximum absolute PC loading of one parameter.

By contrast, groundwater quality in confined aquifers was mainly controlled by three factors. PC1 (factor 1) had strong positive loadings with TDS, Cl^-, Na^+, Mg^{2+}, TH, SO_4^{2-}, Mn, B, HCO_3^-, Fe, I^-, and Ca^{2+}, and a moderate positive loading with COD. As mentioned previously, the water–rock interaction, such as minerals dissolution and ion exchange, was mainly responsible for the enrichment of TDS, Cl^-, Na^+, Mg^{2+}, TH, SO_4^{2-}, HCO_3^-, and Ca^{2+} in confined groundwater (Figure 3). Additionally, the reductive dissolution of Fe/Mn (hydroxy)oxides loaded with I^- in the reductive environment was mainly responsible for the enrichment of groundwater Mn, Fe, and I^- in the Hebei Plain [46]. Therefore, the PC1 was assumed to be representative of the water–rock interaction and redox processes. PC2 (factor 2) showed strong positive loadings with Zn and S^{2-} and a moderate positive loading with pH, indicating that the PC2 was not a natural factor because groundwater zinc generally presents in low concentrations in the presence of sulfide [51]. On the other hand, groundwater with high values of Zn, S^{2-}, and pH in confined aquifers was distributed in the agricultural area. Therefore, it can be concluded that the PC2 was likely indicative of agricultural pollution. PC3 (factor 3) had strong positive and negative loadings with K^+ and F^-, respectively. This likely represented the input of external water because the input of external water, such as the ecological water with low levels of F^- occurred in the Hebei Plain in recent years [15].

5. Conclusions

Hydrogeochemical characteristics and groundwater quality in phreatic and confined aquifers of the Hebei Plain were investigated. Anthropogenic inputs and/or the infiltration of minerals dissolution in the vadose zone resulted in groundwater TH, TDS, Al, Ba, and major ions excluding K^+ in phreatic aquifers higher than in confined aquifers. By contrast, more alkaline conditions were likely responsible for higher concentrations of groundwater F^- in confined aquifers than in phreatic aquifers. From the Piedmont plain to the littoral plain, phreatic aquifers towards the reducing environment, and the enhancement of water–rock interaction, ion exchange process, and evaporation probably resulted in the increase of groundwater TDS, major ions (excluding HCO_3^- and SO_4^{2-}), B, and Mn concentrations. Moreover, the phreatic groundwater chemistry was mainly controlled by rock weathering changing into evaporite dissolution and seawater intrusion from the Piedmont plain to the littoral plain, according to the Gibbs diagram.

In the Hebei Plain, groundwater in confined aquifers is more suitable to be developed for drinking purposes than in phreatic aquifers because the proportion of drinkable groundwater in confined aquifers was 1.6 times that in phreatic aquifers according to this investigation. For phreatic aquifers, groundwater in the Piedmont plain should be protected because of the much higher proportion of drinkable groundwater in the Piedmont plain than in the other two plains. Groundwater quality in phreatic aquifers was mainly controlled by five factors, including the water–rock interaction, the marine geogenic sources, the agricultural pollution, the acidification, and the reductive environment. By contrast, groundwater quality in confined aquifers was mainly controlled by three factors, including the water–rock interaction and redox processes, agricultural pollution, and the input of external water.

This is the first time to depict the distribution of groundwater chemistry and quality in the whole Hebei Plain in recent years. However, knowledge of the origins of various poor-quality groundwaters in this area is still limited. Thus, in the future, we will focus on sources, driving forces, and treatment of poor-quality groundwater in this area.

Author Contributions: Conceptualization, Y.Q. and Q.H.; methodology, C.W.; software, C.G.; validation, Y.Q., C.W. and S.Z.; formal analysis, C.Y.; investigation, Q.H., S.Z. and X.C.; resources, Y.Q.; data curation, Q.H.; writing—original draft preparation, Y.Q.; writing—review and editing, Q.H. and C.W.; visualization, C.G.; supervision, X.C.; project administration, C.Y.; funding acquisition, Y.Q. All authors have read and agreed to the published version of the manuscript.

Funding: This research was supported by the Investigation and Evaluation of the Groundwater Basic Environment around National Control Assessment Points in Hebei Province in China (IHEGJS2021083) and the S&T Program of Hebei in China (21567632H).

Data Availability Statement: The datasets generated and/or analyzed during the current study are not publicly available.

Conflicts of Interest: The authors declare no conflict of interest.

References

1. Lu, X.; Jin, M.; van Genuchten, M.T.; Wang, B. Groundwater Recharge at Five Representative Sites in the Hebei Plain, China. *Groundwater* **2011**, *49*, 286–294. [CrossRef]
2. Huan, H.; Hu, L.; Yang, Y.; Jia, Y.; Lian, X.; Ma, X.; Jiang, Y.; Xi, B. Groundwater nitrate pollution risk assessment of the groundwater source field based on the integrated numerical simulations in the unsaturated zone and saturated aquifer. *Environ. Int.* **2020**, *137*, 105532. [CrossRef]
3. Zhang, M.; Huang, G.; Liu, C.; Zhang, Y.; Chen, Z.; Wang, J. Distributions and origins of nitrate, nitrite, and ammonium in various aquifers in an urbanized coastal area, south China. *J. Hydrol.* **2020**, *582*, 124528. [CrossRef]

4. Ayari, J.; Barbieri, M.; Agnan, Y.; Sellami, A.; Braham, A.; Dhaha, F.; Charef, A. Trace element contamination in the mine-affected stream sediments of Oued Rarai in north-western Tunisia: A river basin scale assessment. *Environ. Geochem. Health* **2021**, *43*, 4027–4042. [CrossRef]
5. Huang, G.; Pei, L.; Li, L.; Liu, C. Natural background levels in groundwater in the Pearl River Delta after the rapid expansion of urbanization: A new pre-selection method. *Sci. Total Environ.* **2022**, *813*, 151890. [CrossRef]
6. Ayari, J.; Barbieri, M.; Barhoumi, A.; Boschetti, T.; Braham, A.; Dhaha, F.; Charef, A. Trace metal element pollution in media from the abandoned Pb and Zn mine of Lakhouat, Northern Tunisia. *J. Geochem. Explor.* **2023**, *247*, 107180. [CrossRef]
7. Bi, P.; Liu, R.; Huang, G.; Li, D. Evaluating natural background levels of heavy metals in shallow groundwater of the Pearl River Delta via removal of contaminated groundwaters: Comparison of three preselection related methods. *Environ. Pollut.* **2023**, *335*, 122382. [CrossRef]
8. Gan, L.; Huang, G.; Pei, L.; Gan, Y.; Liu, C.; Yang, M.; Han, D.; Song, J. Distributions, origins, and health-risk assessment of nitrate in groundwater in typical alluvial-pluvial fans, North China Plain. *Environ. Sci. Pollut. Res.* **2022**, *29*, 17031–17048. [CrossRef]
9. Qian, Y.; Cui, X.; Yue, C.; Guo, C.; Zhen, S.; Wang, W.; Huang, G.; Li, H.; Wang, Y.; Su, C.; et al. *Report on the Investigation and Evaluation of the Basic Environmental Conditions of Groundwater Around the State Control Assessment Points in Hebei Province*; The Institute of Hydrogeology and Environmental Geology, Chinese Academy of Geological Sciences: Shijiazhuang, China, 2022. (In Chinese)
10. Xing, L.; Guo, H.; Zhan, Y. Groundwater hydrochemical characteristics and processes along flow paths in the North China Plain. *J. Asian Earth Sci.* **2013**, *70–71*, 250–264. [CrossRef]
11. Zhang, Y.; Chen, Z.; Sun, J.; Wang, J. Natural background levels of chemical components in groundwater of Hutuo River catchment area, North China Plain. *Environ. Forensics* **2017**, *18*, 62–73. [CrossRef]
12. Zhang, Y.; Wu, Y.; Sun, J.; Hu, S.; Zhang, Y.; Xiang, X. Controls on the spatial distribution of iodine in groundwater in the Hebei Plain, China. *Environ. Sci. Pollut. Res.* **2018**, *25*, 16702–16709. [CrossRef]
13. Zhang, L.; Dong, D.; Lv, S.; Ding, J.; Yan, M.; Han, G. Spatial evolution analysis of groundwater chemistry, quality, and fuoride health risk in southern Hebei Plain, China. *Environ. Sci. Pollut. Res.* **2023**, *30*, 61032–61051. [CrossRef]
14. Liu, F.; Zou, J.; Liu, J.; Zhang, J.; Zhen, P. Factors controlling groundwater chemical evolution with the impact of reduced exploitation. *Catena* **2022**, *214*, 106261. [CrossRef]
15. Hao, C.; Wei, Q.; Dong, H.; Ma, T.; Liu, X.; Li, D.; Dong, J. Geochemical evidence for annual variability of fluoride levels in deep groundwater on land subsidence plain in Cangzhou, Hebei Province, China. *Geochemistry* **2023**, 125985. [CrossRef]
16. Zhan, Y.; Guo, H.; Wang, Y.; Li, R.; Hou, C.; Shao, J.; Cui, Y. Evolution of groundwater major components in the Hebei Plain: Evidences from 30-year monitoring data. *J. Earth Sci.* **2014**, *25*, 563–574. [CrossRef]
17. Huang, G.; Sun, J.; Zhang, Y.; Chen, Z.; Liu, F. Impact of anthropogenic and natural processes on the evolution of groundwater chemistry in a rapidly urbanized coastal area, South China. *Sci. Total Environ.* **2013**, *463–464*, 209–221. [CrossRef]
18. Marandi, A.; Shand, P. Groundwater chemistry and the Gibbs Diagram. *Appl. Geochem.* **2018**, *97*, 209–212. [CrossRef]
19. Liu, F.; Huang, G.; Sun, J.; Jing, J.; Zhang, Y. A new evaluation method for groundwater quality applied in Guangzhou region, China: Using fuzzy method combining toxicity index. *Water Environ. Res.* **2016**, *88*, 99–106. [CrossRef]
20. Güler, C.; Kurt, M.A.; Alpaslan, M.; Akbulut, C. Assessment of the impact of anthropogenic activities on the groundwater hydrology and chemistry in Tarsus coastal plain (Mersin, SE Turkey) using fuzzy clustering, multivariate statistics and GIS techniques. *J. Hydrol.* **2012**, *414–415*, 435–451. [CrossRef]
21. Hu, X.; Shi, L.; Zeng, J.; Yang, J.; Zha, Y.; Yao, Y.; Cao, G. Estimation of actual irrigation amount and its impact on groundwater depletion: A case study in the Hebei Plain, China. *J. Hydrol.* **2016**, *543*, 433–449. [CrossRef]
22. Cao, G.; Zheng, C.; Scanlon, B.R.; Liu, J.; Li, W. Use of flow modeling to assess sustainability of groundwater resources in the North China Plain. *Water Resour. Res.* **2013**, *49*, 159–175. [CrossRef]
23. Chen, Z.; Qi, J.; Xu, J.; Xu, J.; Ye, H.; Nan, Y. Paleoclimatic interpretation of the past 30 ka from isotopic studies of the deep confined aquifer of the North China Plain. *Appl. Geochem.* **2003**, *18*, 997–1009.
24. Zhong, H.; Sun, L.; Fischer, G.; Tian, Z.; Liang, Z. Optimizing regional cropping systems with a dynamic adaptation strategy for water sustainable agriculture in the Hebei Plain. *Agric. Syst.* **2019**, *173*, 94–106. [CrossRef]
25. Zhang, Z.; Shi, D.; Shen, Z.; Zhong, Z.; Xue, Y. Evolution and development of groundwater environment in North China Plain under human activities. *Acta Geosci. Sin.* **1997**, *18*, 337–344.
26. Chen, Z.; Nie, Z.; Zhang, Z.; Qi, J.; Nan, Y. Isotopes and sustainability of ground water resources, North China Plain. *Ground Water* **2005**, *43*, 485–493.
27. Zhang, Z.; Fei, Y.; Chen, Z.; Zhao, Z.; Xie, Z.; Wang, Y. *Investigation and Assessment of Sustainable Utilization of Groundwater Resources in the North China Plain*; Geology Press: Beijing, China, 2009; pp. 28–42. (In Chinese)
28. General administration of quality supervision inspection and quarantine of the people's republic of China (GAQSIQPRC). In *Standard for Groundwater Quality*; Standards Press of China: Beijing, China, 2017.
29. Zhang, F.; Huang, G.; Hou, Q.; Liu, C.; Zhang, Y.; Zhang, Q. Groundwater quality in the Pearl River Delta after the rapid expansion of industrialization and urbanization: Distributions, main impact indicators, and driving forces. *J. Hydrol.* **2019**, *577*, 124004. [CrossRef]
30. Huang, G.; Chen, Z.; Liu, F.; Sun, J.; Wang, J. Impact of human activity and natural processes on groundwater arsenic in an urbanized area (South China) using multivariate statistical techniques. Environ. *Sci. Pollut. Res.* **2014**, *21*, 13043–13054. [CrossRef]

31. Huang, G.; Liu, C.; Zhang, Y.; Chen, Z. Groundwater is important for the geochemical cycling of phosphorus in rapidly urbanized areas: A case study in the Pearl River Delta. *Environ. Pollut.* **2020**, *260*, 114079. [CrossRef]
32. Huang, G.; Hou, Q.; Han, D.; Liu, R.; Song, J. Large scale occurrence of aluminium-rich shallow groundwater in the Pearl River Delta after the rapid urbanization: Co-effects of anthropogenic and geogenic factors. *J. Contam. Hydrol.* **2023**, *254*, 104130. [CrossRef] [PubMed]
33. Zhang, E.; Wang, Y.; Qian, Y.; Ma, T.; Zhang, D.; Zhan, H.; Zhang, Z.; Fei, Y.; Wang, S. Iodine in groundwater of the North China Plain: Spatial patterns and hydrogeochemical processes of enrichment. *J. Geochem. Explor.* **2013**, *135*, 40–53. [CrossRef]
34. Huang, G.; Zhang, M.; Liu, C.; Li, L.; Chen, Z. Heavy metal(loid)s and organic contaminants in groundwater in the Pearl River Delta that has undergone three decades of urbanization and industrialization: Distributions, sources, and driving forces. *Sci. Total Environ.* **2018**, *635*, 913–925. [CrossRef] [PubMed]
35. Xiao, Y.; Hao, Q.; Zhang, Y.; Zhu, Y.; Yin, S.; Qin, L.; Li, X. Investigating sources, driving forces and potential health risks of nitrate and fluoride in groundwater of a typical alluvial fan plain. *Sci. Total Environ.* **2022**, *802*, 149909. [CrossRef] [PubMed]
36. Guo, Y.; Shen, Z.; Zhong, Z. Hydrogeochemical modeling of groundwater chemical environmental evolution in Hebei Plain. *Sci. China Ser. D Earth Sci.* **1997**, *40*, 502–508. [CrossRef]
37. Ravenscroft, P.; McArthur, J.M. Mechanism of regional enrichment of groundwater by boron: The examples of Bangladesh and Michigan, USA. *Appl. Geochem.* **2004**, *19*, 1413–1430. [CrossRef]
38. Hou, Q.; Zhang, Q.; Huang, G.; Liu, C.; Zhang, Y. Elevated manganese concentrations in shallow groundwater of various aquifers in a rapidly urbanized delta, south China. *Sci. Total Environ.* **2020**, *701*, 134777. [CrossRef] [PubMed]
39. Huang, G.; Liu, C.; Sun, J.; Zhang, M.; Jing, J.; Li, L. A regional scale investigation on factors controlling the groundwater chemistry of various aquifers in a rapidly urbanized area: A case study of the Pearl River Delta. *Sci. Total Environ.* **2018**, *625*, 510–518. [CrossRef]
40. Edmunds, W.M.; Cook, J.M.; Darling, W.G.; Kinniburgh, D.G.; Miles, D.L.; Bath, A.H.; Morgan-Jones, M.; Andrews, J.N. Baseline geochemical conditions in the Chalk aquifer, Berkshire, U.K.: A basis for groundwater quality management. *Appl. Geochem.* **1987**, *2*, 251–274. [CrossRef]
41. Sahib, L.Y.; Marandi, A.; Schüth, C. Strontium isotopes as an indicator for groundwater salinity sources in the Kirkuk region, Iraq. *Sci. Total Environ.* **2016**, *562*, 935–945. [CrossRef]
42. Barica, J. Salinization of groundwater in arid zones. *Water Res.* **1972**, *6*, 925–933. [CrossRef]
43. Sarwade, D.V.; Nandakumar, M.V.; Kesari, M.P.; Mondal, N.C.; Singh, V.S.; Singh, B. Evaluation of sea water ingress into an Indian atoll. *Environ. Geol.* **2007**, *52*, 1475–1483. [CrossRef]
44. Huang, G.; Song, J.; Han, D.; Liu, R.; Liu, C.; Hou, Q. Assessing natural background levels of geogenic contaminants in groundwater of an urbanized delta through removal of groundwaters impacted by anthropogenic inputs: New insights into driving factors. *Sci. Total Environ.* **2023**, *857*, 159527. [CrossRef] [PubMed]
45. Mather, J.D.; Porteous, N.C. The geochemistry of boron and its isotopes in groundwaters from marine and non-marine sandstone aquifers. *Appl. Geochem.* **2001**, *16*, 821–834. [CrossRef]
46. Li, J.; Zhou, H.; Qian, K.; Xie, X.; Xue, X.; Yang, Y.; Wang, Y. Fluoride and iodine enrichment in groundwater of North China Plain: Evidences from speciation analysis and geochemical modeling. *Sci. Total Environ.* **2017**, *598*, 239–248. [CrossRef]
47. Huang, G.; Liu, C.; Li, L.; Zhang, F.; Chen, Z. Spatial distribution and origin of shallow groundwater iodide in a rapidly urbanized delta: A case study of the Pearl River Delta. *J. Hydrol.* **2020**, *585*, 124860. [CrossRef]
48. Bi, P.; Huang, G.; Liu, C.; Li, L. Geochemical factors controlling natural background levels of phosphate in various groundwater units in a large-scale urbanized area. *J. Hydrol.* **2022**, *608*, 127594. [CrossRef]
49. Huang, G.; Han, D.; Song, J.; Li, L.; Pei, L. A sharp contrasting occurrence of iron-rich groundwater in the Pearl River Delta during the past dozen years (2006-2018): The genesis and mitigation effect. *Sci. Total Environ.* **2022**, *829*, 154676. [CrossRef]
50. Rye, R.O.; Back, W.; Hanshaw, B.B.; Rightmire, C.T.; Pearson, F.J., Jr. The origin and isotopic composition of dissolved sulfide in groundwater from carbonate aquifers in Florida and Texas. *Geochim. Cosmochim. Acta* **1981**, *45*, 1941–1950. [CrossRef]
51. Hem, J.D. Chemistry and occurrence of cadmium and zinc in surface water and groundwater. *Water Resour. Res.* **1972**, *8*, 661–679. [CrossRef]

water

MDPI

Article

Distribution and Origins of Hardness in Shallow and Deep Groundwaters of the Hebei Plain, China

Yong Qian [1,2,*], Shijun Zhen [1,2], Chen Yue [1,2] and Xiangxiang Cui [1,2]

1 Institute of Hydrogeology and Environmental Geology, Chinese Academy of Geological Sciences, Shijiazhuang 050061, China; zhenshijun@mail.cgs.gov.cn (S.Z.); yuechen24@163.com (C.Y.); cuixiangxiang@mail.cgs.gov.cn (X.C.)
2 Hebei Key Laboratory of Groundwater Remediation, Shijiazhuang 050061, China
* Correspondence: desertqy@163.com

Abstract: Elevated hardness concentrations in groundwater have become a noteworthy concern in recent decades because long-term drinking of groundwater with high levels of hardness is an important factor resulting in chronic kidney diseases. In this study, the distribution and origins of groundwater total hardness (TH) in various sub-plains and different land-use areas of the Hebei Plain (HBP) were investigated. A total of 445 groundwater samples in the HBP were collected once in 2021, and twelve chemical parameters, including TH in groundwater, were analyzed. Results showed that TH-rich (>450 mg/L) shallow groundwater in both the central and littoral plains was more than twice that in the Piedmont plain. Similarly, TH-rich deep groundwater accounted for about 18% in the central plain but was negligible in the Piedmont plain. In the Piedmont plain, TH-rich shallow groundwater in urban areas was twice or more than in other land use types. By contrast, both TH-rich shallow and deep groundwaters in agricultural areas in the central plain were higher than those in rural areas. This was opposite to TH-rich shallow groundwater in the littoral plain. In the Piedmont plain, TH-rich shallow groundwater was mainly attributed to water-rock interaction, groundwater over-extraction, and the infiltration of domestic sewage and animal waste. In the central plain, both TH-rich shallow and deep groundwaters likely ascribed to the evaporite dissolution and seawater intrusion. By contrast, the leaching of agricultural fertilizers resulting in the dissolution of Ca-rich and Mg-rich minerals in the vadose zone was mainly responsible for the occurrence of TH-rich shallow groundwater in the littoral plain. Therefore, in order to limit elevated hardness concentrations in groundwater in the HBP, limiting shallow groundwater extraction and strengthening the supervision of the domestic sewage and animal waste in the Piedmont plain are recommended. Besides, restricting the use of nitrogenous fertilizers in the littoral plain is also recommended.

Keywords: total hardness; TH-rich shallow groundwater; groundwater over-extraction; evaporites dissolution; agricultural areas; Hebei plain

Citation: Qian, Y.; Zhen, S.; Yue, C.; Cui, X. Distribution and Origins of Hardness in Shallow and Deep Groundwaters of the Hebei Plain, China. *Water* **2024**, *16*, 310. https://doi.org/10.3390/w16020310

Academic Editor: Fernando António Leal Pacheco

Received: 12 December 2023
Revised: 5 January 2024
Accepted: 10 January 2024
Published: 17 January 2024

1. Introduction

Groundwater is important for food security, human health, and ecosystems. Globally, groundwater is the main source of drinking water for more than 2.5 billion people, and about 70% of groundwater abstraction is used for agricultural irrigation [1]. Especially in areas with insufficient surface water, such as the Hebei Plain (HBP) and islands, groundwater is often the major water resource for drinking and irrigation purposes [2,3]. For example, groundwater accounted for more than 52% of the total water supply in the Hebei Province of China in 2019 [4]. However, the use of groundwater is always limited by its water quality, which is always controlled by kinds of natural and anthropogenic factors [5,6] because poor quality groundwater is often harmful to human beings and/or unsuitable for irrigation and industrial purposes [7–10]. For example, long-term drinking of groundwater with high levels of hardness is an important factor resulting in chronic kidney diseases that are associated with reduced glomerular filtration rate and/or increased albumin excretion [11]. In addition, elevated hardness groundwater is also unfit for agricultural irrigation because it can render salinity and alkali hazards to soil [12]. Unfortunately, one publication in recent reported that elevated hardness was one of the major indicators for poor quality groundwater in the HBP; more than half of groundwater in phreatic aquifers and about 5% of groundwater in confined aquifers in the HBP showed that total hardness concentration above the allowable limit (450 mg/L) recommended by China [13].

To date, the origins of elevated hardness groundwater in many areas have already been investigated globally. A couple of studies reported that the infiltration of wastewater/leachate with high levels of hardness was often mainly responsible for elevated hardness in groundwater in industrial areas and landfills [14,15]. Similarly, the release of Ca^{2+} and Mg^{2+} from sediments via the ion exchange process under the condition of sewage irrigation was generally the major driving force for increasing groundwater hardness in irrigation areas [16,17]. By contrast, Cloutier et al. pointed out that groundwater with elevated hardness in a sedimentary rock aquifer system in Québec of Canada, was likely attributed to geogenic sources such as the dissolution of dolostone [18]. In addition, Cucchi et al. revealed that the occurrence of elevated hardness groundwater in plains aquifers of northeastern Italy probably ascribed to the over-extraction of groundwater [19]. This was also the main factor for elevated hardness groundwater in the HBP in the last century [20]. Besides, some researchers revealed hydrochemical characteristics related to groundwater hardness via different hydrogeochemical zones in the HBP during 1975–2010 [21,22]. For example, Zhan et al. reported that average concentrations of Ca^{2+}, Mg^{2+}, and total hardness in shallow groundwater in the Piedmont alluvial fan-recharge zone of the HBP during 1975–2005 increased by 42.6%, 40.5%, and 20.6%, respectively; meanwhile, from the piedmont alluvial fan-recharge zone to the coast plain-discharge zone, total hardness concentration in shallow groundwater in the HBP was from less than 450 mg/L increased to more than 900 mg/L [22]. However, human activities in the HBP accompanied by urbanization have intensified in recent decades [23]. This may affect spatial distribution and origins of groundwater hardness in the HBP. Recently, Li et al. investigated groundwater quality in the Hutuo river alluvial plain of the HBP and showed that about 26% of shallow groundwater and 20% of deep groundwater were characterized by high levels of total hardness (>450 mg/L) [24]. Bai et al. investigated groundwater chemistry in Yongqing County in the northern part of the HBP and reported that groundwater total hardness concentration in phreatic aquifers was up to 840 mg/L [25]. Guo et al. surveyed groundwater quality in the Bazhou irrigation district of the HBP and found that the total hardness concentration was up to 1483 mg/L in shallow groundwater but less than 90 mg/L in deep groundwater [26]. However, knowledge of the distribution and origins of hardness-rich shallow and deep groundwaters in the whole HBP is still limited.

Therefore, this study aims to depict the current spatial distribution of groundwater hardness in different land use types in the HBP and discuss the origins of elevated hardness in both shallow and deep groundwaters. Results will be beneficial for groundwater resource management in the HBP.

2. Study Area

2.1. Geographical Conditions

The HBP is part of the North China Plain and occupies an area of about 7.3×10^4 km^2. It is bounded by Henan province and Bohai Bay in the south and east and bounded by Taihang and Yanshan mountains in the west and north, respectively (Figure 1A). It has deep soil, abundant sunshine, and very flat topography [22]. The mean annual precipitation of the HBP is 500–600 mm and about 60–70% concentrated between June and August, whereas the mean annual evaporation rate is in the order of 1000–1300 mm [27,28]. The HBP is dominant in agriculture, with agricultural land more than 70% of the total area, and groundwater is the main water resource for agricultural irrigation [29]. With the development of urbanization, the construction land in the HBP increased by more than 20% during the period of 2010–2020, which was mainly transformed from agricultural land [23].

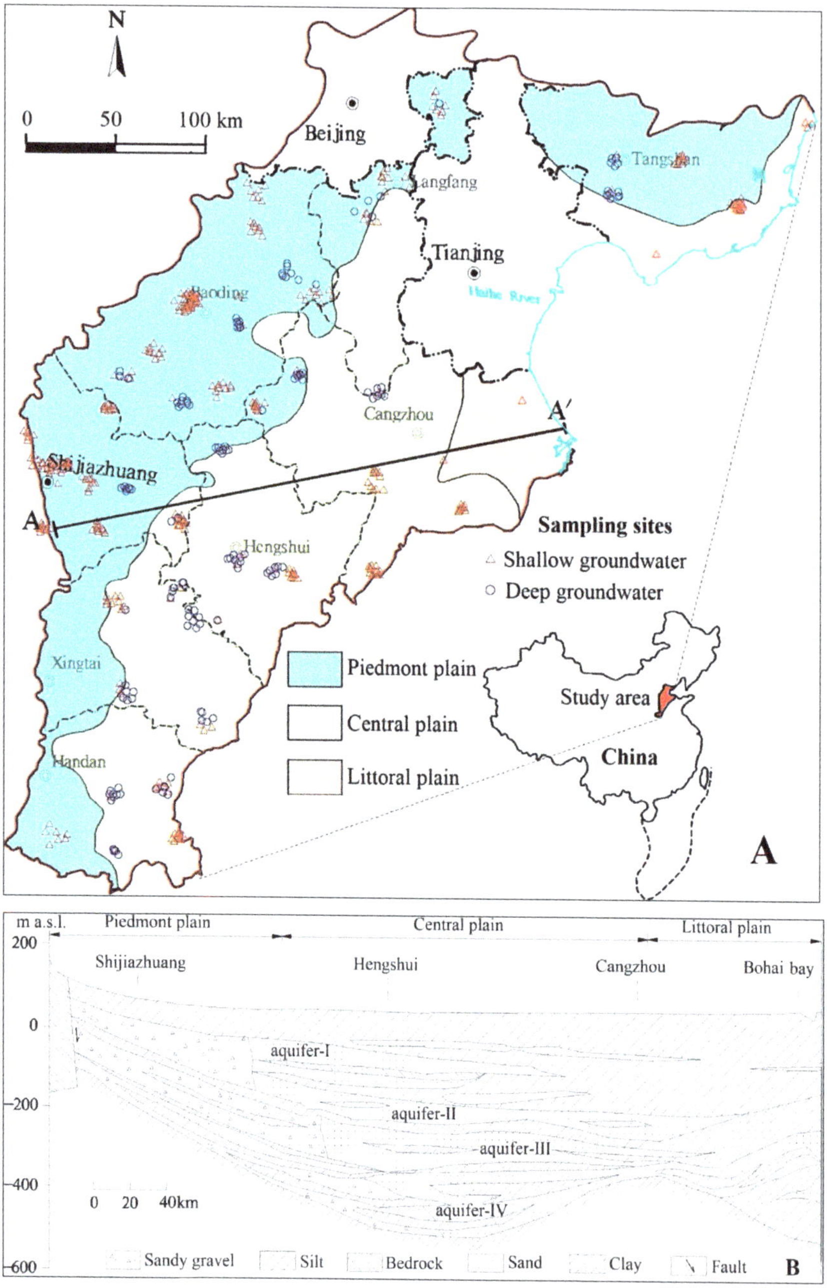

Figure 1. Hydrogeological setting and sampling sites in the Hebei Plain. (**A**) sampling sites, (**B**) cross section.

2.2. Geological and Hydrogeological Conditions

The HBP is a large sedimentary basin. Quaternary sediments originate from the middle and lower reaches of the Yellow River, the Haihe River, the Luanhe River, and their tributaries. Its thickness is approximately 150–500 m. It can be divided into the Piedmont Plain (PP), the Central Plain (CP), and the Littoral Plain (LP) from west to east [30]. They consist of fluvial deposits, alluvial and lacustrine deposits, and alluvial deposits with interbedded marine deposits, respectively. The PP has more plentiful groundwater resources in comparison with CP and LP because it has more recharge sources [31]. The HBP has four aquifer groups (Figure 1B). The aquifer-I at the top of the HBP is formed in the Holocene and a range of 10–50 m below land surface (BLS), with coarse-grained sand in the PP to fine-grained sand in the LP. The aquifer-II is formed in the upper Pleistocene with a range of 120–170 m BLS, with sandy gravel and medium to fine sand [21]. Groundwater in the above two aquifers is denoted as shallow groundwater and is a major water resource for agricultural irrigation. The aquifer-III is formed in the middle Pleistocene with a range of 170–350 m BLS. It consists of sandy gravel in the PP and medium to fine sand in the CP and LP [21]. While the aquifer-IV is formed in the lower Pleistocene and >350 m BLS, it consists of cemented sandy gravel and thin layers of weathered sand. Groundwater in these two aquifers is denoted as deep groundwater [32,33]. Groundwater was mainly recharged by lateral flow from mountain areas and vertical infiltration from rivers and irrigation return [21,33]. On a regional scale, groundwater flows are from the west to the east or the northeast. The groundwater flow velocity is 0.013–0.26 m/d in the PP and decreases to 0.002–0.10 m/d in the CP [34]. Recently, deep groundwater levels in urban areas increased significantly but still decreased in agricultural areas, which are often recharged by the leakage of overlying shallow groundwater [22,35].

3. Materials and Methods

3.1. Groundwater Sampling

A total of 306 shallow groundwater samples and 139 deep groundwater samples in the HBP were collected from July to September 2021. Among them, 202 shallow samples and 61 deep groundwater samples were collected from the PP, 84 shallow samples and 78 deep groundwater samples were collected from the CP, and 20 shallow samples were collected from the LP, respectively. In order to ensure samples representing the in-situ conditions, they were collected after pumping at least 3 well volumes or 30 min. Samples were filtered through 0.45 μm membrane filters to remove suspended solids in the field. All samples were stored at 4 °C until laboratory procedures could be performed.

3.2. Analytical Techniques

The pH was measured in the field by a multi-parameter portable meter (HANNA, HI 98121, Shanghai, China). HCO_3^-, CO_3^{2-}, and the total dissolved solids (TDS) were determined by volumetric and gravimetric methods, respectively. Other major anions (NO_3^-, SO_4^{2-}, Cl^-) were carried out on IC (Shimadzu LC-10ADvp, Kyoto, Japan). Chemical oxygen demand (COD) and total hardness (TH) were determined by potassium dichromate and EDTA titration methods, respectively. Major cations (Ca^{2+}, Na^+, K^+, Mg^{2+}) were measured by ICP-AES (ICAP6300, Thermo, New York, NY, USA). To assure data quality for indicators, each groundwater sample was analyzed in triplicate, sample batches were regularly interspersed with standards and blanks, and all data were corrected for instrument drift. The relative errors were $<\pm5\%$ for all analyzed indicators.

3.3. Principal Components Analysis (PCA)

The PCA reduces high dimensional space into a smaller number of dimensions called principal components (PCs) by linearly combining measurements [36–38]. In this study, the PCA was used to extract the PCs that control the TH in shallow and deep groundwaters in various plains from 12 groundwater chemical parameters [39–41]. The units of these parameters were mg/L, except for pH. Log-transformed data and a standardized data matrix were used in the PCA to give each variable equal weight in the multivariate statistical analysis. Rotation of the PCs was conducted using the Varimax method [42,43]. The number of PCs to keep was based on the Kaiser criterion (eigenvalues > 1). The terms "strong", "moderate", and "weak" (as applied to PC loadings) referred to the absolute loading values of >0.75, 0.75–0.5 and 0.5–0.3, respectively [8]. The software SPSS® Version 23.0 (SPSS Inc., Chicago, IL, USA) was used for the PCA in this study.

4. Results and Discussion

4.1. Characteristics of Groundwater Chemistry

Statistics for hydrochemical parameters in shallow and deep groundwaters in various sub-plains of the HBP are shown in Table 1. Median pH values in both shallow and deep groundwater in various sub-plains of the HBP were approximately 7.3–8.1, indicating that both shallow and deep groundwaters in the HBP were near neutral to weak alkaline. Moreover, deep groundwater showed a little bit more alkaline than shallow groundwater in both PP and CP according to their median pH values. From the PP to the LP, the median COD concentration in shallow groundwater increased gradually, indicating that shallow groundwater towards reducing environments from the PP to the LP. This is likely ascribed to the decrease of shallow groundwater flow velocity from the PP to the LP and the coarse-grained sand in the PP to fine-grained sand in the LP [21,34,37]. Moreover, deep groundwater showed a more oxidizing environment in comparison with shallow groundwater because median COD concentrations in shallow groundwater in both PP and CP were higher than those in deep groundwater. This probably ascribes to the more frequent reductive contamination (e.g., ammonium consuming dissolved oxygen) in shallow groundwater than in deep groundwater in the HBP [13,44]. Median concentrations of TDS and some major ions, such as Na^+, Cl^-, and SO_4^{2-}, in shallow groundwater in both CP and LP were more than twice that in the PP. This likely ascribes to the stronger water-rock interaction and evaporation and the occurrence of seawater intrusion in the latter two sub-plains [13]. These are also likely responsible for the higher median concentrations of K^+, Ca^{2+}, Mg^{2+}, and HCO_3^- in shallow groundwater in the latter two sub-plains than in the PP. In both of PP and CP, median values of TDS and most major ions (excluding K^+ and Na^+) in shallow groundwater were more than 1.5 times that in deep groundwater. This is likely due to the more frequent salt contamination in shallow groundwater than in deep groundwater because salt-contaminated groundwater was commonly characterized by high levels of TDS, Cl^-, and SO_4^{2-} [36,44–46]. By contrast, median concentrations of K^+ and Na^+ in shallow groundwater in the PP were lower than those in deep groundwater. This is likely because of the cation exchange in vadose zones that adsorbed Ca^{2+} and Mg^{2+} are replaced by K^+ and Na^+ [13]. Unlike the above major ions, the median value of NO_3^- in shallow groundwater in the PP was more than twice that in CP and LP. This indicates that shallow groundwater NO_3^- contamination was more often in the PP than in the other two sub-plains because shallow groundwater in the PP was characterized by stronger oxidizing environments and lower COD levels in comparison with that in the other two sub-plains [10,13,47]. This is also probably responsible for the higher median concentration of NO_3^- in deep groundwater in the PP than in the CP. Moreover, in the PP, the median NO_3^- concentration in shallow groundwater was 7 times that in deep groundwater, indicating that shallow groundwater in the PP was more vulnerable to NO_3^- contamination in comparison with deep groundwater.

4.2. Distribution of Groundwater Total Hardness

Groundwater TH concentrations in various sub-plains in the HBP are shown in Figure 2. Shallow groundwater TH concentrations in the PP ranged widely from 15 to 1411 mg/L with a median value of 343 mg/L. Median concentrations of shallow groundwater TH in CP and LP were 2.3 times and 1.8 times that in the PP, respectively. This is probably attributed to the stronger water-rock interaction and evaporation process in CP and LP than in the PP because shallow groundwater in the former two sub-plains presents slower flow velocity and shallower groundwater levels than in the PP [13,34]. This is also likely responsible for the higher median concentration of TH in deep groundwater in the PP than in the CP. In both PP and CP, median concentrations of TH in shallow groundwater were more than twice that in deep groundwater. This is probably due to the cation exchange that adsorbed Ca^{2+} and Mg^{2+} in vadose zones replaced by Na^+ and/or K^+, and finally infiltrating into shallow groundwater rather than deep groundwater via the water flow [13,48,49]. In addition, groundwater TH concentrations in areas with different land uses in the HBP were also investigated. In the PP, the median concentration of shallow groundwater TH in urban areas was more than 1.5 times that in rural and agricultural areas. In contrast, median concentrations of deep groundwater TH in rural and agricultural areas were 1.3 times and 1.1 times that in urban areas. Moreover, median TH concentrations in shallow groundwater in urban, rural, and agricultural areas were 4.1 times, 2.1 times, and 2.4 times those in deep groundwater, respectively. These likely indicate that anthropogenic inputs of TH in urban areas were stronger than those in rural and agricultural areas, and anthropogenic inputs of TH had contributed to shallow groundwater rather than deep groundwater because the intensity of human activities in urban areas was commonly stronger than that in other areas [39,43]. Besides, in the CP, median TH concentrations in both shallow and deep groundwaters in rural areas were lower than those in agricultural areas. By contrast, in the LP, the median TH concentration in shallow groundwater in rural areas was approximately twice that in agricultural areas.

Table 1. Descriptive statistics of chemical parameters in shallow and deep groundwaters in the Hebei Plain.

Items	Shallow Groundwater									Deep Groundwater					
	Piedmont Plain			Central Plain			Littoral Plain			Piedmont Plain			Central Plain		
	Min.	Med.	Max.	Min.	Med.	Max.	Min.	Med.	Max.	Min.	Med.	Max.	Min.	Med.	Max.
pH	5.0	7.5	13.4	6.2	7.4	8.2	6.7	7.3	9.2	6.2	7.8	10.3	7.1	8.1	8.3
COD	0.3	0.7	8.4	0.3	1.4	2.8	0.5	1.6	14.1	0.2	0.6	2.0	0.4	0.8	4.0
TDS	140	507	3688	224	1857	12,180	284	1116	23,550	185	324	1857	325	714	9570
K^+	0.3	1.4	13.8	0.4	1.8	12.4	0.6	5.0	51.0	0.3	1.5	3.8	0.3	1.2	7.5
Na^+	6	33	724	35	384	2670	53	100	6465	7	52	489	74	206	1940
Ca^{2+}	5	82	257	9	111	509	32	147	571	4	34	166	5	32	282
Mg^{2+}	1	33	217	5	131	833	6	55	1219	1	13	135	2	24	759
SO_4^{2-}	4	61	1999	30	438	5967	21	178	3665	5	37	804	46	179	3964
CO_3^{2-}	<DL	<DL	18	<DL	<DL	12	<DL	<DL	59	<DL	<DL	12	<DL	<DL	30
HCO_3^-	109	319	915	135	651	1118	227	449	604	103	229	877	57	226	1172
Cl^-	6	51	966	19	342	2040	21	213	13,830	2	19	466	12	132	2066
NO_3^--N	<DL	5.6	44.5	<DL	0.4	36.2	<DL	2.0	70.1	<DL	0.8	23.5	<DL	0.4	6.3

Note: pH without the unit while other parameters with the unit of mg/L; <DL: below detection limits.

Compared to the allowable limit (450 mg/L) of TH in groundwater recommended by China [50], about 33.2% of shallow groundwater in the PP exceeded the allowable limit of TH and is denoted as TH-rich groundwater in this study. TH-rich shallow groundwater in CP and LP were 2.4 times and 2.1 times that in the PP, respectively. Similarly, TH-rich deep groundwater in the CP was 11.2 times that in the PP. Moreover, TH-rich shallow groundwater in PP and CP was 20.8 times and 4.4 times TH-rich deep groundwater, respectively. In the PP, the proportion of TH-rich shallow groundwater in urban areas was 72%, and was 2.0 times and 2.7 times that in rural and agricultural areas. By contrast, TH-rich deep groundwater in the PP only occurred in agricultural areas rather than urban and rural areas. In the CP, both TH-rich shallow and deep groundwaters in agricultural areas were higher than those in rural areas. In contrast, TH-rich shallow groundwater in rural areas in the LP was 100% and was 1.4 times that in agricultural areas.

As shown in Figure 3, TH-rich shallow groundwater in the PP was commonly dominated by Ca^{2+} in cations and HCO_3^- in anions, respectively, and Ca-HCO_3 facies accounted for 80.6% of TH-rich shallow groundwater in the PP. By contrast, in the CP, TH-rich shallow groundwater was generally dominated by Na^+ in cations, and Na-SO_4 facies (36.4%), Na-HCO_3 facies (24.2%), and Na-Cl facies (19.7%) were major hydrochemical facies in TH-rich shallow groundwater. Similarly, TH-rich deep groundwater in the CP was also dominated by Na^+ in cations, and Na-Cl facies (42.9%) and Na-SO_4 facies (35.7%) were major hydrochemical facies. In the LP, TH-rich shallow groundwater was commonly dominated by Ca^{2+} in cations and Cl^- in anions, respectively, and Ca-HCO_3 facies (42.9%) and Na-Cl facies (28.6%) were major hydrochemical facies in TH-rich shallow groundwater. In addition, in the PP, TH-rich shallow groundwater dominated by Mg^{2+} in urban, rural, and agricultural areas was 0%, 10.3%, and 40%, respectively. In both CP and LP, TH-rich shallow groundwater dominated by Mg^{2+} in agricultural areas accounted for 21.0% and 11.1%, respectively, but none in rural areas. Besides, TH-rich deep groundwater dominated by Mg^{2+} only occurred in agricultural areas (7.7%) in the CP.

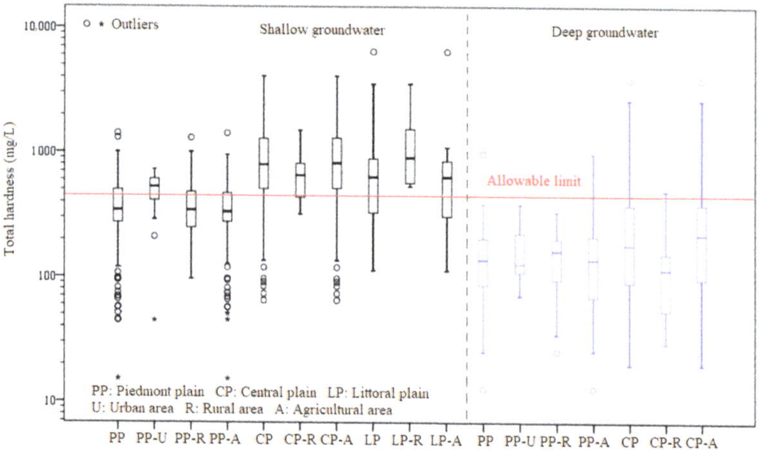

Figure 2. Distributions of total hardness in shallow and deep groundwaters in various sub-plains and different land-use types in the Hebei Plain.

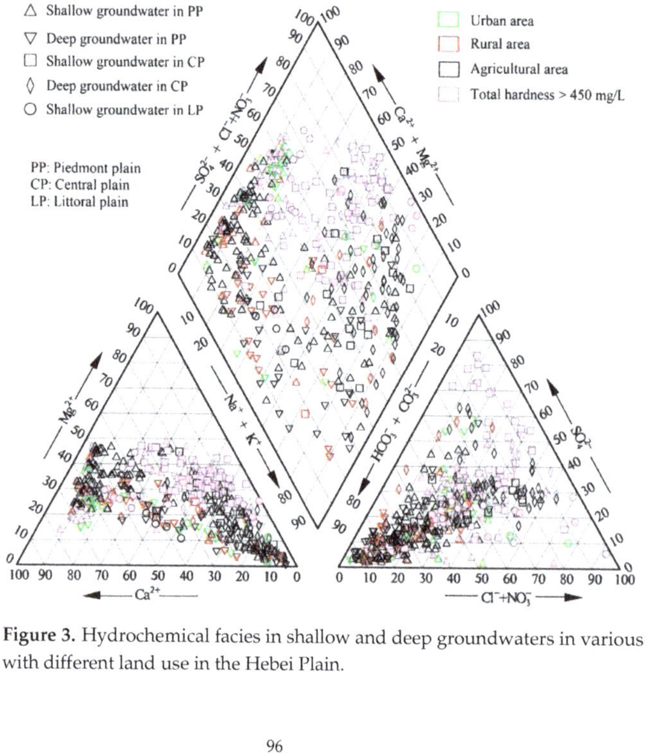

Figure 3. Hydrochemical facies in shallow and deep groundwaters in various sub-plains and areas with different land use in the Hebei Plain.

4.3. *Origins of Groundwater Total Hardness*

4.3.1. Gibbs Diagram

In this study, the Gibbs Diagram was used to analyze hydrogeochemical processes that control groundwater TH levels in the HBP. As seen in Figure 4, Na/(Na + Ca) ratios in 75.7% of shallow groundwater and 83.6% of TH-rich shallow groundwater in the PP were below 0.5 and accompanied by low levels of TDS (<1200 mg/L). This indicates that rock weathering was the dominant hydrochemical process for TH-rich shallow groundwater in the PP [49]. In the PP, the mean TDS concentration in TH-rich shallow groundwater with Na/(Na + Ca) ratios < 0.5 was more than twice that in shallow groundwater with TH < 450 mg/L and Na/(Na + Ca) ratios < 0.5. This indicates that the salinisation with cation exchange was probably another important hydrochemical process for TH-rich shallow groundwater in the PP [51]. On the other hand, the vadose zone thickness covering shallow groundwater in the PP was thicker because shallow groundwater in the PP was over-extracted, and its water table has been deeper continuously for the past 40 years [52]. In this case, more Ca^{2+} and Mg^{2+} would be released from the vadose zone into shallow groundwater by the cation exchange of Na^{+} and/or K^{+} via water flow and finally result in elevated TH and TDS concentrations in shallow groundwater [19,20]. This inference was supported by the evidence that shallow groundwater TH concentrations in the PP had a significantly positive correlation with the depth of the shallow groundwater table (Figure 5). Therefore, it can be concluded that the hydrochemical process of TH-rich shallow groundwater in the PP was mainly controlled by both rock weathering and salinisation with cation exchange resulting from groundwater over-extraction. By contrast, TH-rich shallow groundwater in the other two sub-plains was unaffected by the salinisation with cation exchange because the over-extraction of shallow groundwater in these two sub-plains was insignificant, and their shallow groundwater TH concentrations had insignificant correlations with the depth of shallow groundwater table (Figure 5).

In the CP, Na/(Na + Ca) ratios and TDS concentrations in TH-rich shallow groundwater were higher in comparison with that in the PP. For instance, Na/(Na + Ca) ratios in 89.4% of TH-rich shallow groundwater in the CP were above 0.5, with a mean TDS concentration of 3013 mg/L (Figure 4). This infers that the dissolution of evaporites with cation exchange and/or the evaporation process was probably mainly controlled by hydrochemistry of TH-rich shallow groundwater in the CP [51] because shallow groundwater in the CP was shown shallower depth of water table and longer residence time in comparison with that in the PP [33]. However, the depth of the shallow groundwater table in the CP was generally deeper than 3 m [44]. This indicates that the evaporation process had little influence on the hydrochemistry of TH-rich shallow groundwater in the CP because groundwater evaporation is negligible when the groundwater table is deeper than 3 m [52]. Thus, evaporite dissolution rather than the evaporation process was likely the major hydrochemical process for TH-rich shallow groundwater in the CP. It was also probably the major hydrochemical process for TH-rich deep groundwater in the CP because Na/(Na + Ca) ratios in 92.9% of TH-rich shallow groundwater in the CP were above 0.5 with a mean TDS concentration of 2937 mg/L (Figure 4). In addition, about 16.7% of TH-rich shallow groundwater and 21.4% of TH-rich deep groundwater in the CP were characterized by Na/(Na + Ca) ratios > 0.8 and TDS concentrations > 3000 mg/L (Figure 4). This indicates that seawater intrusion was likely another important factor controlling the hydrochemistry of TH-rich shallow and deep groundwaters in the CP [44]. As a consequence, evaporite dissolution and seawater intrusion were two major hydrochemical processes controlling both TH-rich shallow and deep groundwaters in the CP.

In the LP, more than 70% of TH-rich shallow groundwater was characterized by Na/(Na + Ca) ratios < 0.5 and TDS concentrations < 1500 mg/L (Figure 4). This indicates that rock weathering was likely the major factor for TH-rich shallow groundwater in the LP [53]. Besides, about 21.4% of TH-rich shallow groundwater in the LP was characterized by Na/(Na + Ca) ratios > 0.9 and TDS concentrations > 7000 mg/L (Figure 4), indicating that seawater intrusion was another important factor controlling hydrochemistry of TH-rich shallow groundwater in the LP. Thus, rock weathering and seawater intrusion were two major hydrochemical processes controlling TH-rich shallow groundwater in the LP.

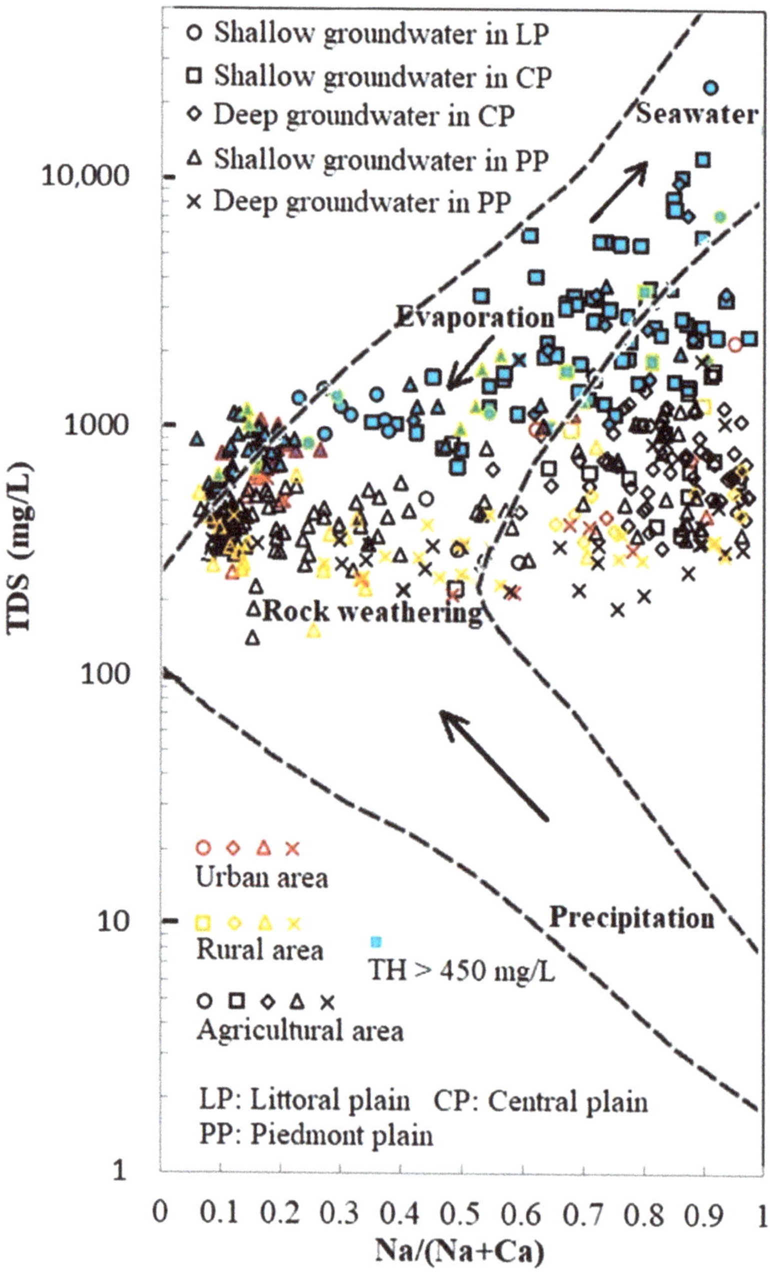

Figure 4. Gibbs diagram of shallow and deep groundwaters in various sub-plains and areas with different land use in the Hebei Plain.

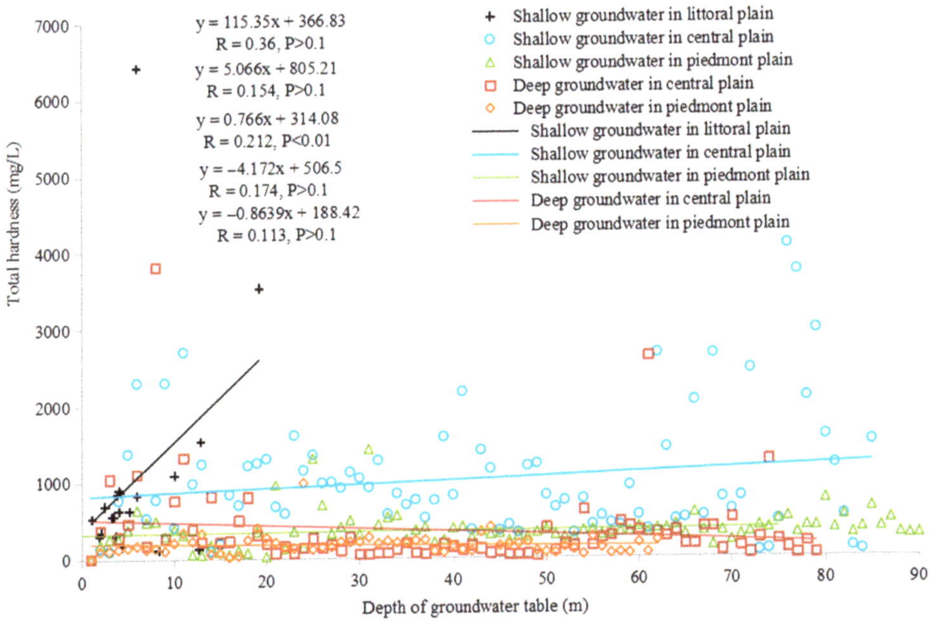

Figure 5. Relationships between total hardness concentrations and groundwater tables in shallow and deep groundwaters in various sub-plains in the Hebei Plain.

4.3.2. Principal Components Analysis

In this study, factors (PCs) controlling groundwater chemistry in various sub-plains of the HBP, where TH-rich groundwater often occurs, were extracted by the PCA. As shown in Table 2, shallow groundwater chemistry in the PP was mainly controlled by four factors. PC1 was shown strong positive loadings (SPL) (>0.75) of TDS, Na^+, SO_4^{2-}, and Mg^{2+}, and moderate positive loadings (MPL) (0.5–0.75) of Cl^- and TH. This indicates that the PC1 was likely indicative of the water-rock interaction and salinisation (resulted from groundwater over-extraction), because TH-rich shallow groundwater in the PP was mainly controlled by rock weathering and salinisation (as mentioned previously) that resulted in elevated TDS, Na^+, SO_4^{2-}, Mg^{2+}, Cl^-, and TH concentrations in shallow groundwater in the PP [13,20]. Besides, TH-rich shallow groundwater in the PP was also affected by another factor (PC2) because it also had a MPL of TH. Specifically, the PC2 had SPL of NO_3^- and Ca^{2+} and MPL of K^+ and TH. This factor probably represented the dissolution of Ca-rich minerals accompanied by the infiltration of domestic sewage and animal waste because domestic sewage and animal waste are an enrichment of N and K and are main sources for high levels of NO_3^- and K^+ in shallow groundwater in the PP [10,36,47], and NO_3^- contamination results in the dissolution of calcium-rich minerals in vadose zone via water flow [44]. As a consequence, water-rock interaction, groundwater over-extraction (salinisation), and the infiltration of domestic sewage and animal waste (dissolution of calcium-rich minerals) were mainly responsible for the occurrence of TH-rich shallow groundwater in the PP.

In the CP, shallow groundwater chemistry was mainly controlled by two PCs. PC1 showed SPL of TDS, TH, Mg^{2+}, SO_4^{2-}, Na^+, Cl^-, and Ca^{2+}. This indicates that TH-rich shallow groundwater in the CP was likely attributed to the evaporite dissolution and seawater intrusion because they were two major factors controlling TH-rich shallow groundwater in the CP (as mentioned above), and the dissolution of evaporites (e.g., dolomite and bitter salt) would enrich shallow groundwater TDS, TH, Mg^{2+}, SO_4^{2-}, and Ca^{2+} concentrations [46], besides, seawater intrusion would result in high levels of TDS, Na^+, and Cl^- in shallow groundwater [36,45]. Similarly, they were also mainly responsible for the occurrence of TH-rich deep groundwater in the CP, because PC1 in deep groundwater in the CP also had SPL of TDS, Mg^{2+}, SO_4^{2-}, TH, Na^+, Cl^-, and Ca^{2+}, and a MPL of HCO_3^- (Table 2).

Table 2. Principal component (PC) loadings for chemical parameters in shallow and deep groundwaters in various sub-plains in the Hebei Plain.

Items	SG in PP				Items	SG in CP		Items	SG in LP			Items	DG in CP		
	PC1	PC2	PC3	PC4		PC1	PC2		PC1	PC2	PC3		PC1	PC2	PC3
TDS	***0.94***	0.25	0.17	0.11	TDS	***0.98***	0.09	Mg^{2+}	***0.99***	−0.06	−0.04	TDS	***1.00***	0.04	−0.03
Na^+	***0.86***	−0.35	−0.13	0.13	TH	***0.97***	0.18	TDS	***0.99***	−0.08	−0.06	Mg^{2+}	***0.98***	0.06	−0.01
SO_4^{2-}	***0.84***	0.23	−0.13	−0.01	Mg^{2+}	***0.97***	0.16	Cl^-	***0.99***	0.02	−0.07	SO_4^{2-}	***0.98***	0.03	0.01
Mg^{2+}	***0.77***	0.15	0.49	0.10	SO_4^{2-}	***0.94***	−0.06	TH	***0.99***	0.07	0.02	TH	***0.98***	0.11	0.00
Cl^-	***0.71***	−0.01	0.17	0.01	Na^+	***0.94***	0.08	Na^+	***0.99***	−0.11	−0.10	Na^+	***0.97***	−0.03	−0.07
TH	***0.62***	0.60	0.46	0.06	Cl^-	***0.82***	0.12	COD	***0.95***	0.24	−0.03	Cl^-	***0.95***	0.10	−0.03
NO_3^-	−0.03	0.88	0.07	0.00	Ca^{2+}	***0.77***	0.16	K^+	***0.85***	−0.02	−0.08	Ca^{2+}	***0.84***	0.28	0.04
Ca^{2+}	0.31	0.84	0.31	−0.01	K^+	0.34	−0.32	SO_4^{2-}	***0.70***	−0.56	−0.08	HCO_3^-	***0.68***	−0.13	−0.17
K^+	−0.14	0.65	−0.17	0.59	HCO_3^-	0.08	0.81	pH	0.24	−0.86	−0.21	K^+	0.13	0.92	−0.01
pH	0.06	−0.23	−0.78	0.03	pH	−0.14	−0.78	HCO_3^-	0.14	0.74	−0.18	COD	0.17	0.88	−0.16
HCO_3^-	0.50	−0.10	0.63	0.33	COD	0.24	0.72	Ca^{2+}	0.66	0.67	0.32	NO_3^-	−0.07	0.33	0.12
COD	0.17	−0.01	0.13	0.91	NO_3^-	0.01	0.42	NO_3^-	−0.10	0.04	0.98	pH	−0.03	0.03	0.98
Eigenvalue	4.2	2.6	1.7	1.3		6.1	2.2		7.5	2.1	1.2		7.0	1.8	1.0
EV (%)	35.2	21.5	14.1	11.1		50.6	18.1		62.7	17.7	9.7		58.0	15.4	8.6
CV (%)	35.2	56.7	70.8	81.9		50.6	68.7		62.7	80.3	90.1		58.0	73.3	81.9

Notes: SG: Shallow groundwater; DG: Deep groundwater; PP: Piedmont plain; CP: Central plain; LP: Littoral plain; EV: Explained variance; CV: Cumulative variance; Bold and italics numbers: maximum absolute PC loading of one parameter.

By contrast, in the LP, shallow groundwater chemistry was mainly controlled by three PCs. PC2 showed SPL of Ca^{2+}, TH, and NO_3^-, and a MPL of Mg^{2+} (Table 2). This was likely indicative of the dissolution of Ca-rich and Mg-rich minerals (e.g., dolomite) resulting from the NO_3^- contamination due to agricultural fertilizers because the mean concentration of shallow groundwater NO_3^- in agricultural areas in the LP was more than twice that in urban and rural areas, and NO_3^- contamination results in the dissolution of Ca-rich and Mg-rich minerals in vadose zone via water flow [44]. Therefore, the occurrence of TH-rich shallow groundwater in the LP is mainly ascribed to the leaching of agricultural fertilizers.

5. Conclusions

In this study, distributions of TH in shallow and deep groundwaters in various sub-plains in the HBP were investigated. Elevated groundwater TH concentrations in both CP and LP were more severe than that in the PP because TH-rich shallow groundwater in both CP and LP was more than twice that in the PP, and TH-rich deep groundwater accounted for about 18% in the CP but was negligible in the PP. Besides, TH-rich groundwater in the CP occurred more frequently in agricultural areas than in rural areas. This was opposite to TH-rich shallow groundwater in the LP. By contrast, TH-rich shallow groundwater in the PP was more frequent in urban areas than in other land use types.

In addition, the origins of TH-rich shallow and deep groundwaters in various sub-plains in the HBP were also investigated by the Gibbs diagram and PCA. In the PP, TH-rich shallow groundwater was mainly attributed to water-rock interaction, groundwater over-extraction (salinisation), and the infiltration of domestic sewage and animal waste (dissolution of calcium-rich minerals). In the CP, both TH-rich shallow and deep groundwaters likely ascribed to the evaporite dissolution and seawater intrusion. By contrast, the leaching of agricultural fertilizers resulting in the dissolution of Ca-rich and Mg-rich minerals in the vadose zone was mainly responsible for the occurrence of TH-rich shallow groundwater in the LP.

Therefore, in the PP, local governments should limit shallow groundwater extraction and strengthen the supervision of domestic sewage and animal waste to restrict shallow groundwater TH concentrations. Besides, in the LP, they should strengthen the supervision of the use of N-fertilizers in agricultural activities.

Author Contributions: Conceptualization, X.C.; Methodology, S.Z.; Software, C.Y.; Validation, C.Y.; Investigation, S.Z. and X.C.; Writing—original draft, Y.Q.; Writing—review & editing, Y.Q. All authors have read and agreed to the published version of the manuscript.

Funding: This research was supported by National Geological Survey Project (DD20230423) and Hebei Province High level Talent Team Construction Project (225A4204D).

Data Availability Statement: The datasets generated and/or analyzed during the current study are not publicly available.

Conflicts of Interest: The authors declare no conflict of interest.

References

1. Mukherjee, A.; Scanlon, B.; Aureli, A.; Langan, S.; Guo, H.; McKenzie, A. *Global Groundwater: Source, Scarcity, Sustainability, Security and Solutions*, 1st ed.; Elsevier: Amsterdam, The Netherlands, 2020; pp. 3–18.
2. Hu, X.; Shi, L.; Zeng, J.; Yang, J.; Zha, Y.; Yao, Y.; Cao, G. Estimation of actual irrigation amount and its impact on groundwater depletion: A case study in the Hebei Plain, China. *J. Hydrol.* **2016**, *543*, 433–449. [CrossRef]
3. Alberti, L.; Antelmi, M.; Oberto, G.; La Licata, I.; Mazzon, P. Evaluation of Fresh Groundwater Lens Volume and Its Possible Use in Nauru Island. *Water* **2022**, *14*, 3201. [CrossRef]
4. Water Resources Department of Hebei Province, 2020. Hebei Water Resources Bulletin of 2019. Available online: http://slt.hebei.gov.cn/resources/43/202010/1603098695816085596.pdf (accessed on 1 December 2023).
5. Antelmi, M.; Mazzon, P.; Höhener, P.; Marchesi, M.; Alberti, L. Evaluation of MNA in A Chlorinated Solvents-Contaminated Aquifer Using Reactive Transport Modeling Coupled with Isotopic Fractionation Analysis. *Water* **2021**, *13*, 2945. [CrossRef]
6. Zhang, F.; Huang, G.; Hou, Q.; Liu, C.; Zhang, Y.; Zhang, Q. Groundwater quality in the Pearl River Delta after the rapid expansion of industrialization and urbanization: Distributions, main impact indicators, and driving forces. *J. Hydrol.* **2019**, *577*, 124004. [CrossRef]
7. Huang, G.; Zhang, M.; Liu, C.; Li, L.; Chen, Z. Heavy metal(loid)s and organic contaminants in groundwater in the Pearl River Delta that has undergone three decades of urbanization and industrialization: Distributions, sources, and driving forces. *Sci. Total Environ.* **2018**, *635*, 913–925. [CrossRef] [PubMed]
8. Huang, G.; Liu, C.; Li, L.; Zhang, F.; Chen, Z. Spatial distribution and origin of shallow groundwater iodide in a rapidly urbanized delta: A case study of the Pearl River Delta. *J. Hydrol.* **2020**, *585*, 124860. [CrossRef]
9. Huang, G.; Han, D.; Song, J.; Li, L.; Pei, L. A sharp contrasting occurrence of iron-rich groundwater in the Pearl River Delta during the past dozen years (2006–2018): The genesis and mitigation effect. *Sci. Total Environ.* **2022**, *829*, 154676. [CrossRef]
10. Zhang, M.; Huang, G.; Liu, C.; Zhang, Y.; Chen, Z.; Wang, J. Distributions and origins of nitrate, nitrite, and ammonium in various aquifers in an urbanized coastal area, south China. *J. Hydrol.* **2020**, *582*, 124528. [CrossRef]
11. Liyanage, D.; Diyabalanage, S.; Dunuweera, S.; Rajapakse, S.; Rajapakse, R.; Chandrajith, R. Significance of Mg-hardness and fluoride in drinking water on chronic kidney disease of unknown etiology in Monaragala, Sri Lanka. *Environ. Res.* **2021**, *203*, 111779. [CrossRef]
12. Haritash, A.K.; Kaushik, C.P.; Kaushik, A.; Kansal, A.; Yadav, A.K. Suitability assessment of groundwater for drinking, irrigation and industrial use in some North Indian villages. *Environ. Monit. Assess.* **2008**, *145*, 397–406. [CrossRef]
13. Qian, Y.; Hou, Q.; Wang, C.; Zhen, S.; Yue, C.; Cui, X.; Guo, C. Hydrogeochemical Characteristics and Groundwater Quality in Phreatic and Confined Aquifers of the Hebei Plain, China. *Water* **2023**, *15*, 3071. [CrossRef]
14. Vadillo, I.; Andreo, B.; Carrasco, F. Groundwater Contamination by Landfill Leachates in a Karstic Aquifer. *Water Air Soil Pollut.* **2005**, *162*, 143–169. [CrossRef]
15. Shankar, B.S.; Balasubramanya, N.; Maruthesha Reddy, M.T. Impact of industrialization on groundwater quality-a case study of Peenya industrial area, Bangalore, India. *Environ. Monit. Assess.* **2008**, *142*, 263–268. [CrossRef]
16. Kass, A.; Gavrieli, I.; Yechieli, Y.; Vengosh, A.; Starinsky, A. The impact of freshwater and wastewater irrigation on the chemistry of shallow groundwater: A case study from the Israeli Coastal Aquifer. *J. Hydrol.* **2005**, *300*, 314–331. [CrossRef]
17. Yesilnacar, M.I.; Gulluoglu, M.S. Hydrochemical characteristics and the effects of irrigation on groundwater quality in Harran Plain, GAP Project, Turkey. *Environ. Geol.* **2007**, *54*, 183–196. [CrossRef]
18. Cloutier, V.; Lefebvre, R.; Therrien, R.; Savard, M.M. Multivariate statistical analysis of geochemical data as indicative of the hydrogeochemical evolution of groundwater in a sedimentary rock aquifer system. *J. Hydrol.* **2008**, *353*, 294–313. [CrossRef]
19. Cucchi, F.; Franceschini, G.; Zini, L. Hydrogeochemical investigations and groundwater provinces of the Friuli Venezia Giulia Plain aquifers, northeastern Italy. *Environ. Geol.* **2007**, *55*, 985–999. [CrossRef]
20. Bi, E.; Mu, H.; Chen, Z.; Wang, Z. Impacts of human activities on the evolution of groundwater quality in Hebei Plain. *Acta Geosci. Sin.* **2001**, *22*, 365–368.
21. Xing, L.; Guo, H.; Zhan, Y. Groundwater hydrochemical characteristics and processes along flow paths in the North China Plain. *J. Asian Earth Sci.* **2013**, *70*, 250–264. [CrossRef]
22. Zhan, Y.; Guo, H.; Wang, Y.; Li, R.; Hou, C.; Shao, J.; Cui, Y. Evolution of groundwater major components in the Hebei Plain: Evidences from 30-year monitoring data. *J. Earth Sci.* **2014**, *25*, 563–574. [CrossRef]

23. Zhao, L.; Su, M.; Wang, X.; Li, X.; Chang, X.; Zhang, P. Spatial-temporal evolution and prediction of habitat quality in Beijing–Tianjin–Hebei region based on land use change. *Land* **2023**, *12*, 667. [CrossRef]
24. Li, Y.; Zhang, Z.; Fei, Y.; Chen, H.; Qian, Y.; Dun, Y. Investigation of quality and pollution characteristics of groundwater in the Hutuo River Alluvial Plain, North China Plain. *Environ. Earth Sci.* **2016**, *75*, 581. [CrossRef]
25. Bai, X.; Tian, X.; Li, J.; Wang, X.; Li, Y.; Zhou, Y. Assessment of the Hydrochemical Characteristics and Formation Mechanisms of Groundwater in A Typical Alluvial-Proluvial Plain in China: An Example from Western Yongqing County. *Water* **2022**, *14*, 2395. [CrossRef]
26. Guo, H.; Li, M.; Wang, L.; Wang, Y.; Zang, X.; Zhao, X.; Wang, H.; Zhu, J. Evaluation of Groundwater Suitability for Irrigation and Drinking Purposes in an Agricultural Region of the North China Plain. *Water* **2021**, *13*, 3426. [CrossRef]
27. Wang, B.; Jin, M.; Nimmo, J.R.; Yang, L.; Wang, W. Estimating groundwater recharge in Hebei Plain, China under varying land use practices using tritium and bromide tracers. *J. Hydrol.* **2008**, *356*, 209–222. [CrossRef]
28. Lu, X.; Jin, M.; van Genuchten, M.T.; Wang, B. Groundwater Recharge at Five Representative Sites in the Hebei Plain, China. *Groundwater* **2011**, *49*, 286–294. [CrossRef] [PubMed]
29. Chu, Y.; Shen, Y.; Yuan, Z. Water footprint of crop production for different crop structures in the Hebei southern plain, North China. *Hydrol. Earth Syst. Sci.* **2017**, *21*, 3061–3069. [CrossRef]
30. Chen, Z.; Qi, J.; Xu, J.; Xu, J.; Ye, H.; Nan, Y. Paleoclimatic interpretation of the past 30 ka from isotopic studies of the deep confined aquifer of the North China Plain. *Appl. Geochem.* **2003**, *18*, 997–1009.
31. Zhong, H.; Sun, L.; Fischer, G.; Tian, Z.; Liang, Z. Optimizing regional cropping systems with a dynamic adaptation strategy for water sustainable agriculture in the Hebei Plain. *Agric. Syst.* **2019**, *173*, 94–106. [CrossRef]
32. Zhang, Z.; Shi, D.; Shen, Z.; Zhong, Z.; Xue, Y. Evolution and development of groundwater environment in North China Plain under human activities. *Acta Geosci. Sin.* **1997**, *18*, 337–344.
33. Chen, Z.; Nie, Z.; Zhang, Z.; Qi, J.; Nan, Y. Isotopes and sustainability of ground water resources, North China Plain. *Ground Water* **2005**, *43*, 485–493.
34. Zhang, Z.; Fei, Y.; Chen, Z.; Zhao, Z.; Xie, Z.; Wang, Y. *Investigation and Assessment of Sustainable Utilization of Groundwater Resources in the North China Plain*; Geology Press: Beijing, China, 2009; pp. 28–42. (In Chinese)
35. Yang, H.; Cao, W.; Zhi, C.; Li, Z.; Bao, X.; Ren, Y.; Liu, F.; Fan, C.; Wang, S.; Wang, Y. Evolution of groundwater level in the North China Plain in the past 40 years and suggestions on its overexploitation treatment. *Geol. China* **2021**, *48*, 1142–1155.
36. Huang, G.; Liu, C.; Sun, J.; Zhang, M.; Jing, J.; Li, L. A regional scale investigation on factors controlling the groundwater chemistry of various aquifers in a rapidly urbanized area: A case study of the Pearl River Delta. *Sci. Total Environ.* **2018**, *625*, 510–518. [CrossRef] [PubMed]
37. Huang, G.; Pei, L.; Li, L.; Liu, C. Natural background levels in groundwater in the Pearl River Delta after the rapid expansion of urbanization: A new pre-selection method. *Sci. Total Environ.* **2021**, *813*, 151890. [CrossRef] [PubMed]
38. Hou, Q.; Zhang, Q.; Huang, G.; Liu, C.; Zhang, Y. Elevated manganese concentrations in shallow groundwater of various aquifers in a rapidly urbanized delta, south China. *Sci. Total Environ.* **2019**, *701*, 134777. [CrossRef]
39. Huang, G.; Liu, C.; Zhang, Y.; Chen, Z. Groundwater is important for the geochemical cycling of phosphorus in rapidly urbanized areas: A case study in the Pearl River Delta. *Environ. Pollut.* **2020**, *260*, 114079. [CrossRef] [PubMed]
40. Huang, G.; Hou, Q.; Han, D.; Liu, R.; Song, J. Large scale occurrence of aluminium-rich shallow groundwater in the Pearl River Delta after the rapid urbanization: Co-effects of anthropogenic and geogenic factors. *J. Contam. Hydrol.* **2023**, *254*, 104130. [CrossRef] [PubMed]
41. Bi, P.; Huang, G.; Liu, C.; Li, L. Geochemical factors controlling natural background levels of phosphate in various groundwater units in a large-scale urbanized area. *J. Hydrol.* **2022**, *608*, 127594. [CrossRef]
42. Bi, P.; Liu, R.; Huang, G.; Li, D. Evaluating natural background levels of heavy metals in shallow groundwater of the Pearl River Delta via removal of contaminated groundwaters: Comparison of three preselection related methods. *Environ. Pollut.* **2023**, *335*, 122382. [CrossRef]
43. Huang, G.; Song, J.; Han, D.; Liu, R.; Liu, C.; Hou, Q. Assessing natural background levels of geogenic contaminants in groundwater of an urbanized delta through removal of groundwaters impacted by anthropogenic inputs: New insights into driving factors. *Sci. Total Environ.* **2023**, *857*, 159527. [CrossRef]
44. Qian, Y.; Cui, X.; Yue, C.; Guo, C.; Zhen, S.; Wang, W.; Huang, G.; Li, H.; Wang, Y.; Su, C.; et al. *Report on the Investigation and Evaluation of the Basic Environmental Conditions of Groundwater around the State Control Assessment Points in Hebei Province*; The Institute of Hydrogeology and Environmental Geology, Chinese Academy of Geological Sciences: Shijiazhuang, China, 2022. (In Chinese)
45. Huang, G.; Sun, J.; Zhang, Y.; Chen, Z.; Liu, F. Impact of anthropogenic and natural processes on the evolution of groundwater chemistry in a rapidly urbanized coastal area, South China. *Sci. Total Environ.* **2013**, *463*, 209–221. [CrossRef] [PubMed]
46. Liu, R.; Xie, X.; Qian, Y.; Hou, Q.; Han, D.; Song, J.; Huang, G. Groundwater sulfate in the Pearl River Delta driven by urbanization: Spatial distribution, sources and factors. *Appl. Geochem.* **2023**, *156*, 105766. [CrossRef]
47. Gan, L.; Huang, G.; Pei, L.; Gan, Y.; Liu, C.; Yang, M.; Han, D.; Song, J. Distributions, origins, and health-risk assessment of nitrate in groundwater in typical alluvial-pluvial fans, North China Plain. *Environ. Sci. Pollut. Res.* **2021**, *29*, 17031–17048. [CrossRef]

48. Edmunds, W.; Cook, J.; Darling, W.; Kinniburgh, D.; Miles, D.; Bath, A.; Morgan-Jones, M.; Andrews, J. Baseline geochemical conditions in the Chalk aquifer, Berkshire, U.K.: A basis for groundwater quality management. *Appl. Geochem.* **1987**, *2*, 251–274. [CrossRef]
49. Marandi, A.; Shand, P. Groundwater chemistry and the Gibbs Diagram. *Appl. Geochem.* **2018**, *97*, 209–212. [CrossRef]
50. General administration of quality supervision inspection and quarantine of the people's republic of China (GAQSIQPRC). In *Standard for Groundwater Quality*; Standards Press of China: Beijing, China, 2017.
51. Sahib, L.Y.; Marandi, A.; Schüth, C. Strontium isotopes as an indicator for groundwater salinity sources in the Kirkuk region, Iraq. *Sci. Total Environ.* **2016**, *562*, 935–945. [CrossRef] [PubMed]
52. Barica, J. Salinization of groundwater in arid zones. *Water Res.* **1972**, *6*, 925–933. [CrossRef]
53. Gibbs, J.R. Mechanisms controlling world water chemistry. *Science* **1970**, *170*, 1088–1090. [CrossRef]

water

MDPI

Article

Distribution of Groundwater Hydrochemistry and Quality Assessment in Hutuo River Drinking Water Source Area of Shijiazhuang (North China Plain)

Ziting Yuan, Yantao Jian, Zhi Chen, Pengfei Jin, Sen Gao, Qi Wang, Zijun Ding, Dandan Wang and Zhiyuan Ma *

Hebei Key Laboratory of Environment Monitoring and Protection of Geological Resources, Hebei Geo-Environment Monitoring Institute, Shijiazhuang 050022, China; isyuanziting@outlook.com (Z.Y.); jianyantao@126.com (Y.J.); dire2018@163.com (Z.C.); jinfeixssh@163.com (P.J.); gaosen8888@163.com (S.G.); seven18831991984@outlook.com (Q.W.); mr.dingzj@outlook.com (Z.D.); iswangdandan@foxmail.com (D.W.)

* Correspondence: mzy15031186956@foxmail.com

Abstract: The Hutuo River Drinking Water Source Area is an important water source of Shijiazhuang (North China Plain). Knowing the characteristics of groundwater chemistry/quality is essential for the protection and management of water resources. However, there are few studies focused on the groundwater chemistry evolution over the drinking water area. In this study, total of 160 groundwater samples were collected in November 2021, and the spatial distribution of groundwater chemistry and related controlling factors were analyzed using hydrological and multivariate analysis. The entropy-weighted water quality index (EWQI) was introduced to assess the groundwater quality. The results show that the hydrogeochemical types of groundwater are Ca-HCO_3 (78.1%), mixed Ca-Mg-Cl (20%), and Ca-Cl (1.9%) in the area. Graphical and binary diagrams indicate that groundwater hydrochemistry is mainly controlled by water–rock interaction (i.e., rock weathering, mineral dissolution, and ion exchange). Five principal components separated from the principal component analysis represent the rock–water interaction and agricultural return, redox environment, geogenic sources, the utilization of agricultural fertilizer, the weathering of aluminum silicates, and dissolution of carbonates, respectively. More than 70% of the samples are not recommended for irrigation due to the presence of high salt content in groundwater. EWQI assessment demonstrates that the quality of the groundwater is good. The outcomes of this study are significant for understanding the geochemical status of the groundwater in the Hutuo River Drinking Water Source Area, and helping policymakers to protect and manage the groundwater.

Keywords: groundwater quality; hydrochemistry; multivariate analysis; water sources; North China Plain

Citation: Yuan, Z.; Jian, Y.; Chen, Z.; Jin, P.; Gao, S.; Wang, Q.; Ding, Z.; Wang, D.; Ma, Z. Distribution of Groundwater Hydrochemistry and Quality Assessment in Hutuo River Drinking Water Source Area of Shijiazhuang (North China Plain). *Water* **2024**, *16*, 175. https://doi.org/10.3390/w16010175

Academic Editor: Domenico Cicchella

Received: 9 November 2023
Revised: 26 December 2023
Accepted: 27 December 2023
Published: 3 January 2024

1. Introduction

Groundwater is the most important water source for drinking, industrial, and agricultural water supplies in arid and semi-arid areas, which is essential for the sustainability of the ecosystem and human daily life [1]. Eighty percent of human diseases are induced by water, according to the World Health Organization [2]. Groundwater interacts with the surrounding environment during its process of formation and transportation, the physical and chemical interactions including dissolution/precipitation, adsorption/desorption, oxidation/reduction, and acid–base balance. The hydrochemical composition of groundwater and its distribution reflects the existence of the long-term interaction between groundwater and the environmental media [3]. Clarifying the distribution pattern of groundwater chemistry can help to understand the principal factors and processes that dominate the hydrochemical characteristics of groundwater. The results can provide a hydrochemical background for protecting the groundwater source of the North China Plain and other places.

The chemical evolution assessment is important for the establishment of suitable management policies and groundwater quality improvement. The hydrochemical composition characteristic of groundwater is controlled by sedimentary conditions and human activities [4]. However, it is difficult to figure out the contributions of geogenic processes and anthropogenic activities. Many analytical methods (multivariate statistical analysis, hydrogeochemical facies, binary diagrams, correlation analysis, etc.) are used to characterize the evolution processes [5–7]. Numerous indices have been developed to evaluate water pollution. Among them, the water quality index (WQI) is one of the most widely used indices to determine the overall groundwater quality [8]. The combination of entropy weights and the conventional WQI called the entropy water quality index (EWQI) is proposed; it can improve the reliability of the assessment results [9]. EWQI has been used by numerous studies for water quality assessment [10,11].

Shijiazhuang is one of the biggest cities located in the middle west part of the North China Plain, with a huge population [12]. Groundwater is the dominant source of drinking and irrigation water in Shijiazhuang [13]. However, the quantity and quality of groundwater are threatened with the development of urbanization and industrialization [14]. The demand for groundwater is increasing and the quality of groundwater is deteriorating due to the pollution emissions. The groundwater level has decreased sharply due to drought and over-exploitation, accompanied by the excessive nitrate loading and expansion of saline–alkaline land area [15–17]. Human activities lead to a dramatic increase in the groundwater NO_3^- concentration in pluvial fans of the Hutuo River, which reaches 124.4 mg/L [17]. The decreasing groundwater level will affect the redox environment, and may further affect the chemistry composition of groundwater. The dominant water types are Ca-HCO_3, HCO_3-Ca-Mg, and HCO_3-Na-K, as investigated 15 years ago. However, they have been transformed from HCO_3-type to HCO_3-Cl, HCO_3-SO_4, and SO_4-HCO_3 types in recent years [18]. Due to the variety of geological media and input from various anthropogenic sources, the distribution variability of groundwater chemistry and associated interactions remain unclear in this area.

The Hutuo River Drinking Water Source Area is located in the northern region of Shijiazhuang, and upstream of the water source are the Gangnan and Huangbizhuang reservoirs, which provide water for the Fourth and Seventh Water Plants of Shijiazhuang. The Hutuo River Drinking Water Source Area is an important water source in the North China Plain, and it is the water intake for the Shijiazhuang Drinking Water Company. However, there are few studies focused on groundwater chemistry evolution on the scale of drinking water areas. The hydrochemistry, origin analysis, and quality assessment of the groundwater in the Hutuo River Drinking Water Source Area have not been investigated comprehensively. Therefore, it is essential to analyze the geochemical characteristics in groundwater and the associated quality risks in water source areas. This study aims to: (1) measure the prevailing levels of 19 chemical variables based on national standards; (2) clarify the mutual associations among the measured parameters and the processes controlling the groundwater chemistry by statistical methods; (3) evaluate the suitability of the groundwater in the Hutuo River Drinking Water Source Area for irrigation/drinking. The study can provide a scientific support for the establishment or optimization of groundwater quality/quantity monitoring and management systems.

2. Materials and Methods

2.1. Site Description

Shijiazhuang (37°27′–38°47′ N, 113°30′–115°20′ E), is the capital of Hebei Province and located in the Mideast part of North China Plain, covering an area of 13,504 km^2. The average annual temperature of the study area is 13 °C. The annual precipitation of Shijiazhuang is 481.9 mm, and the rainfall is mainly concentrated in summer. The annual evaporation rate is 1557.8 mm/a, which is significantly higher than the mean annual precipitation. The Taihang Mountains are located in the western part of Shijiazhuang. The topography is low in the east and high in the west, with a slope of approximately 0.5~1.0‰.

Quaternary deposits are distributed on the surface of the region, with sediments gradually thickening from west to east and particles becoming coarse to fine. The primary rivers are the Hutuo River and Ye River. According to the Adjustment Plan for Emergency Reserve Drinking Water Source Areas in the Main Urban Area of Shijiazhuang, the drinking water source areas of the Hutuo River are divided into first-level protected area (63.95 km^2) and second-level protected area (196.84 km^2) (Figure 1). According to the Technical Specification for Classification of Drinking Water Source Protection Areas of China (HJ 338-2018), sewage outlets and the cultivation and stocking of poultry and livestock are prohibited in the first-grade protection zone. The Hutuo River Drinking Water Source Area was established in 1971 with 77 water source wells, the designed water intake quantity is 9.49×10^{10} tons, and the actual water intake quantity is 8.79×10^7 tons.

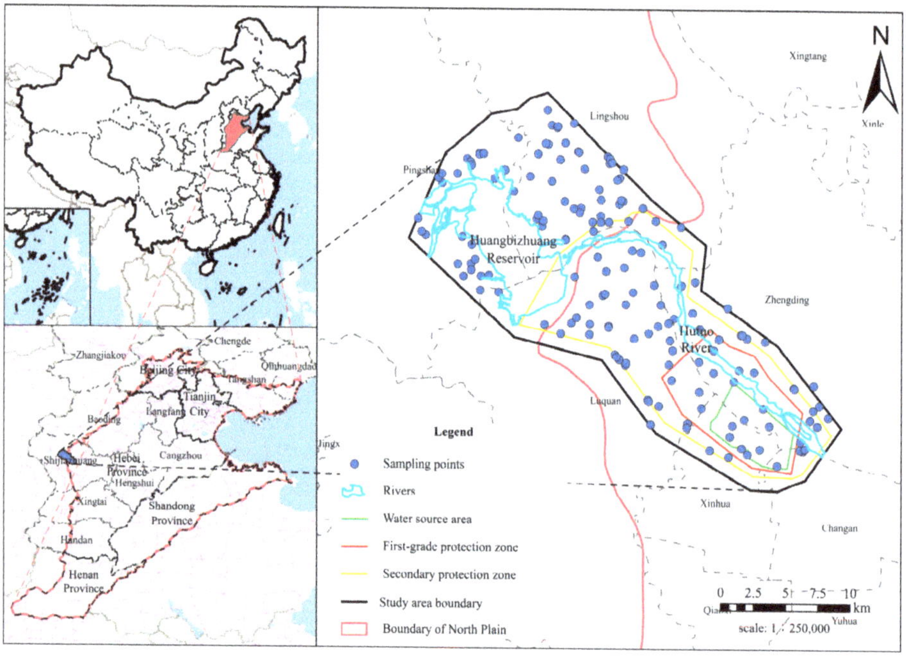

Figure 1. Geographical map of the study area.

The survey area belongs to the top and middle sections of the Hutuo River alluvial fan in the Ziya River system. It is divided into the hydrogeological structure zoning of fissure karst and the hydrogeological structure zoning of loose rock pore aquifer within a depth of 120 m. Fissure karst water is distributed in the northwest of the survey area and is hidden fissure karst water. Loose rock pore water is distributed throughout the region, and the main water source for the Hutuo River system groundwater type water source is the second water bearing rock group of loose rock pore water.

2.2. Sample Collection and Analysis

In this study, 160 groundwater samples were collected in November 2021 from the Hutuo River Drinking Water Source Area. The location of sampling sites was recorded by real-time kinematic devices. For all the samples, well water was pumped several times before water sampling to ensure that freshwater samples were collected. Samples were collected from private wells with sampling depths below 0.5 m from the water table. The shallow groundwater table is buried at a depth of 10–35 m. Polyethylene bottles were cleaned three times by the groundwater before being collected. A quantity of 250 mL of groundwater was collected for analysis of iron (Fe^{2+}), manganese (Mn^{2+}), zinc (Zn^{2+}), and aluminum (Al^{3+}), and 1500 mL for analyses of pH, total hardness (TH), total dissolved solids (TDS), sodium (Na^{+}), arsenic (As^{3+}), chloride (Cl^{-}), fluorine (F^{-}), sulfate (SO_4^{2+}), nitrate (NO_3^{-}), nitrite (NO_2^{-}), calcium (Ca^{2+}), magnesium (Mg^{2+}), potassium (K^{+}). All the bottles were filled with groundwater and sealed with parafilm to avoid interference from atmospheric gases. In addition, to ensure the analysis reliability, HNO_3 (65%) was added to adjust the pH below 2 in the bottle. Afterward, groundwater samples were stored at 4 °C before laboratory analysis. All samples were analyzed at the Hebei Provincial Geological Experimental Testing Center, Baoding.

Seventeen chemical parameters were measured. The pH was measured using a portable acidity meter (Zhengzhou Baojing Electronic Technology Co., Ltd., Zhengzhou, China, PHS-3C). The total hardness was calculated using the gravimetric titration method. The total dissolved solids was estimated by a weighing method. Heavy metal ions (Zn^{2+}, Al^{3+}, Mn^{2+}) were identified using Agilent Inductively Coupled Plasma Mass Spectrometry (ICP-MS, 7700XI); the detection limits of Zn^{2+}, Al^{3+}, Mn^{2+} were 0.67, 1.15, 0.12 μg/L, respectively. A detection limit of 0.3 μg/L was determined by an Atomic Fluorescence Photometer (Beijing Haiguang Instrument Co., Ltd., Beijing, China, AFS-8510). Major cations were detected by the Agilent Inductively Coupled Plasma Spectrometer (Agilent-725). The detection limits were 120 μg/L for Na^{+}, 20 μg/L for Ca^{2+}, 3 μg/L for Mg^{2+}, 50 μg/L for K^{+}, 10 μg/L for Fe^{2+}. Quantification of anions (F^{-}, SO_4^{2-}, Cl^{-}, NO_3^{-}) was carried out by ion chromatography (Dionex, Sunnyvale, CA, USA, ICS-1100), and the detection limits were 6 μg/L for F^{-}, 18 μg/L for SO_4^{2-}, 7 μg/L for Cl^{-}, 4 μg/L for NO_3^{-}. NO_2^{-} was detected with a spectrophotometer (T6), and the detection limit was 3 μg/L. HCO_3^{-} and CO_3^{2-} were determined by a titration method (Method for Analysis of Groundwater Quality, DZ/T0064.49-2021) using an acid burette. Each sample was analyzed in triplicate, and standards and blanks were interspersed regularly in sample batches to ensure the data quality. All data were corrected for instrument drift. The relative errors of these parameters were < ±10%. In addition, the charge balance error percentage (%CBE) was checked. The %CBE of all water samples was < ±5%.

2.3. Statistical Analysis

The principal components (PCs) were calculated by the multivariate principal component analysis (PCA), and PCs were used to characterize the study object by dimensional reduction. In the PCA, PCs with eigenvalues greater than 1 are significant and can be used to describe the impact of hydrochemical processes and human activities on groundwater chemical components. The PCA and Pearson correlation were conducted through IBM SPSS Statistics 24. The Pearson correlation is calculated by:

$$r = \frac{\sum_{i=1}^{n}(X_i - X)(Y_i - Y)}{\sqrt{\sum_{i=1}^{n}(X_i - X)^2}\sqrt{\sum_{i=1}^{n}(Y_i - Y)^2}}.$$

2.4. Groundwater Quality Assessment

Sodium hazard can be described by the sodium adsorption ratio (SAR). It is an important index and can be calculated by the relative concentrations of Ca^{2+}, Na^{+}, and Mg^{2+} in the irrigation groundwater. The formula can be expressed as follows:

$$SAR = \frac{[Na^{+}]}{\sqrt{\frac{1}{2}\left(\left[Ca^{2+}\right] + \left[Mg^{2+}\right]\right)}}$$

Based on the United States Department of Agriculture (USDA), groundwater with a SAR less than 10 (more than 26) is considered ideal (inappropriate) for irrigation. Groundwater with a high SAR value and low/moderate salinity will decrease the aeration and infiltration rate of the soil, affect the normal condition of plant growth, and lead to crop yield reduction [19].

Residual sodium carbonate (RSC) can be calculated by the relative content of Ca^{2+}, Mg^{2+}, $CO_3{}^{2-}$, and $HCO_3{}^{-}$ in the groundwater. The formula can be expressed as follows:

$$RSC = \left(CO_3{}^{2-} + HCO_3{}^{-}\right) - \left(Ca^{2+} + Mg^{2+}\right)$$

Groundwater with an RSC value of less than 2.5 meq/L has been considered satisfactory for plant irrigation. Groundwater with a high RSC value would lead to rapid salinization and codification of the soil profile, thereby harming the growth of crops [20].

Salinity hazard can be described by potential salinity, which reflects the relative content of Cl^{-} and $SO_4{}^{2-}$ in the groundwater.

$$PS = Cl^{-} + \sqrt{SO_4{}^{2-}}$$

where ion concentrations are in meq/L. Groundwater with PS values more than 3 and 15 are considered conditioned and not recommended, respectively, which may cause accumulation of salinity in soil [21].

The entropy weight quality index (EWQI) method was applied to determine the quality of the groundwater. EWQI calculation (Figure S1) has been used by numerous studies due to its simple calculation process and the integration of multiple chemical parameters [22–24]. In this study, the groundwater quality indices including pH, TH, TDS, Fe^{2+}, Na^{+}, Al^{3+}, Mn^{2+}, Zn^{2+}, As^{3+}, F^{-}, $SO_4{}^{2-}$, Cl^{-}, $NO_3{}^{-}$, and $NO_2{}^{-}$ were selected to identify the groundwater quality of Hutuo River Drinking Water Source Area. The EWQI can be calculated through two processes (Figure S1) [25]. The first step is the calculation of the rating scale (q_j) of EWQI for the chemical parameters of groundwater. C_i is the concentration of groundwater quality, S_i represents the values of grade III standards in the groundwater quality standard limits of the People's Republic of China. The q_j of pH values was evaluated based on the standard pH of 6.5–8.5. The second step is the standardization of matrix; C_{min} and C_{max} are the minimum and maximum of the chemical parameters, respectively.

Groundwater can be classified into 5 levels according to EWQI value [26,27], including Level 1: excellent (25 > EWQI), Level 2: good (50 > EWQI > 25), Level 3: fair (100 > EWQI > 50), Level 4: poor (150 > EWQI > 100), Level 5: very poor (EWQI > 150).

3. Results and Discussion

3.1. Groundwater Hydrochemistry

The statistical analysis results of the groundwater chemical parameters in the Hutuo River Drinking Water Source Area are illustrated in Table 1. The pH values range from 6.92 to 8.20 in groundwater samples. Only one groundwater sample has an acidic pH of 6.92, and all the other groundwater samples are slightly alkaline to neutral. The total hardness (TH) range is 204–1668 mg/L, with a mean concentration of 554.48 mg/L. The results reveal that 71.25% of the groundwater samples exceeded the allowable limit of TH (450 mg/L). TH is mainly originated from the dissolution of minerals and illustrates the total content of Mg^{2+} and Ca^{2+} [28]. Total dissolved solids (TDS) values of groundwater range from 271 to 2371 mg/L. The wide variation in TDS content indicates that various factors are correlated with the chemical composition of groundwater. The TDS values of the samples above 1000 mg/L are 13.75%, indicating that these groundwater samples are unsuitable for domestic consumption, leading to inferior palatability and gastrointestinal irritation. According to salinity classification [29], the groundwater in the Hutuo River Drinking Water Source Area is categorized into no-saline/fresh and low-saline water.

Table 1. Statics of chemical parameters of groundwater (unit: mg/L, except pH).

Parameters	Max	Min	Mean	SD	CV	Low Quartile	Upper Quartile	Allowable Limits [a]	Percentage Exceeding the Standard
pH	8.20	6.92	7.49	0.17	0.02	7.39	7.59	6.5–8.5	0.00
TH	1668.00	204.00	551.48	200.72	0.36	427.00	629.00	450.00	71.25
TDS	2371.00	271.00	747.29	284.70	0.38	574.00	872.00	1000.00	13.75
Fe^{2+}	0.33	0.00	0.01	0.03	6.57	0.00	0.00	0.30	0.63
Na^{+}	154.00	8.87	39.50	19.65	0.50	28.40	45.20	200.00	0.00
Al^{3+}	0.05	0.00	0.01	0.01	1.09	0.00	0.01	0.20	0.00
Mn^{2+}	0.07	0.00	0.00	0.01	2.56	0.01	0.00	0.10	0.00
Zn^{2+}	2.44	0.00	0.05	0.21	3.80	0.04	0.03	1.00	0.63
As^{3+}	0.00	0.00	0.00	0.00	1.38	0.00	0.00	0.01	0.00
F^{-}	0.62	0.05	0.26	0.10	0.37	0.20	0.32	1.00	0.00
SO_4^{2-}	552.00	11.00	191.84	93.03	0.48	141.00	238.00	250.00	20.00
Cl^{-}	404.00	8.98	72.54	50.75	0.70	43.90	82.90	250.00	2.50
NO_3^{-}-N	104.00	1.79	17.23	13.27	0.77	7.65	22.40	20.00	30.63
NO_2^{-}-N	1.55	0.00	0.01	0.12	8.58	0.00	0.01	1.00	0.63
Ca^{2+}	528.00	64.60	161.41	62.26	0.39	122.00	183.00	-	0.00
Mg^{2+}	97.00	9.68	36.59	14.32	0.39	28.60	42.30	-	0.00
K^{+}	23.60	0.36	2.15	2.19	1.02	0.97	2.60	-	0.00
CO_3^{2-}	6.00	0.00	0.04	0.47	0.08	0.00	0.00	-	0.00
HCO_3^{-}	578.00	186.00	315.86	68.88	4.59	258.00	360.00	-	0.00

Note: [a] Allowable limits represent the values of grade III standards in the groundwater quality standard limits of the People's Republic of China.

In the Hutuo River Drinking Water Source Area, the cationic dominance in the groundwater can be described in the following order: $Ca^{2+} > Na^{+} > Mg^{2+} > K^{+}$. Ca^{2+} is the dominant cation over the area, ranging from 64.6 to 528 mg/L, with an average concentration of 161.4 mg/L (Table 1). Ca^{2+} may originate from the dissolution of Ca-rich minerals such as $CaCO_3$ and $CaMg(CO_3)_2$ during recharge. The second dominant cation is Na^{+}; its concentration in the study area ranges from 8.87 to 154 mg/L. Na^{+} plays a crucial role as a salinity indicator in groundwater; it originates from various processes including ion exchange in halite and clay minerals, and plagioclase feldspar weathering [24]. The Na^{+} contents in all samples meet grade III of the groundwater quality standard limits of the People's Republic of China. In this study area, the Mg^{2+} and K^{+} content varied from 9.68 to 97 mg/L and 0.36 to 23.6 mg/L.

Over the Hutuo River Drinking Area, the order of anionic dominance in the groundwater is $HCO_3^{-} > SO_4^{2-} > Cl^{-} > NO_3^{-} > F^{-} > NO_2^{-}$. HCO_3^{-} is the dominant anion, varying from 186 to 578 mg/L, with an average of 186 mg/L. SO_4^{2-} ranges from 11 to 552 mg/L, with a mean concentration of 191.84 mg/L. The SO_4^{2-} contents in 20% of the samples surpass the permissible limit (250 mg/L). The Cl^{-} is from 8.98 to 404 mg/L (mean 72.54 mg/L), and the contents of Cl^{-} are more than its desirable limit (250 mg/L) in 75% of the total groundwater samples, which may cause a salty taste. The NO_3^{-}-N is between 1.79 to 104 mg/L (mean 17.32 mg/L), and the proportion of NO_3^{-}-N surpasses the permissible limit for grade III standards in the groundwater quality standard limits in 30.63% of the samples. Due to the highest standard-exceeding ratio of NO_3^{-}, the source and health risk of NO_3^{-} should be investigated and analyzed further. The maximum, minimum, and average concentrations of F^{-} are 0.62, 0.05, and 0.26 mg/L, respectively, and all samples have concentrations lower than the standard limits.

3.2. Factors Affecting the Groundwater Chemistry

The Piper diagram is a widely used graphical tool for recognizing groundwater chemistry reactions and the evolution of groundwater [30,31]. The Piper diagram contains three main areas, and the cation and anion proportions of groundwater are plotted to evaluate the different types of the samples. As shown in Figure 2, the groundwater sample points in the Hutuo River Drinking Water Source Area are mainly distributed in Zone 1, Zone 2, and Zone 3. The results suggest that the groundwater belongs to the Ca-HCO_3 (78.1%), mixed Ca-Mg-Cl (20%), and Ca-Cl (1.9%) classifications. The anions and cations of groundwater samples are primarily concentrated in Zone E and Zone A, demonstrating that the dominant anions and cations detected over the study area are HCO_3^- + CO_3^{2-} and Ca^{2+}, respectively. The results suggest that carbonate mineral weathering (such as calcite and dolomite) may be the main factor controlling groundwater chemistry in the Hutuo River Drinking Water Source Area. A total of 78.1% of the samples are classified as the HCO_3 water type, indicating the increased possibility of the ion exchange process, silicates, and carbonates weathering in groundwater. The other groundwater samples are classified as the Cl-water type, which implies evaporated dissolution and anthropogenic inputs in groundwater. Alkaline earth metals (Ca^{2+} and Mg^{2+}) are the most important for most groundwater samples instead of alkaline metals (Na^+ and K^+); the high concentration of alkaline earth metals may be due to dolomite weathering. Samples clustered in Zone 1, Zone 2, Zone E, and Zone B illustrate the dominant role of carbonate and silicated weathering in the hydrochemistry of groundwater [32].

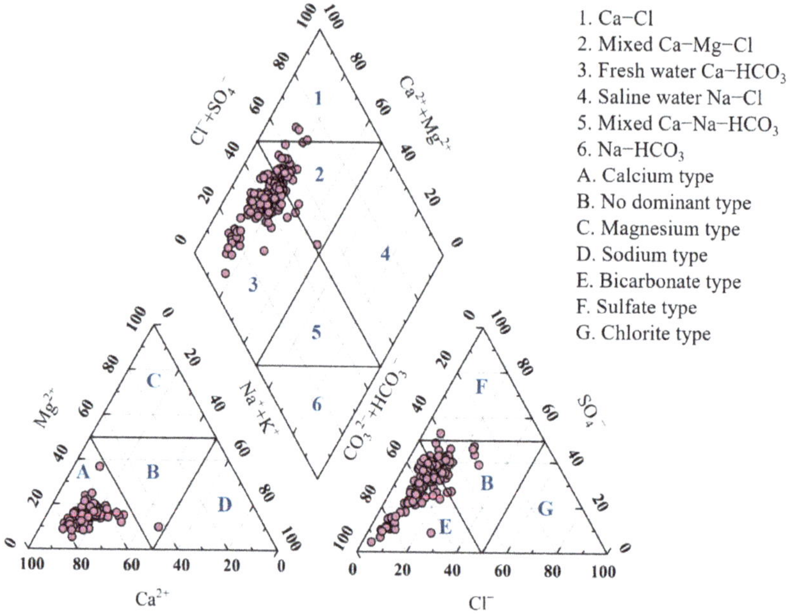

Figure 2. Piper diagram for classification of groundwater types of the groundwater samples.

Except for the impact of rock weathering, the hydrochemistry of groundwater is easily affected by climate factors. The Gibbs diagrams are divided into three parts; groundwater samples located in three regions represent different evolution processes of groundwater, including rock, precipitation, and evaporation dominance [33]. Apart from one sample, most of the groundwater samples concentrate in the rock dominance area (Figure 3), suggesting that water–rock interaction is the dominant factor and influences the formation of groundwater hydrochemistry. As shown in Figure 3a, one groundwater sample is located in the evaporation dominance area, indicating that this groundwater sample is controlled by the evaporation factor. The evaporation process would upsurge salinity by increasing the content of Na^+ and Cl^-, and TDS would be intensified subsequently.

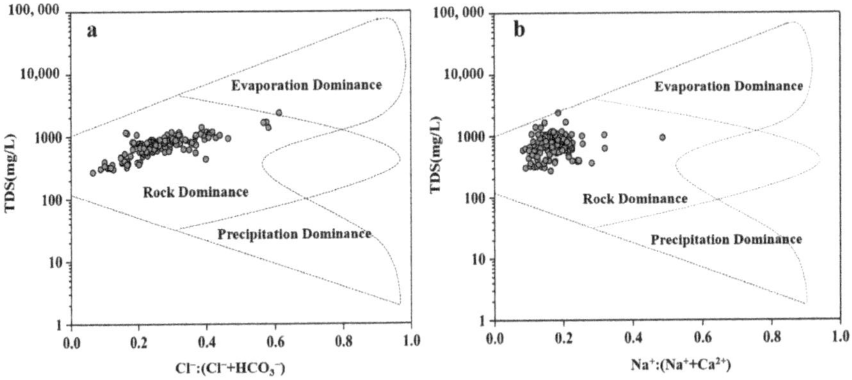

Figure 3. Gibbs diagrams. (**a**) TDS versus $Cl^-/(Cl^- + HCO_3^-)$. (**b**) TDS versus $Na^+/(Na^+ + Ca^{2+})$.

Reactions between groundwater and the aquifer media, such as rock weathering, evaporation, and ion exchange, would affect the hydrochemistry properties of groundwater significantly, and can be used for identification of the source of groundwater solutes. Several scatter plots are used to identify these reactions. As shown in Figure 4a, most of the samples plot along a 1:1 line for the bivariate diagram of Na^+ and Cl^-, which indicates that Na^+ may originate from the dissolution of halite [27]. The halite rock might be a prime loading source of Cl^- due to the clustering of groundwater samples above the dissolution line (1:1), whereas the deviation of the sample points toward the Na^+ axis also suggests that there are alternative sources of Na^+, such as ion exchange and silicate rock weathering [34].

As shown in Figure 4b, the groundwater samples are mainly clustered above the y = x relationship line, which indicates that the dissolution of gypsum is not the main origin of Ca^{2+} and SO_4^{2-} [35]. In a plot of Ca^{2+} versus HCO_3^- (Figure 4c), only seven groundwater sample points fall within the 1:1 and 1:2 theoretical lines. Some of the points are located along the 1:1 line, proving that the dissolution of carbonated rocks is one of the Ca^{2+} sources [36]. Most of the points are located above the 1:1 line, which indicates multiple origins of Ca^{2+}. The excess of Ca^{2+} over HCO_3^- indicates that the dissolution of gypsum might increase the concentration of Ca^{2+} in groundwater.

As shown in Figure 4d, some of the sample points scatter along the 1:1 line, which shows that the dissolution of calcite, dolomite, anhydrite, and gypsum are the dominant processes that affect the hydrochemistry of the groundwater [37]. The majority of the points are clustered above the equiline; the dominance of $Ca^{2+} + Mg^{2+}$ over $HCO_3^- + SO_4^{2-}$ can be an indicator of reverse ion exchange [38]. The phenomenon of cation exchange in the groundwater is further described by the negative slope of the trendline on the interionic plot $Na^+ + K^+ - Cl^-$ vs. $(Ca^{2+} + Mg^{2+}) - (HCO_3^- + SO_4^{2-})$ [24,39] (Figure 4e). The scattering of sample points not only demonstrates the loading of anthropogenic input in groundwater but also indicates the presence of reverse ion exchange, which is predicted by $(Ca^{2+} + Mg^{2+})$ versus $(HCO_3^- + SO_4^{2-})$ [40].

3.3. Principal Components Analysis Results

Principal components analysis (PCA) is utilized to clarify the components that control the groundwater chemistry characteristics. Principal components (PCs) are calculated from the seventh parameter. The Kaiser–Meyer–Olkin (KMO) value is 0.705, and the significance value of Bartlett's sphericity test is lower than 0.001. As shown in Figure 5, five PCs with eigenvalues > 1 explain 74.67% of the total variance in the original dataset. The results reflect the correlation between parameters and PCs, which are divided into three parts according to their absolute values: strong, medium, and weak, with values > 0.75, 0.75–0.5, and 0.5–0.3, respectively [41]. The parameters can be interpreted by each PC, and each PV has a geochemical understanding of the original datum.

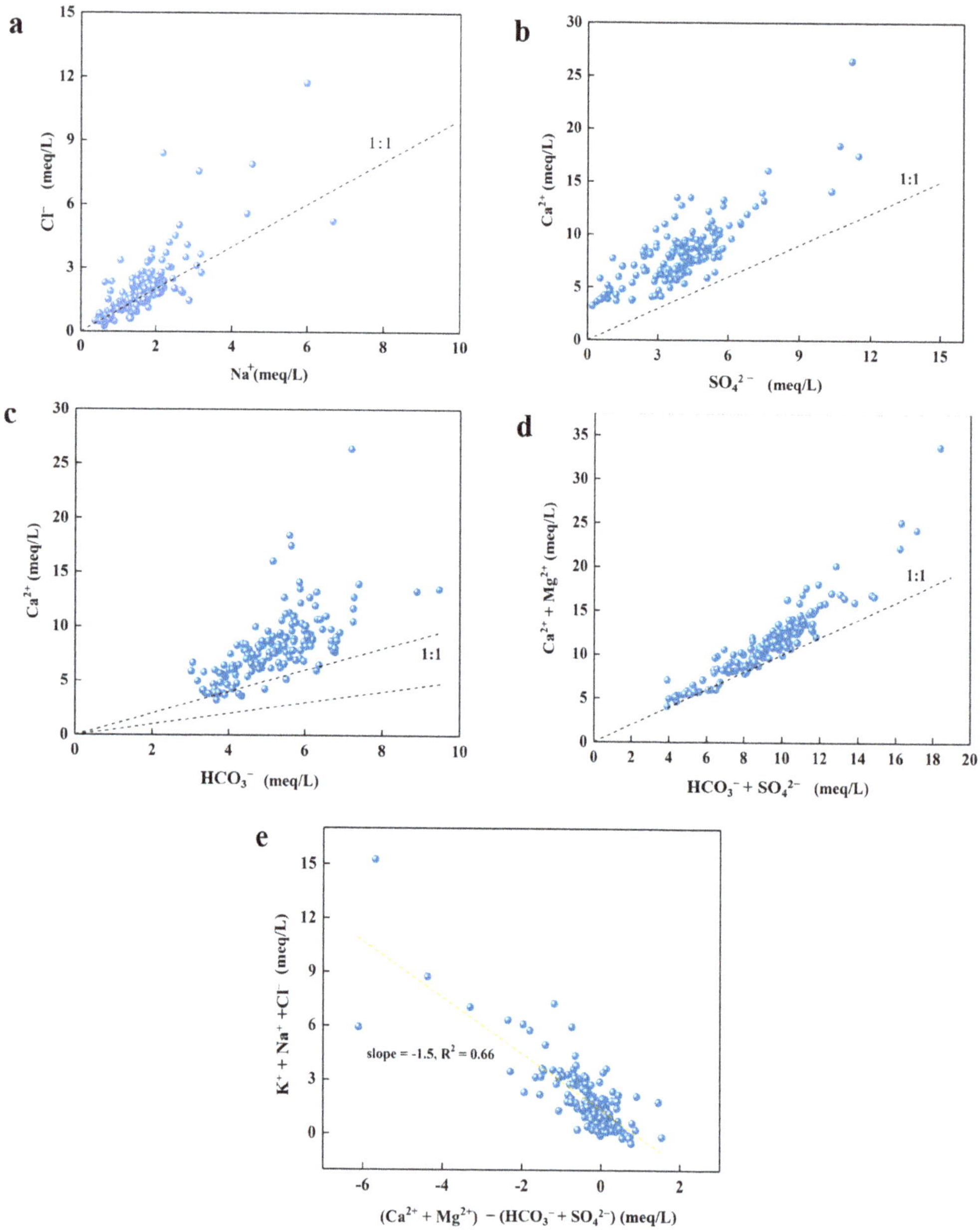

Figure 4. Relationship between the concentration of (**a**) Cl vs. Na^+, (**b**) Ca^{2+} vs. $SO_4{}^{2-}$, (**c**) Ca^{2+} vs. $HCO_3{}^-$, (**d**) (Ca^{2+} + Mg^{2+}) vs. ($HCO_3{}^-$ + $SO_4{}^{2-}$), (**e**) (K^+ + Na^+ − Cl^-) vs. (Ca^{2+} + Mg^{2+} − $HCO_3{}^-$ − $SO_4{}^{2-}$).

The Pearson correlation coefficient is introduced to quantify the relationship between the chemical parameters (Figure 6). The correlation coefficient of 1 indicates a good positive correlation, a perfect negative relation between two parameters is implied by −1, while zero suggests no relationship between two parameters [19].

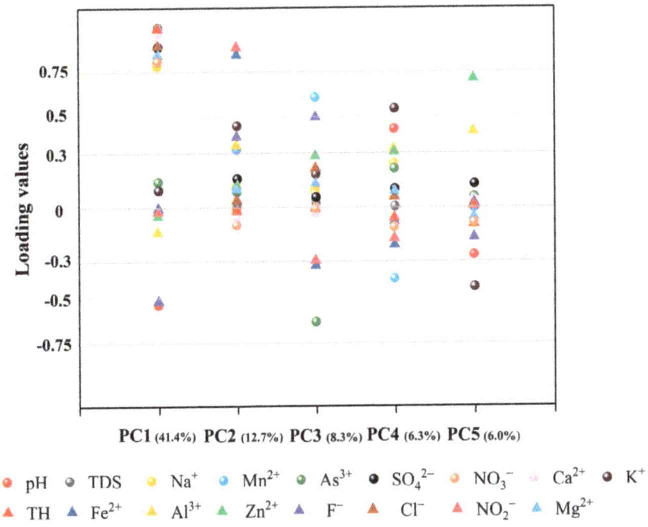

Figure 5. The principal component loadings of groundwater in the study area.

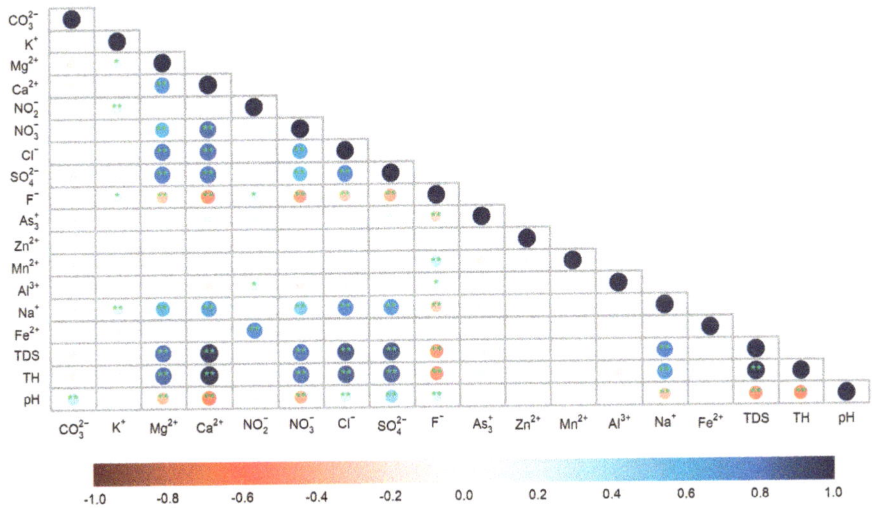

Figure 6. The Pearson correlation coefficient of major geochemical parameters. The correlations with p-values < 0.05, and 0.01 are shown with * and **, respectively. Except for HCO_3^-, all the parameters have passed the normality tests.

Principal component 1 (PC1) is dominated by TDS, TH, SO_4^{2-}, Cl^-, NO_3^-, Ca^{2+}, and Mg^{2+} (loadings between 0.78 and 0.99), and negatively correlated with pH (loading is −0.54). Thus, PC1 is considered as a salinity factor. As shown in Figure 6, TDS is positively correlated with Na^+, Ca^{2+}, Mg^{2+}, SO_4^{2-}, Cl^-, and NO_3^-; thus the variation of TDS in groundwater can reflect the regional impact of the salinity parameter and the corresponding modification [42]. Salinity is directly impacted by the contents of Na^+, Ca^{2+}, Mg^{2+}, SO_4^-, and Cl^-; the ions are contributed by rock–water interaction, especially the dissolution of halite, carbonated rocks, and silicate rocks as described above.

Among the 160 groundwater samples, there are 49 samples with a concentration of NO_3^- higher than the maximum permissible limit of the groundwater quality standard grade III limits of the People's Republic of China (20 mg/L). Nitrogen is of great significance in the biogeochemical cycle. Nitrate contamination is widespread in China, and the residents are exposed to the risk of high nitrate concentration in groundwater [43]. The source identification of NO_3^- can facilitate the improvement in groundwater quality, such as reduction in the usage of chemical fertilizers, improvement in the treatment processes of wastewater, and the remediation of the contaminated groundwater [44]. The nitrate in groundwater may be from multiple natural factors and anthropogenic activities, but it is difficult to quantify nitrate sources in groundwater [45]. The correlation between NO_3^-/Na^+ and Cl^-/Na^+ makes a distinction of nitrate sources between agricultural activities and urban sewage; different NO_3^- sources have different NO_3^-/Cl^- ratios [23]. As shown in Figure 7, the values of NO_3^-/Cl^- vary from 0.002 to 0.38; the low NO_3^-/Cl^- and elevated Cl^- suggest that NO_3^- in groundwater is significantly affected by the combination of both. Agricultural activities play the dominant role in the hydrochemistry formation process of groundwater [46]. A combination of agricultural components and communal effluent is in accordance with the dense population and farmland in the study area. The usage of agricultural fertilizer should be controlled. Thus, rock–water interaction and agricultural return control the PC1.

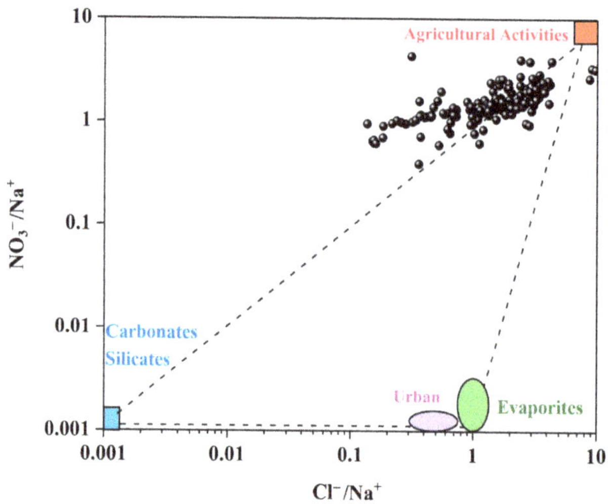

Figure 7. The binary diagram of NO_3^-/Na^+ vs. Cl^-/Na^+ (as molar ratios).

PC2 is distinguished by the reduced species Fe^{2+} and NO_2^-, which may indicate a redox environment. Fe enrichment in groundwater usually indicates the reducing condition, where denitrification is inhibited [47]. PC3 shows maximum loading for Mn^{2+} (0.609) and F^- (0.498). It is reported that the dissolution of F-bearing carbonate mineral and Mn-hydroxides are the origin of F^- in groundwater [48]; the F^- and Mn^{2+} in groundwater mainly originate from geogenic sources [49].

PC4 is dominated by K^+ (with a loading of 0.545). The site investigation indicates that the Hutuo River Drinking Water Source Area is located on the North China Plain, where agricultural activities are intensive and fertilizer is over-used. Fertilization with KNO_3 and $NaNO_3$ leads to deterioration of groundwater quality, and K^+ is frequently detected with Na^+ and other ions in groundwater [50,51]. The correlation of K^+ and Na^+ ($p < 0.01$) shown in Figure 6 verifies the utilization of agricultural fertilizer (Figure 6).

PC5 shows maximum loadings for Zn^{2+} (0.709) and Al^{3+} (0.416), and it is noted that both of them are not significantly correlated with other parameters (Figure 6). Therefore, the increase in these metals may be attributed to the aluminum silicate weathering and dissolution of carbonates (i.e., Smithsonite) [52]. The agricultural pollution sources may cause serious pollution to regional groundwater, and the exceeding of several water quality parameters indicates that the groundwater quality in the Hutuo River Drinking Water Source Area needs appropriate assessment.

3.4. Groundwater Quality Assessment

To further analyze the contributing parameters in the geochemical composition and their corresponding entropy weights, 14 parameters with different entropy weights and concentrations are employed to calculate the EWQI value. The entropy weights (w_j) and information entropy (e_j) values for the qualitative classification are provided in Table S1; the parameter with a higher entropy value would have a greater influence on groundwater quality. According to the calculated results, Fe^{2+} and NO_2^- have the highest weights of 0.28 and 0.26, respectively, while pH shows the lowest weight of 0.004. The EWQI results revealed that 97.5%, 1.88%, and 0.63% of the samples are classified as excellent, good, and fair, respectively (Figure 8).

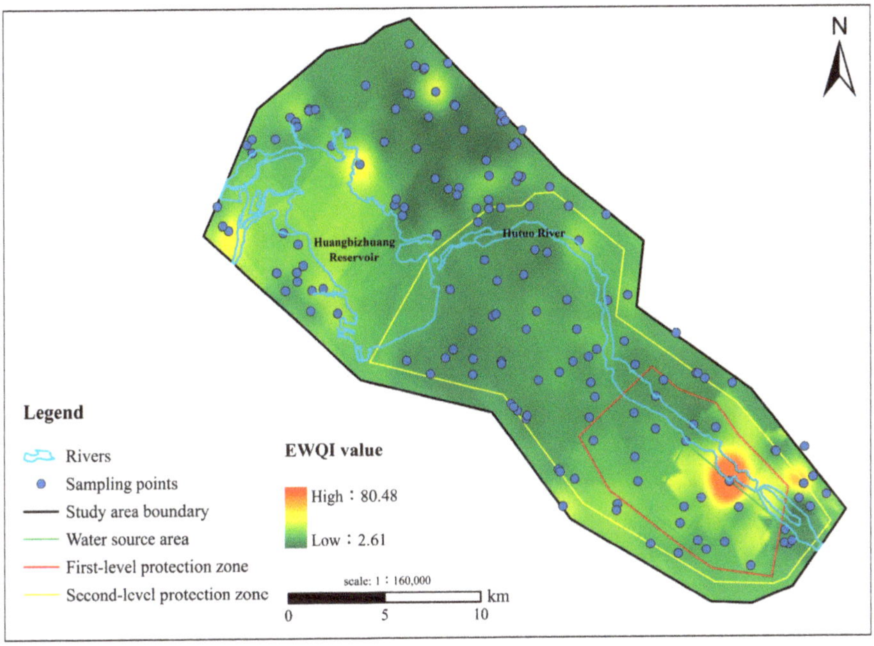

Figure 8. Groundwater quality based on the EWQI method.

The evaluation of different parameters indicates that the components of some groundwater samples are higher than grade III of the groundwater quality standard limits of the People's Republic of China. As shown in Table 1, the TH values of 71.25% of the samples exceed the limit, and the NO_3^- values for 30.63% of the samples are higher than the limit. The EWQI calculations show that 99.37% of the samples are classified into excellent and good quality (grades I and II), indicating that the quality of groundwater in the Hutuo River Drinking Water Source Area is good and suitable for drinking [23]. These results demonstrate that only using limited parameters cannot estimate the real groundwater quality comprehensively.

To estimate whether the groundwater is ideal for agricultural irrigation in the Hutuo River Drinking Water Source Area, the sodium adsorption ratio (SAR), residual sodium carbonate (RSC), and potential salinity (PS) are calculated. Groundwater with HCO_3^- concentration above 90 mg/L is not ideal for irrigation, as it would exaggerate the sodic condition of the soil [24].

15. Zhang, Y.; Chen, Z.; Huang, G.; Yang, M. Origins of groundwater nitrate in a typical alluvial-pluvial plain of North China plain: New insights from groundwater age-dating and isotopic fingerprinting. *Environ. Pollut.* **2023**, *316*, 120592. [CrossRef]
16. Nakayama, T.; Yang, Y.; Watanabe, M.; Zhang, X. Simulation of groundwater dynamics in the North China Plain by coupled hydrology and agricultural models. *Hydrol. Process.* **2006**, *20*, 3441–3466. [CrossRef]
17. Zhang, Q.; Wang, H.; Wang, L. Tracing nitrate pollution sources and transformations in the over-exploited groundwater region of north China using stable isotopes. *J. Contam. Hydrol.* **2018**, *218*, 1–9. [CrossRef]
18. Zhang, Q.; Wang, H.; Wang, Y.; Yang, M.; Zhu, L. Groundwater quality assessment and pollution source apportionment in an intensely exploited region of northern China. *Environ. Sci. Pollut. Res.* **2017**, *24*, 16639–16650. [CrossRef]
19. Batool, M.; Toqeer, M.; Shah, M.H.H. Assessment of water quality, trace metal pollution, source apportionment and health risks in the groundwater of Chakwal, Pakistan. *Environ. Geochem. Health* **2023**, *45*, 4327–4352.
20. Prasad, A.; Kumar, D.; Singh, D.V. Effect of residual sodium carbonate in irrigation water on the soil sodication and yield of palmarosa (Cymbopogon martinni) and lemongrass (Cymbopogon flexuosus). *Agric. Water Manag.* **2001**, *50*, 161–172. [CrossRef]
21. Alfredo Ramos-Leal, J.; Lopez-Alvarez, B.; Santacruz-De Leon, G.; Almanza-Tovar, O.; Moran-Ramirez, J.; Padilla-Reyes, D.A.; Gonzalez-Acevedo, Z.I. Quality indices of groundwater for agricultural use in the region of Tierra Nueva, San Luis Potosi, Mexico. *Arab. J. Geosci.* **2016**, *9*, 1–17.
22. Yang, Y.; Li, P.; Elumalai, V.; Ning, J.; Xu, F.; Mu, D. Groundwater Quality Assessment Using EWQI With Updated Water Quality Classification Criteria: A Case Study in and Around Zhouzhi County, Guanzhong Basin (China). *Expo. Health* **2022**, *15*, 825–840. [CrossRef]
23. Amiri, V.; Sohrabi, N.; Li, P.; Amiri, F. Groundwater Quality for Drinking and Non-Carcinogenic Risk of Nitrate in Urban and Rural Areas of Fereidan, Iran. *Expo. Health* **2022**, *15*, 807–823. [CrossRef]
24. Marghade, D.; Malpe, D.B.; Duraisamy, K.; Patil, P.D.; Li, P. Hydrogeochemical evaluation, suitability, and health risk assessment of groundwater in the watershed of Godavari basin, Maharashtra, Central India. *Environ. Sci. Pollut. Res.* **2021**, *28*, 18471–18494. [CrossRef] [PubMed]
25. Wu, J.; Li, P.; Qian, H.; Chen, J. On the sensitivity of entropy weight to sample statistics in assessing water quality: Statistical analysis based on large stochastic samples. *Environ. Earth Sci.* **2015**, *74*, 2185–2195. [CrossRef]
26. Amiri, V.; Bhattacharya, P.; Nakhaei, M. The hydrogeochemical evaluation of groundwater resources and their suitability for agricultural and industrial uses in an arid area of Iran. *Groundw. Sustain. Dev.* **2021**, *12*, 100527. [CrossRef]
27. Li, P.; Wu, J.; Tian, R.; He, S.; He, X.; Xue, C.; Zhang, K. Geochemistry, Hydraulic Connectivity and Quality Appraisal of Multilayered Groundwater in the Hongdunzi Coal Mine, Northwest China. *Mine Water Environ.* **2018**, *37*, 222–237. [CrossRef]
28. Li, P.; Wu, J.; Qian, H.; Zhang, Y.; Yang, N.; Jing, L.; Yu, P. Hydrogeochemical Characterization of Groundwater in and Around a Wastewater Irrigated Forest in the Southeastern Edge of the Tengger Desert, Northwest China. *Expo. Health* **2016**, *8*, 331–348. [CrossRef]
29. Robinove, C.J.; Langford, R.H. *Saline-Water Resources of North Dakota*; Water Supply Paper; US Government Printing Office: Washington, DC, USA, 1958.
30. Marghade, D.; Malpe, D.B.; Rao, N.S. Applications of geochemical and multivariate statistical approaches for the evaluation of groundwater quality and human health risks in a semi-arid region of eastern Maharashtra, India. *Environ. Geochem. Health* **2021**, *43*, 683–703. [CrossRef]
31. Li, P.Y.; He, X.D.; Guo, W.Y. Spatial groundwater quality and potential health risks due to nitrate ingestion through drinking water: A case study in Yan'an City on the Loess Plateau of northwest China. *Hum. Ecol. Risk Assess.* **2019**, *25*, 11–31. [CrossRef]
32. Appelo, C.A.J.; Postma, D. *Geochemistry, Groundwater and Pollution*, 2nd ed.; CRC Press: Boca Raton, FL, USA, 2005.
33. Gibbs, R.J. Mechanisms controlling world water chemistry. *Science* **1970**, *170*, 1088–1090. [CrossRef] [PubMed]
34. Kumar, S.; Venkatesh, A.S.; Singh, R.; Udayabhanu, G.; Saha, D. Geochemical signatures and isotopic systematics constraining dynamics of fluoride contamination in groundwater across Jamui district, Indo-Gangetic alluvial plains, India. *Chemosphere* **2018**, *205*, 493–505. [CrossRef] [PubMed]
35. Liu, J.T.; Peng, Y.M.; Li, C.S.; Gao, Z.J.; Chen, S.J. An investigation into the hydrochemistry, quality and risk to human health of groundwater in the central region of Shandong Province, North China. *J. Clean. Prod.* **2021**, *282*, 125416. [CrossRef]
36. Wu, J.; Wang, L.; Wang, S.; Tian, R.; Xue, C.; Feng, W.; Li, Y. Spatiotemporal variation of groundwater quality in an arid area experiencing long-term paper wastewater irrigation, northwest China. *Environ. Earth Sci.* **2017**, *76*, 460. [CrossRef]
37. Barzegar, R.; Moghaddam, A.A.; Tziritis, E.; Fakhri, M.S.; Soltani, S. Identification of hydrogeochemical processes and pollution sources of groundwater resources in the Marand plain, northwest of Iran. *Environ. Earth Sci.* **2017**, *76*, 297. [CrossRef]
38. Elango, L.; Kannan, R. Chapter 11 Rock–water interaction and its control on chemical composition of groundwater. *Dev. Environ. Sci.* **2007**, *5*, 229–243.
39. Fijani, E.; Moghaddam, A.A.; Tsai, F.T.C.; Tayfur, G. Analysis and Assessment of Hydrochemical Characteristics of Maragheh-Bonab Plain Aquifer, Northwest of Iran. *Water Resour. Manag.* **2017**, *31*, 765–780. [CrossRef]
40. Adimalla, N. Groundwater Quality for Drinking and Irrigation Purposes and Potential Health Risks Assessment: A Case Study from Semi-Arid Region of South India. *Expo. Health* **2019**, *11*, 109–123. [CrossRef]
41. Zhang, H.; Han, X.; Wang, G.; Mao, H.; Chen, X.; Zhou, L.; Huang, D.; Zhang, F.; Yan, X. Spatial distribution and driving factors of groundwater chemistry and pollution in an oil production region in the Northwest China. *Sci. Total Environ.* **2023**, *875*, 162635. [CrossRef]

42. Krishan, G.; Bhagwat, A.; Sejwal, P.; Yadav, B.K.; Kansal, M.L.; Bradley, A.; Singh, S.; Kumar, M.; Sharma, L.M.; Muste, M. Assessment of groundwater salinity using principal component analysis (PCA): A case study from Mewat (Nuh), Haryana, India. *Environ. Monit. Assess.* **2023**, *195*, 37. [CrossRef]
43. Gu, B.; Ge, Y.; Chang, S.X.; Luo, W.; Chang, J. Nitrate in groundwater of China: Sources and driving forces. *Glob. Environ. Chang. Hum. Policy Dimens.* **2013**, *23*, 1112.
44. Amiri, V.; Nakagawa, K. Using a linear discriminant analysis (LDA)-based nomenclature system and self-organizing maps (SOM) for spatiotemporal assessment of groundwater quality in a coastal aquifer. *J. Hydrol.* **2021**, *603*, 127082. [CrossRef]
45. Linhoff, B. Deciphering natural and anthropogenic nitrate and recharge sources in arid region groundwater. *Sci. Total Environ.* **2022**, *848*, 157345. [CrossRef] [PubMed]
46. Huang, X.; Jin, M.; Ma, B.; Liang, X.; Cao, M.; Zhang, J.; Zhang, Z.; Su, J. Identifying nitrate sources and transformation in groundwater in a large subtropical basin under a framework of groundwater flow systems. *J. Hydrol.* **2022**, *610*, 127943. [CrossRef]
47. Puig, R.; Soler, A.; Widory, D.; Mas-Pla, J.; Domenech, C.; Otero, N. Characterizing sources and natural attenuation of nitrate contamination in the Baix Ter aquifer system (NE Spain) using a multi-isotope approach. *Sci. Total Environ.* **2017**, *580*, 518–532. [CrossRef]
48. Gao, X.; Luo, W.; Luo, X.; Li, C.; Zhang, X.; Wang, Y. Indigenous microbes induced fluoride release from aquifer sediments. *Environ. Pollut.* **2019**, *252*, 580–590. [CrossRef]
49. Yin, S.; Xiao, Y.; Han, P.; Hao, Q.; Gu, X.; Men, B.; Huang, L. Investigation of Groundwater Contamination and Health Implications in a Typical Semiarid Basin of North China. *Water* **2020**, *12*, 1137. [CrossRef]
50. Griffioen, J. Potassium adsorption ratios as an indicator for the fate of agricultural potassium in groundwater. *J. Hydrol.* **2001**, *254*, 244–254. [CrossRef]
51. Ye, X.; Zhang, Q.; Liu, J.; Li, X.; Xu, C.-y. Distinguishing the relative impacts of climate change and human activities on variation of streamflow in the Poyang Lake catchment, China. *J. Hydrol.* **2013**, *494*, 83–95. [CrossRef]
52. Rajmohan, N.; Niyazi, B.A.M.; Masoud, M.H.Z. Trace metals pollution, distribution and associated health risks in the arid coastal aquifer, Hada Al-Sham and its vicinities, Saudi Arabia. *Chemosphere* **2022**, *297*, 134246. [CrossRef]
53. Li, P.; Li, X.; Meng, X.; Li, M.; Zhang, Y. Appraising Groundwater Quality and Health Risks from Contamination in a Semiarid Region of Northwest China. *Expo. Health* **2016**, *8*, 361–379. [CrossRef]

water

Article

Hydrogeochemical Characteristics and Evolution of Karst Groundwater in Heilongdong Spring Basin, Northern China

Ming Gao [1,2,3], Xiangquan Li [1,3], Jiazhong Qian [2], Zhenxing Wang [1,3,*], Xinwei Hou [1,3], Changchang Fu [1,3,*], Jianfei Ma [1,3], Chunchao Zhang [1,3] and Jinqiu Li [1,3]

1 Institute of Hydrogeology and Environmental Geology, Chinese Academy of Geological Sciences, Shijiazhuang 050061, China
2 School of Resources and Environmental Engineering, Hefei University of Technology, Hefei 230009, China
3 Key Laboratory of Groundwater Sciences and Engineering, Ministry of Natural Resources, Shijiazhuang 050061, China
* Correspondence: wangzhxing9987@126.com (Z.W.); fuchangchang1988@163.com (C.F.)

Abstract: Understanding the impact of natural processes and anthropogenic activities on geochemical evolution is vital for groundwater protection and utilization. This research was devoted to identifying the water quality status and the main controlling factors of the hydrochemical evolution of karst groundwater by combining hydrogeochemical indicators with multi-isotope analysis techniques in the Heilongdong Spring Basin, North China. The results showed that the karst groundwater in the area was of meteoric origin, and the dissolution of carbonate minerals was dominant in water–rock interactions. Meanwhile, the positive and negative cation exchange occurred in the process. The main hydrochemical types of karst groundwater were HCO_3-Ca·Mg and HCO_3-Ca in the recharge area, while the predominant hydrochemical types were the HCO_3·SO_4-Ca·Mg and HCO_3·SO_4-Ca in the runoff and discharge area. Under the influence of coal mining and other factors, the average concentrations of major ions kept rising in the runoff area where coal mines were distributed, and the SO_4^{2-} concentrations of the karst groundwater changed the most in the study area. In addition, sewage from agricultural production and domestic sources had also negatively impacted the quality of regional groundwater in the runoff and discharge area, as evidenced by the increasing NO_3^- and Cl^- contents in the Quaternary sediment groundwater, Permian bedrock groundwater and a small portion of karst groundwater. These results were helpful to explain the mechanism of gradual hydrogeochemical changes and provided a scientific basis for the effective management and utilization of karst groundwater.

Keywords: karst groundwater; hydrogeochemistry; stable isotopes; water–rock interaction; coal mining; anthropogenic activities

Citation: Gao, M.; Li, X.; Qian, J.; Wang, Z.; Hou, X.; Fu, C.; Ma, J.; Zhang, C.; Li, J. Hydrogeochemical Characteristics and Evolution of Karst Groundwater in Heilongdong Spring Basin, Northern China. *Water* **2023**, *15*, 726. https://doi.org/10.3390/w15040726

Academic Editor: Domenico Cicchella

Received: 15 December 2022
Revised: 7 February 2023
Accepted: 8 February 2023
Published: 12 February 2023

1. Introduction

Water is essential to life, yet 844 million people worldwide lack access to basic drinking water. Most of them are individuals who reside in the rural areas of developing countries, such as in Subsaharan Africa [1,2]. Since the last century, the population has increased threefold, and the consumption of water has increased sixfold. Moreover, due to global climate change, the supply of water has decreased, resulting in prolonged drought and increased pollution [3]. We are facing a global water crisis. Previous studies have shown that groundwater is more adaptable to climate change than surface water, and it is more reliable and easier to obtain than surface water, because it can be directly exploited by users [4]. However, climate change can also affect groundwater quality by reducing the aquifer recharge and increasing artificial pressures [5].

Karst groundwater is one of the most important and well-used groundwater sources around the world. More than a quarter of the world's population lives in areas where karst groundwater is used as a source drinking water [6,7]. In some areas, it is even the only available water resource [8]. Most of Northern China has a large area of arid or semiarid climate with a relative shortage of surface water resources, while groundwater resources are more extensively exploited and utilized [9–11]. The area of carbonate rocks in Northern China can reach 68.5×10^4 km^2, with a total exposed area of 7.78×10^4 km^2 [12,13]. Karst groundwater with abundant and good water quality is the significant water supply for Northern China, supporting a huge population and providing an important guarantee for local industry, agriculture and civil use. While the karst groundwater is convenient for human beings, people gradually realize the serious consequences caused mainly by the long-term abstraction of karst groundwater. The excessive exploitation of karst groundwater has caused a series of ecological and environmental problems in Northern China, such as the decline in the regional groundwater level, attenuation of karst springs and deterioration of water quality [9,14,15].

Hydrochemical and isotopic indicators are effective tools that are frequently used to solve multiple problems in hydrology and hydrogeology [11,16–18]. The combination of hydrogeochemical methods such as the Gibbs diagram, Piper diagram and ionic ratios have been widely used in the analysis of hydrochemical evolution process, which provides a way to deeply understand the control mechanism of various geochemical reactions occurring in groundwater. Additionally, as an effective tracer, stable isotopes are useful in studying the origin of groundwater as well as calculating the mixing proportion among different water sources [19,20]. The $\delta^{18}O$ and δ^2H isotopes have frequently been used to identify the origin and recharge mechanism of groundwater. The stable isotopes of dissolved sulfate ($\delta^{34}S$ and $\delta^{18}O$) are excellent tracers for investigating the source of sulfate in groundwater or surface water. Therefore, the combination of hydrochemical and isotopic methods can determine the geochemical factors and mechanisms controlling the hydrochemical composition of groundwater.

The Heilongdong Spring Basin (HSB), located in Southwestern Hebei Province, is a typical representative of karst groundwater systems in Northern China. The abundant and good quality of karst groundwater in this region has long been an important source of water for domestic use, agricultural irrigation and industrial production in Handan, Northern China. The region is also the main coal mining distribution area and agricultural cultivation area in Northern China. In the past decades, the demand for groundwater has increased with the development of industry and agriculture. However, the local groundwater quality has shown a continuous deterioration due to the long-term discharge of coal mining water, industrial wastewater and domestic sewage, as well as the extensive use of agricultural fertilizers and pesticides.

Since the 1970s, many scholars have successively carried out research on the evaluation of groundwater circulation and evolution processes and made a lot of achievements in the area. Guo et al. [21] used four parameter ionic ratios to reveal the hydrochemical characteristics in the Fengfeng mining area and conducted reverse hydrogeochemical process modeling. Hao et al. [22] found that the hydraulic gradient of karst groundwater increased after coal mining activities and the groundwater quality became deteriorated in the Fengfeng coal mining area. Hao et al. [23] revealed the ^2H and ^{18}O drift were attributed to water–rock reactions between the Ordovician limestone, carbonate and silicate minerals in the Fengfeng coal mining area. However, most of the previous studies have focused on regional hydrogeological investigations and the hydrochemical and isotopic characteristics of a small local scale or in a particular year [7]. The knowledge of the hydrogeochemical evolution characteristics of the whole spring basin and the relationship between different aquifers is still insufficient or unclear. The present study systematically studied the geochemical evolution of karst groundwater in the whole HSB and obtained a new understanding.

To comprehensively understand the hydrochemical characteristics and evolution processes of karst groundwater in the HSB, it is necessary to clarify the impact of natural and anthropogenic activities on karst groundwater chemistry. Therefore, the main objectives of this study are to: (1) ascertain the sources of karst groundwater and clarify the hydraulic connection between karst aquifers and other aquifers and (2) compare the hydrochemical characteristics of karst groundwater in different periods and reveal the hydrochemical formation mechanisms controlling the chemical composition, especially SO_4^{2-} and NO_3^- in karst groundwater.

As a typical representative of the karst groundwater system in Northern China, the Heilongdong Spring Area is facing huge pollution risks with the rapid development of coal mining and agricultural activities, which makes it practically significant to evaluate the spatiotemporal variations in karst groundwater. The findings will provide a scientific basis for promoting the protection and utilization of karst groundwater resources in the HSB and a reference for the management of other karst groundwater resources in mining areas.

2. Study Area

2.1. General Setting

The HSB is located in the southwestern part of Hebei Province, which belongs to Handan City in Northern China. To the west of the study area is a series of mountains, while the east is mainly hilly and plain. The elevation gradually decreases from west to east, with the Heilongdong (HLD) spring groups at the lowest altitude in the whole region. The total area of the HSB is 2002 km^2.

The study area has significant seasonal variations with a mean annual temperature of 13.9 °C. The mean annual precipitation is 545 mm, while the mean annual evaporation is 1895.7 mm. The rainy season (July to September) accounts for 66.4% of the annual precipitation. Zhanghe River, Fuyang River and Nanming River are the main rivers in the basin. Additionally, Yuecheng (YC) Reservoir and Dongwushi (DWS) Reservoir are two large reservoirs in the region.

The Heilongdong spring groups are located at the intersection of Gushan Mountain and Fuyang River. They are composed of more than 60 springs, which are concentrated in an area of about 2 km^2. The outcrop elevation of the springs is 122.8~132.4 m. Fuyang River is mainly formed by the spring groups' discharge. The HLD springs' flow ranged from 7 m^3/s to 9 m^3/s before the 1980s. After the 1980s, due to the increase in human exploitation and the decrease in precipitation, the HLD springs dried up successively in 1987. Since then, the springs have only flowed intermittently in the rainy years or seasons (usually from July to September). It was not until 2016 that the spring flow completely resumed, and the current spring flow was 4.76 m^3/s in 2019.

2.2. Geological and Hydrogeological Setting

The boundaries of the HSB are defined by groundwater divides, water-blocking faults and impermeable igneous rocks. Therefore, the study area is a relatively independent hydrogeological unit. As a whole, it is an east-inclining monoclinic structure with a stratigraphic dip angle roughly of 10°~20°. The faults, with NE, NW, near SN and EW strikes, are distributed throughout the region. The main strata exposed in the study are of Sinian (Z), Cambrian (∈), Ordovician (O), Carboniferous (C), Permian (P), Triassic (T) and Quaternary sediments (Q), in which the main coal-bearing strata are Carboniferous and Permian.

The key aquifers (simplified) from bottom to top in the area are the Cambrian limestone aquifer, Ordovician limestone aquifer, Carboniferous thin-limestone aquifer, Permian sandstone aquifer and Quaternary sediment aquifer (Figure 1). The Cambrian and Ordovician limestone aquifers, with large thicknesses and high water abundance, are the main aquifer in the study area. The Cambrian and Ordovician strata are exposed in the western and central mountainous areas yet are covered by Quaternary and coal measure strata in the middle and east of the basin, which are mainly composed of limestone and dolomite [24]. In addition, the karstic fissures, honeycombed karstic pores and layered gypsum lenses are also developed in the Ordovician strata, with a total thickness of 470~584 m [23]. The gypsum layers in the Ordovician strata are about 2~3 mm thick [25]. Karst groundwater (KGW), which is extremely abundant in the Cambrian and Ordovician limestone aquifers, is an important source of drinking water for local residents. The Carboniferous thin limestone and the Permian sandstone aquifer, located above the Ordovician limestone aquifer and overlain by the Quaternary sediment aquifer, have poor water yields. The Carboniferous and Permian strata are composed of multilayer coal seams, and the main lithology is sandstone, thin limestone and multilayer coal seams with shale interbedded. The vertical distance between the lower coal seam of the Carboniferous strata and the underlying Ordovician limestone aquifer is generally 20~50 m. Due to the widespread distribution of structural fractures, karst collapse columns and fractures formed in the process of coal mining, there is a direct or indirect hydraulic connection between the coal seam and the underlying Ordovician limestone aquifer as well as the overlying Permian sandstone aquifer. The Quaternary sedimentary aquifer is dominated by sand, gravel and clay with a thickness of 0~60 m, which is mainly distributed around the piedmont area along the main rivers in the central and eastern regions.

Karst groundwater is mostly recharged by atmospheric precipitation from the outcrop areas of limestone in the western and central mountainous areas. Under natural conditions, the flow direction of karst groundwater is generally from the west, north, southwest and northeast to the Heilongdong spring groups. However, due to the overexploitation of groundwater extraction and mine drainage, the regional karst groundwater level has dropped dramatically since the 1980s, resulting in several karst groundwater depression cones around some coal mines and well fields [7]. With the increasing amount of groundwater pumping and mine drainage, karst groundwater gradually changed from spring drainage to groundwater exploitation and mine drainage.

Since the middle route of the South-to-North Water Transfer Project (SNWTP) was opened in 2014, the water-using structure of Handan has changed significantly. The water source of the SNWTP gradually replaced the local groundwater resources, and the exploitation of the Yangjiaopu (YJP) well field was greatly reduced. At the same time, the local government implemented a lot of measures such as some old mine closures, groundwater-pumping reduction and mine drainage utilization, resulting in a rise in the regional groundwater table to varying degrees.

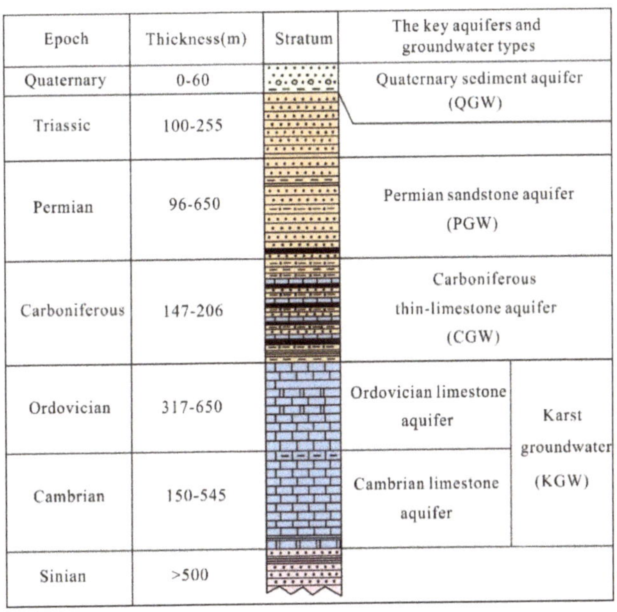

Figure 1. Stratigraphic column and the key aquifers of the study area.

2.3. Coal Mining

The coal mines are located in the northern, southern and eastern areas of the spring area. The main coal-bearing strata are the upper Carboniferous Taiyuan Formation and the lower Permian Shanxi formation, with a total thickness of 170~250 m. There are 22 state-owned and local-owned coal mines (M1–M22) in the spring basin. Specifically, 7 coal mines were closed before 2015 (M1–M7) and 4 coal mines were closed from 2015 to 2018 (M8–M11); the other 11 coal mines still stay in production (M12–M22). Moreover, the production mines are mainly located in the eastern and southern areas of the spring area (Figure 2).

After decades of mining, the shallow coal resources have been nearly exhausted, and most of the production mines have been transferred to the deep coal seams. As mentioned above, the vertical distance between the deep coal seams of Carboniferous strata and the underlying Ordovician limestone aquifer is only 20~50 m. So, the risk of water inrush is increasing due to the more abundant and higher water pressure of the Ordovician karst groundwater [26].

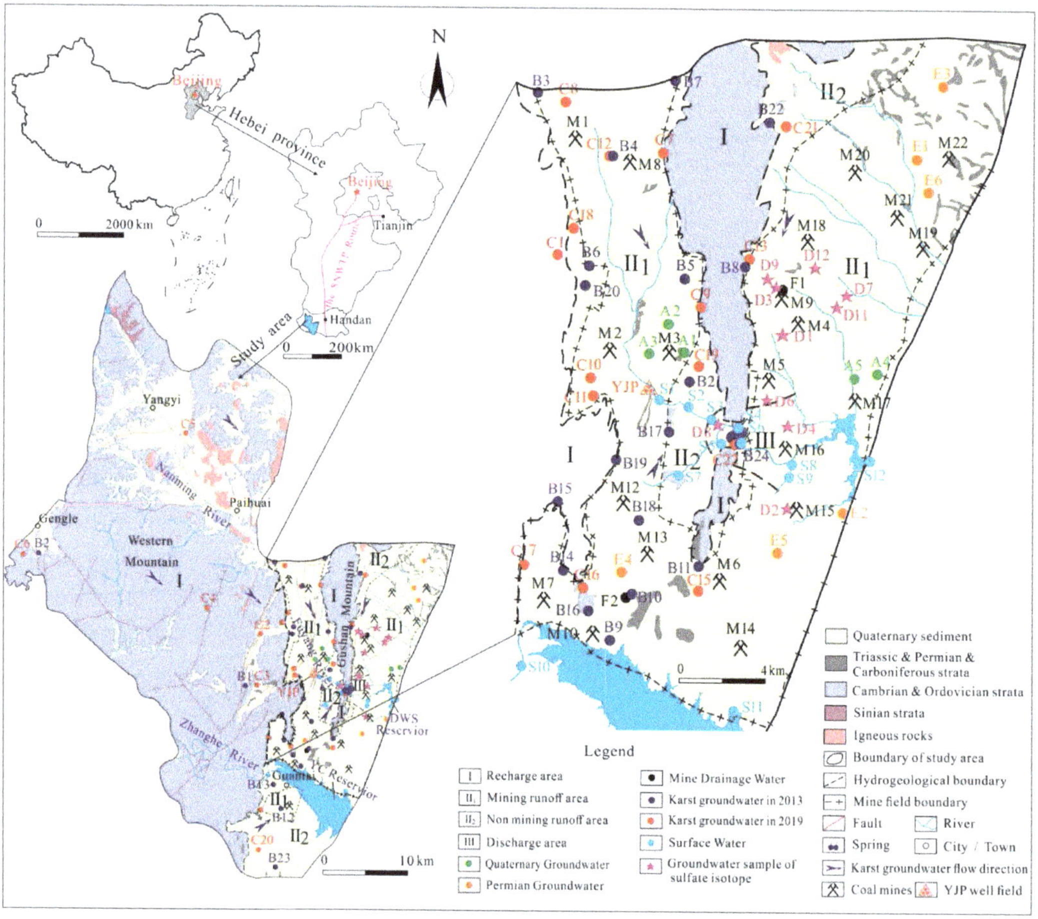

Figure 2. The hydrogeological sketch of Heilongdong Spring Basin.

3. Materials and Methods

3.1. Sample Collection and Analysis

A total of 59 water samples from wells and springs were taken from late May to mid-September in 2013 and 2019, respectively. Specifically, 24 samples were taken in 2013, all of which were karst groundwater (B1–B24). Among them, there were 2 samples (B1–B2) in the recharge area (Zone I), 17 samples (B3–B19) in the mining runoff area (Zone II_1), 4 samples (B20–B23) in the nonmining runoff area (Zone II_2) and 1 sample (B24) in the discharge area (Zone III).

To compare the differences in the major chemical compositions from different aquifers in different years, 35 samples were taken in 2019, including 22 samples of karst groundwater (C1–C22), 5 samples of Quaternary sediment groundwater (A1–A5), 6 samples of Permian groundwater (E1–E6) and 2 samples of mine drainage water (F1–F2). Among the karst groundwater samples, there were 6 samples (C1–C6) in the recharge area (Zone I), 11 samples (C7–C17) in the mining runoff area (Zone II_1), 4 samples (C18–C21) in the nonmining runoff area (Zone II_2) and 1 sample (C22) in the discharge area (Zone III). In addition, 12 samples of surface water (S1–S12) were collected in 2019, including 9 samples of Fuyang River water (S1–S9). The other 3 water samples were from Zhanghe River, Yuecheng (YC) Reservoir and Dongwushi (DWS) Reservoir (S10–S12) [7].

Most of the karst groundwater samples were collected from the public and private supply wells ranging in depth from 170 to 600 m, and almost all the Quaternary sediment and Permian groundwater were pumped from the private wells (ranging from 5 to 100 m). Surface water samples were collected from rivers and reservoirs. Mine drainage water and spring water samples were taken directly from the mine drainage outlet and karst spring outlet, respectively.

Only major ions and trace element analyses were carried out in all the samples of karst groundwater taken in 2013. δ^2H and $\delta^{18}O$ isotope tests were carried out in all the samples of groundwater (except C6, C14 and C20) in addition to the major ions and trace element analysis in 2019. The above 3 samples failed the tests for the δ^2H and $\delta^{18}O$ isotope due to damage during transportation. Moreover, 10 samples were chosen from the 35 samples for the $\delta^{34}S_{SO4}$ analysis in 2019, including 7 samples (C1, C4, C6, C7, C8, C11, C22) of Ordovician groundwater and 3 samples (E2, E5, E6) of Permian sandstone groundwater. In addition, 10 samples were collected for $\delta^{34}S_{SO4}$ analysis in August 2015, including 7 samples (D1-D4, D6-D8) of Ordovician groundwater and 3 samples (D9, D11, D12) of Carboniferous thin limestone water [20]. Specifically, among the karst groundwater samples, there were 3 samples (C1, C4, C6) in the recharge area (Zone I), 9 samples (C7, C8, C11, D1-D4, D6-D7) in the mining runoff area (Zone II_1) and 2 samples (C22, D8) in the discharge area (Zone III). All the other samples of Permian sandstone groundwater (E2, E5, E6) and Carboniferous thin limestone water (D9, D11, D12) were taken from the mining runoff area (Zone II_1).

Prior to sampling, each well was pumped at least three well volumes of groundwater with a low-rate submersible pump to remove the stagnant water, and each sampling bottle was rinsed three times with the sample water before collection. Then, the pH, electrical conductivity (EC) and water temperature were measured in situ using the HANNA (HI98194) multiparameter instrument, with a precision of ± 0.02 for pH, ± 0.15 °C for temperature and ± 1 μs/cm for EC.

All the water samples were filtered through 0.45 μm membrane filters and poured into 1.5 L and 250 mL polyethylene bottles for analyses of major and trace elements. The 250 mL samples were acidified in situ by adding double distilled nitric acid to Ph < 2, while the 1.5L samples were untreated. All the samples were stored at 4 °C until analysis (within one week). In addition, groundwater samples for the stable isotope (δ^2H and $\delta^{18}O$) analysis were collected in 50 mL glass bottles, which were sealed with airtight caps. To measure the values of $\delta^{34}S$ of the dissolved $SO_4{}^{2-}$ in the groundwater, the water samples were separately collected in a 5 L bucket and acidified with HCl to pH 2–3; then, they were reacted with $BaCl_2$ to form $BaSO_4$ precipitates and were sent immediately to the laboratory.

3.2. Analytical Methods

K^+, Na^+, Ca^{2+}, Mg^{2+}, Cl^-, $SO_4{}^{2-}$, $NO_3{}^-$ and other chemical indicators were tested at the Groundwater Mineral Water and Environmental Monitoring Center of the Institute of Hydrogeology and Environmental Geology, Chinese Academy of Geological Sciences.

The TDS was determined using the gravimetric method. Major anions, except $HCO_3{}^-$, were analyzed using a Thermo Scientific Dionex ICS-4000 (precision, ±1%). $HCO_3{}^-$ and $CO_3{}^{2-}$ were titrated with 0.05 mol/L HCl in situ on the sampling day using phenolphthalein and Methyl orange as indicators. Major cations and minor elements were analyzed with an Inductively Coupled Plasma Optical Emission Spectrometer (ICP-OES, PerkinElmer Optima 8300) (precision, ±1%). The reliability of the chemical data was checked using the Charge Balance Error (% CBE). The results showed that all the samples were within CBE values of ±10%, indicating that the measurement accuracy was reasonably good in the study area.

The hydrogen (δ^2H) and oxygen ($\delta^{18}O$) compositions in the water samples were identified using isotopic mass spectrometry (Picarro L2130-I) at the Institute of Hydrogeology and Environmental Geology, Chinese Academy of Geological Sciences. The stable isotope ratios are expressed in the standard δ-notation and calculated with respect to the Vienna Standard Mean Ocean Water (VSMOW, ‰) with the precision of ±1.0‰ (δ^2H) and ± 0.1‰ ($\delta^{18}O$).

The ^{34}S isotope of $SO_4{}^{2-}$ was analyzed with a thermal conversion elemental analyzer (TCEA) coupled with an isotope ratio mass spectrometer (Finnigan MAT253) at the Analytical Laboratory of the Beijing Research Institute of Uranium Geology (ALBRIUG). The notation was expressed in terms of δ (‰) relative to the Vienna Canyon Diablo Troilite (V-CDT) standard, and the working standards of $\delta^{34}S_{SO4}$ were GBW 04414 and GBW 04415. The results of the repeated analysis of internal standards showed that the precision was better than ± 0.2‰ for $\delta^{34}S_{SO4}$.

Quality assurance (QA) and quality control (QC) measures for field sampling, storage, transportation and laboratory analysis were implemented. To ensure the data quality, each sample was analyzed in triplicate, and each batch was interspersed with the standard and blank sample in proportion to evaluate the experimental methods. In addition, about 20% of the samples in each batch were randomly selected as validation set samples to ensure that the relative standard deviations were within 10%.

4. Results

4.1. Statistics of Hydrochemical Compositions

The water samples, taken in 2013 and 2019, were classified into five categories: Karst groundwater in the Cambrian and Ordovician limestone aquifers (KGW), groundwater in the Quaternary sediment aquifer (QGW), groundwater in the Permian sandstone aquifer (PGW), mine drainage water (MDW) and surface water (SW). Additionally, karst groundwater samples were further divided into four groups based on the sampling location, which were taken in the recharge area (Zone I), the mining runoff area (Zone II_1), the nonmining runoff area (Zone II_2) or the discharge area (Zone III). Subsequently, the samples of karst groundwater were divided into KGW1 ~ KGW4 in 2013 and KGW5 ~ KGW8 in 2019, as listed in Tables 1 and 2.

As shown in Table 1, all the pH values were slightly higher than 7.0, indicating their alkaline nature. The TDS ranged from 313.40~599.60mg/L, with an average of 405.95mg/L. The order of abundance of cation concentrations (expressed in mg/L) were $Ca^{2+} > Mg^{2+} > Na^+ > K^+$, while those of the anions (expressed in mg/L) were $HCO_3{}^- > SO_4{}^{2-} > Cl^- > NO_3{}^-$. From Table 2, it can be observed that the range of TDS was 301.50 ~ 798.50 mg/L, with an average of 466.96 mg/L in 2019. The order of abundance of major cation and anion concentrations remained the same, but the concentration of each ion was slightly higher than that in 2013.

In addition, the concentrations of TDS, Ca^{2+}, Mg^{2+}, Cl^-, $SO_4{}^{2-}$ and other major ions increased in the runoff area (Zone II_1, II_2), especially in the Zone II_1 runoff area where coal mines were distributed. The average SO_4^{2-} concentration was the largest, and the average concentration changed the most. Compared with 2013, the average SO_4^{2-} concentration increased by 77.78% in Zone II_1.

Table 1. Statistical summary of hydrochemical parameters in 2013.

	K^+	Na^+	Ca^{2+}	Mg^{2+}	Cl^-	SO_4^{2-}	HCO_3^-	NO_3^-	TDS	pH
KGW Total (n = 24)										
Min	0.92	6.03	68.79	14.95	8.75	47.94	207.5	12.8	313.4	7.55
Max	2.94	32.72	120.1	31.7	49.18	207.3	294.9	57.6	599.6	8.52
Mean	1.37	14.61	91.01	23.46	22.77	95.34	241.91	21.63	405.95	8.05
Std	0.44	7.16	16.85	3.73	8.45	41.82	23	8.74	76.24	0.39
KGW1(n = 2)										
Min	1.3	6.03	82.73	19.96	8.75	64.96	237.8	12.8	336.9	7.55
Max	1.45	7.77	97.67	24.95	22.76	77.45	294.9	29.36	404.9	7.66
Mean	1.38	6.9	90.2	22.46	15.76	71.21	266.35	21.08	370.9	7.61
Std	0.11	1.23	10.56	3.53	9.91	8.83	40.38	11.71	48.08	0.08
KGW2 (n = 17)										
Min	0.92	8.28	70.76	19.29	12.56	47.94	207.5	13.48	317.1	7.57
Max	2.94	32.72	120.1	31.7	49.18	207.3	270.2	27.52	599.6	8.52
Mean	1.39	15.07	88.94	23.83	22.95	94.97	238.53	19.95	402.38	8.15
Std	0.51	7.46	15.04	3.35	8.3	43.06	20.1	3.8	76.57	0.38
KGW3 (n = 4)										
Min	0.99	8.62	68.79	14.95	17.95	49.88	209	17.8	313.4	7.64
Max	1.44	23.02	120	26.31	40.27	141.9	269	57.6	522.2	8.49
Mean	1.2	15.76	95.66	21.43	25.2	94.91	238.2	29.84	416.4	7.97
Std	0.24	7.05	27.75	5.53	10.3	45.3	26.28	18.79	94.07	0.36
KGW4 (n = 1)	1.52	17.59	109.2	27.31	24.05	151.6	265.4	18.48	494.9	7.61

Note: units: concentrations are expressed in milligram per liter (mg/L), except for pH. KGW1: karst groundwater in recharge area; KGW2: karst groundwater in mining runoff area; KGW3: karst groundwater in no mining runoff area; KGW4: karst groundwater in discharge area.

It can also be seen that the major ion concentrations of groundwater in different aquifers varied considerably (Table 2). The highest contents of most major ions (Na^+, Ca^{2+}, Cl^-, SO_4^{2-}, NO_3^-) were observed in QGW, followed by the contents in PGW, and the contents in KGW were the lowest. Moreover, it is worth noticing that the average concentration of SO_4^{2-} in the mine drainage water was very high, which was close to the average concentration of QGW. While the average concentration of NO_3^- in the mine drainage water was the lowest, reflecting the ionic components in the mine drainage water may mainly come from the dissolution of sulfate or other minerals, and they were hardly affected by the external environment and human activities, such as domestic wastes and fertilizer application.

The mean concentrations of ions in the surface water (SW) were similar to those in KGW, while they were significantly different from those in QGW and PGW, implying the close hydraulic connection between surface water (SW) and karst groundwater (KGW).

To sum up, KGW had the best quality groundwater, followed by the PGW and MDW, and the QGW was the worst. It reflected that the effect of human activities such as industrial and agricultural activities had a greater impact on shallow groundwater (QGW and PGW) and a relatively little impact on deep groundwater (KGW). Furthermore, from the spatiotemporal perspective, the quality of karst groundwater was deteriorating.

In recent years, the mean concentrations of TDS, Na^+, Cl^-, SO_4^{2-} and other major ions increased in the runoff area with coal mine distribution (Zone II_1), even exceeding those in the discharge area (Zone III), while the concentrations of the major ions had little change in the recharge area (Zone I) and the discharge area (Zone III). This reflected the increase in ion concentrations in karst groundwater, which may be related to human activities such as coal mining.

Table 2. Statistical summary of hydrochemical parameters in 2019.

	K^+	Na^+	Ca^{2+}	Mg^{2+}	Cl^-	SO_4^{2-}	HCO_3^-	NO_3^-	TDS	pH
KGW Total (n = 22)										
Min	0.92	5.12	70.44	16.44	5.59	35.94	225.8	1.25	301.5	7.3
Max	3.7	57.75	170.2	46.69	54.97	313.8	335	64.62	798.5	7.74
Mean	1.62	16.42	103.7	26.02	25.72	125.72	265.27	22.98	466.96	7.52
Std	0.68	11.82	26.49	7.57	13.24	80.91	27.59	11.44	145.54	0.13
KGW5 (n = 6)										
Min	0.92	5.12	70.44	16.44	5.59	35.94	225.8	18.35	301.5	7.45
Max	2.76	10.03	96.61	24.39	19.56	78.03	256.9	31.42	394.6	7.74
Mean	1.69	7.32	80.61	19.84	13.39	58.43	244.92	22.19	336.68	7.63
Std	0.78	1.64	9.24	2.97	5.88	14.83	13.63	4.79	31.71	0.1
KGW6 (n = 11)										
Min	1.15	10.45	86.58	22.16	15.72	80.51	238	1.25	386.6	7.3
Max	3.7	57.75	170.2	46.69	54.97	313.8	335	64.62	798.5	7.64
Mean	1.76	21.75	117.33	29.83	31.45	168.84	277.8	22.61	545.08	7.46
Std	0.73	14.14	27.37	8.37	14.08	89.33	28.62	15.91	155.45	0.08
KGW7 (n = 4)										
Min	0.95	9.9	82.78	21.21	19.91	64.17	233.1	19.92	378.3	7.38
Max	1.57	22.61	133.6	30.82	40.27	177.5	286.8	29.18	582.9	7.73
Mean	1.2	14.87	102.06	24.5	27.9	108.83	257.65	25.98	446.98	7.54
Std	0.26	5.77	22.63	4.45	8.75	51.03	27.11	4.17	95.82	0.17
KGW8 (n = 1)	1.36	18.5	98.8	27.42	27.94	122.8	280.1	19.72	469.2	7.35
QGW (n = 5)										
Min	0.26	53.95	240	24.08	89.28	377.5	282	34.82	1031	7.21
Max	2.23	78.22	323.9	57.5	156.1	656.7	341.7	190.1	1465	7.4
Mean	1.05	63.49	279.84	43.1	119.34	473.4	311.2	100.52	1253.8	7.31
Std	0.82	8.95	36.49	13.26	28.26	124.14	23.78	60.45	189.21	0.07
PGW (n = 6)										
Min	0.95	32.53	112.2	19.76	37.02	120.5	227.3	12.09	563.4	7.21
Max	2.56	120.4	358.2	71.07	186.9	694.9	402.7	198	1647	7.62
Mean	1.52	57.74	233.95	43.89	109.79	378.15	319.13	86.69	1095.4	7.4
Std	0.63	32.3	95.72	18.07	70.6	208.87	65.82	68.53	437.38	0.14
MDW (n = 2)										
Min	3.76	29.06	185.5	44.34	29.8	368	294.1	10.08	846.6	7.3
Max	4.03	48.75	218	57.12	47.85	536.4	317.9	17.4	1096	7.55
Mean	3.9	38.91	201.75	50.73	38.83	452.2	306	13.74	971.3	7.43
Std	0.19	13.92	22.98	9.04	12.76	119.08	16.83	5.18	176.35	0.18
SW * (n = 12)										
Min	0.98	7.13	30.75	6.33	9.82	28.85	198.07	5.81	223.65	8
Max	6.44	64.57	155.38	40.59	58.91	312.84	349.53	22.31	784.56	8.6
Mean	3.96	31.12	94.94	26.16	32.83	151.15	268.36	17.46	491.8	8.2
Std	1.55	15.55	33.19	9.53	11.77	88.71	47.41	5.27	168.52	0.2

Note: KGW5: karst groundwater in recharge area; KGW6: karst groundwater in mining runoff area; KGW7: karst groundwater in no mining runoff area; KGW8: karst groundwater in discharge area; QGW: Quaternary groundwater; PGW: Permian groundwater; MDW: mine drainage water; SW: surface water; n: number of samples; * data from [7].

4.2. Hydrochemical Types

The Piper diagram is often used to show the relative contents of major ions in water samples, and it can be applied to analyze the evolution laws of groundwater [10,27]. With regard to cations, almost all points were close to the lower left triangle, indicating that the Ca^{2+} type was dominant. For anions, most of the karst groundwater (KGW) and surface water (SW) samples were distributed in the middle and lower corner of the right triangle, showing HCO_3^- as the dominant anion, while the majority of samples of the Quaternary sediment groundwater (QGW) and Permian sandstone groundwater (PGW) showed the dominance of the SO_4^{2-} type. Moreover, as shown in Figure 3, most of the samples (including the KGW and SW samples) were found to concentrate in the lower left side of the diamond-shaped field, indicating the hydrochemical type of the HCO_3-Ca·Mg,

HCO_3-Ca and $HCO_3{\cdot}SO_4$-Ca·Mg facies, while the majority of the QGW and PGW samples were distributed in the upper left side of the diamond-shaped area, which was dominated by $SO_4{\cdot}HCO_3$-Ca·Mg, $SO_4{\cdot}HCO_3$-Ca and SO_4-Ca water types. This was also consistent with the actual situation. The western and central mountain areas were the recharge areas of karst groundwater. The hydrochemical type was relatively simple. The ion concentrations and hydrochemical types increased from the recharge area to the discharge area, indicating that the rock–water interaction was strengthened. Moreover, the QGW and PGW samples were mostly distributed in the runoff and discharge areas with a shallower groundwater depth in the central and eastern parts of the area. Therefore, they had a relatively close hydraulic connection.

Further analysis showed that the KGW samples from the recharge area were closer to the left lower part of the diamond-shaped field, which were mainly HCO_3-Ca·Mg and HCO_3-Ca water types, while the hydrochemical types of the KGW samples in the runoff area (Zone II) and discharge area (Zone III) were similar to that of the SW samples with $HCO_3{\cdot}SO_4$-Ca·Mg water types.

Additionally, it was worth noting that the hydrochemical composition of the mine drainage water (MDW) samples was similar to that of some karst groundwater (Zone II_1) and Permian sandstone groundwater samples with $HCO_3{\cdot}SO_4$-Ca·Mg and $SO_4{\cdot}HCO_3$-Ca·Mg water types. This might be related to coal seam mining. The coal seams of Carboniferous (C) and Permian (P) contained sulfur elements. Due to the coal mining activities, the upper Permian sandstone aquifer (PGW) and the lower Ordovician limestone aquifer (KGW) were connected, resulting in the mean concentrations of SO_4^{2-} and other major ions being higher in the mining runoff area (Zone II_1).

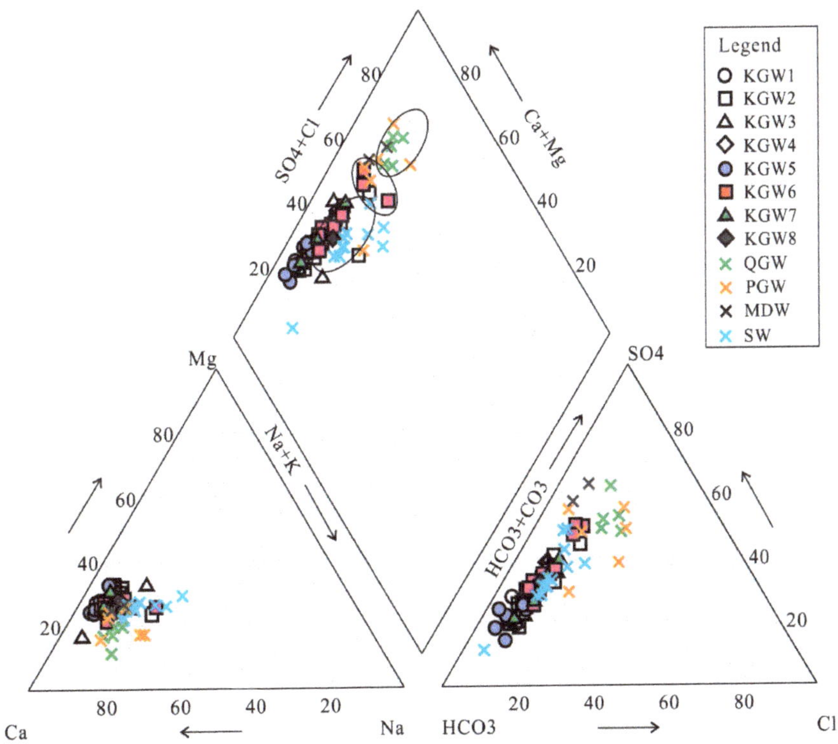

Figure 3. The Piper diagram of water samples in 2013 and 2019.

5. Discussion

5.1. Groundwater Isotopic Composition and Sources

Hydrogen (δ^2H) and oxygen (δ^{18}O) isotopes are effective tools for analyzing hydrological processes across various scales [11]. Table 3 summarizes the isotopic composition of the water samples taken in the HSB. The relationship between δ^2H and δ^{18}O for all water samples is plotted in Figure 4. Due to the lack of precipitation isotope data in the study area, the data of the adjacent Shijiazhuang station were selected. The local meteoric water line was fitted as $\delta^2H = 6.4\delta^{18}O-3.8$. The slope and intercept of the local meteoric water line (LMWL) were lower than that of the global meteoric water line (GMWL: $\delta^2H= 8\delta^{18}O + 10$) [28], which probably resulted from secondary evaporation during rainfall.

Table 3. δ^2H and δ^{18}O isotope composition of all the water samples taken in the HSB.

	QGW (n = 5)			PGW (n = 6)			KGW (n = 19)			MDW (n = 2)			SW (n = 12)		
	Min	Max	Mean	Min	Max	Mean	Min	Max	Mean	Min	Max	Mean	Min	Max	Mean
δ^2H (‰)	−62.0	−57.0	−59.2	−65.0	−64.0	−64.3	−69.0	−64.0	−67.3	−67.0	−66.0	−66.5	−65.4	−57.6	−63.6
δ^{18}O (‰)	−8.4	−7.7	−8.0	−8.8	−8.5	−8.7	−9.6	−8.6	−9.3	−9.1	−8.9	−9.0	−8.9	−7.5	−8.6

Note: QGW: Quaternary groundwater; PGW: Permian groundwater; KGW: Karst groundwater; MDW: mine drainage water; SW: surface water.

As can be seen in Figure 4, all samples were distributed near GMWL and LMWL, indicating that the groundwater in the study area was mainly derived from meteoric water. In addition, all the samples were distributed in parallel to the LMWL, but the intercept was lower. This phenomenon can be interpreted as the evaporated soil moisture mixing with the subsequent rainfall, which infiltrated the soil and impelled the residual water downward by regular rainfall events [7].

The δ^2H and δ^{18}O values of the QGW samples ranged from −62.0 to −57.0 and −8.4~−7.7, respectively. The PGW samples ranged from −65.0 to −64.0 and −8.8~−8.5, respectively. The KGW samples ranged from −69.0 to −64.0 and −9.6~−8.6, respectively, while the MDW samples ranged from −67.0 to −66.0 and −9.1~−8.9, respectively (Table 3).

It can also be seen that the QGW samples were concentrated in the upper right corner (group 1), which reflected the fact that the QGW potentially underwent strong evaporation, resulting in higher δ^2H and δ^{18}O values, while the δ^2H and δ^{18}O values for most of the KGW samples were significantly low, implying the KGW was probably formed by precipitation under wet and cold climate or high-altitude conditions [10,18]. In addition, the δ^2H and δ^{18}O values of the PGW and MDW samples were closer to those of the KGW samples, reflecting that some of the KGW samples had a relatively close hydraulic connection with the PGW and MDW samples.

It is worth noting that most of the surface water (SW) samples were located in group 2, except the YC Reservoir (S11) and DWS Reservoir (S12). The river water samples were isotopically similar to the spring water and some of the KGW samples. This showed that both the reservoir water and the Quaternary groundwater (QGW) had experienced strong evaporation. The concentration of heavy isotopes in the reservoir water was higher than that in the river water. The river water was mainly recharged by the karst groundwater rather than precipitation.

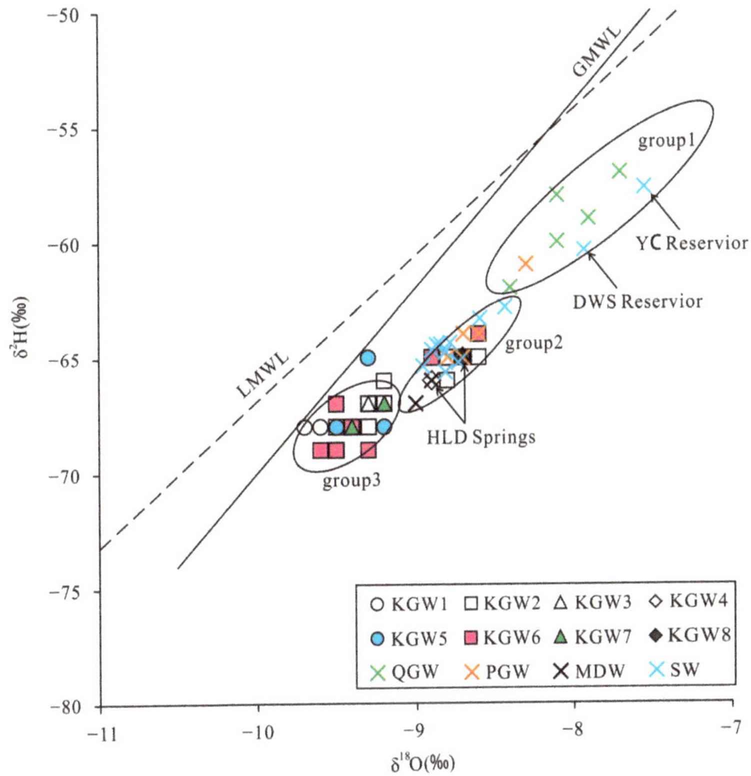

Figure 4. The relationship between δ^2H and $\delta^{18}O$ for all water samples.

5.2. Rock Weathering and Evaporation

A Gibbs diagram can be used to ascertain the influence of rock weathering, precipitation and evaporation on water chemistry [29]. As shown in Figure 5, almost all the groundwater samples were affected by rock weathering. In addition, the values of $\rho(Na^+)/\rho(Na^+ + Ca^{2+})$, $\rho(Cl^-)/\rho(Cl^- + HCO_3^-)$ and TDS in most of the QGW and PGW samples were relatively high, which indicated that the hydrochemical compositions of QGW and PGW were affected by evaporation. It is also worth noting that the TDS of the karst groundwater in the recharge area (Zone I) had little change with the increase in $\rho(Na^+)/\rho(Na^+ + Ca^{2+})$ and $\rho(Cl^-)/\rho(Cl^- + HCO_3^-)$ values, indicating that cation exchange also played an essential role under the background of rock dominance, while the TDS concentrations of the karst groundwater in the runoff area (Zone II) and discharge area (Zone III) had a rising trend with the increase in $\rho(Na^+)/\rho(Na^+ + Ca^{2+})$ and $\rho(Cl^-)/\rho(Cl^- + HCO_3^-)$ values, implying that the increase in TDS concentrations was potentially associated with human activities.

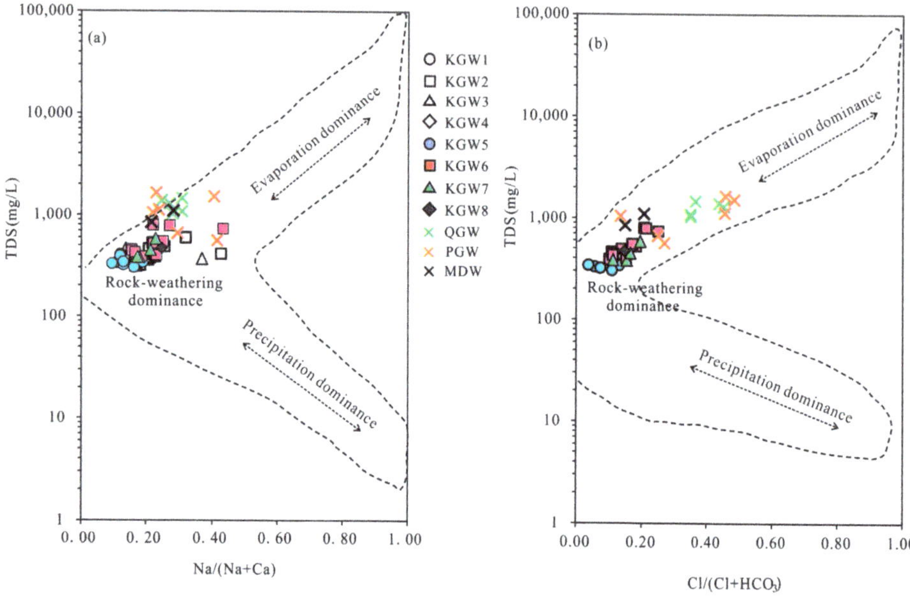

Figure 5. Gibbs diagrams of groundwater samples in the HSB: (**a**) TDS vs. $\rho(Na^+)/\rho(Na^+ + Ca^{2+})$, (**b**) TDS vs. $\rho(Cl^-)/\rho(Cl^- + HCO_3^-)$.

5.3. Correlation of Hydrochemical Composition

The consistency of the sources of the hydrochemical composition in groundwater can be inferred with a correlation analysis [10,30]. The correlation matrix of the major water chemical index was determined using SPSS software (Table 4). The results showed that TDS had a high positive correlation ($r \geq 0.70$, $p = 0.01$) with Na^+, Ca^{2+}, Mg^{2+}, Cl^-, SO_4^{2-} and HCO_3^-, which proved that the above ions were the main factors affecting the TDS value. Na^+-Cl^-, Ca^{2+}-SO_4^{2-}, Ca^{2+}-HCO_3^- and Ca^{2+}-Mg^{2+} also had a high positive correlation ($r \geq 0.70$, $p = 0.01$), indicating that the dissolution of carbonate, sulfate and halite occurred in the study area. In addition, the NO_3^- had a positive strong correlation with Na^+ and Cl^- ($r \geq 0.70$, $p = 0.01$) and moderate correlations with Ca^{2+} and SO_4^{2-} ($0.50 \leq r < 0.70$, $p = 0.01$). As mentioned above, the NO_3^- concentration was relatively high and had a relatively high correlation with Na^+ and HCO_3^-, indicating that NO_3^- in groundwater might be related to anthropogenic activities such as agricultural irrigation and domestic pollution.

Table 4. Correlation matrix of all the samples.

	K	Na	Ca	Mg	Cl	SO_4	HCO_3	NO_3	TDS	pH
K	1.000	0.118	0.106	0.413 **	−0.020	0.274 *	0.244	−0.222	0.161	−0.176
Na	0.118	1.000	0.866 **	0.762 **	0.876**	0.849 **	0.626 **	0.716 **	0.907 **	−0.434 **
Ca	0.106	0.866 **	1.000	0.861 **	0.923**	0.954 **	0.788 **	0.676 **	0.992 **	−0.524 **
Mg	0.413 **	0.762 **	0.861 **	1.000	0.742**	0.932 **	0.819 **	0.395 **	0.896 **	−0.429 **
Cl	−0.020	0.876 **	0.923 **	0.742 **	1.000	0.817 **	0.646 **	0.773 **	0.920 **	−0.452 **
SO_4	0.274 *	0.849 **	0.954 **	0.932 **	0.817**	1.000	0.793 **	0.494 **	0.970 **	−0.504 **
HCO_3	0.244	0.626 **	0.788 **	0.819 **	0.646**	0.793 **	1.000	0.290 *	0.786 **	−0.525 **
NO_3	−0.222	0.716 **	0.676 **	0.395 **	0.773**	0.494 **	0.290 *	1.000	0.668 **	−0.368 **
TDS	0.161	0.907 **	0.992 **	0.896 **	0.920**	0.970 **	0.786 **	0.668 **	1.000	−0.519 **
pH	−0.176	−0.434 **	−0.524 **	−0.429 **	−0.452**	−0.504 **	−0.525 **	−0.368 **	−0.519 **	1.000

Note: ** correlation is significant at the 0.01 level (two-tailed); * correlation is significant at the 0.05 level (two-tailed).

5.4. Water–Rock Interaction

The ratios of ion concentrations can effectively reflect the formation mechanisms and sources of the major ions in groundwater [11,31].

The scatter diagram of Cl^- versus Na^+ (Figure 6a) showed that almost all the water samples were close to the 1:1 line. It indicated that the dissolution of halite might be the main source of Cl^- and Na^+ in different types of groundwater. The ratio of Na^+/Cl^- ranged from 0.47 to 1.80 with a mean value of 0.99. The results followed the expected 1:1 trend line, which further confirmed that Na^+ and Cl^- were derived primarily from halite dissolution. Halite could dissolve continuously into the groundwater.

In the plot of $SO_4{}^{2-}$ and Ca^{2+}, if gypsum dissolution is the only source of $SO_4{}^{2-}$ in the groundwater, $Ca^{2+}/SO_4{}^{2-}$ will follow the 1:1 line. It can be seen in Figure 6b that the majority of the samples were distributed above the 1:1 line, and the Ca^{2+} concentration was higher than $SO_4{}^{2-}$, suggesting that Ca^{2+} was not completely from gypsum dissolution and that there were other sources such as carbonate (dolomite, calcite) mineral dissolution, reverse cation exchange or anthropogenic input.

As shown in Figure 6c, there were no significant correlations between $SO_4{}^{2-}$ and Cl^- in some of the PGW and KGW samples (Zone I), while there were significant positive correlations between $SO_4{}^{2-}$ and Cl^- in the majority of QGW, PGW and KGW samples from the runoff area (Zone II). It indicated that $SO_4{}^{2-}$ and Cl^- in these samples probably had the same sources and might have been related to anthropogenic activities, such as coal mining development, agricultural irrigation and industrial production. In addition, it was worth noting that the slope of $Cl^-/SO_4{}^{2-}$ in the QGW was higher than that in the KGW, implying that anthropogenic activities had a greater impact on shallow groundwater, resulting in higher concentrations of $SO_4{}^{2-}$ and Cl^-.

The relationship between ($HCO_3{}^-$+$SO_4{}^{2-}$) and (Ca^{2+}+Mg^{2+}) (expressed in meq/L) in groundwater samples will follow the 1:1 line if these ions are strongly controlled by the carbonate and gypsum equilibrium [18]. As shown in the scatter plot (Figure 6d), nearly all the KGW samples were distributed along the equiline (1:1 line), reflecting that Ca^{2+}, Mg^{2+}, $HCO_3{}^-$ and $SO_4{}^{2-}$ were mainly derived from the dissolution of carbonate and gypsum, whereas most of the QGW and PGW samples fell above the 1:1 line, indicating that there were extra sources of Ca^{2+} and Mg^{2+} in the QGW and PGW, such as the dissolution of silicate minerals.

From the relationship between $HCO_3{}^-$ and (Cl^-+$SO_4{}^{2-}$) (Figure 6e), it can be seen that most of the PGW, QGW and KGW samples had different influence mechanisms in the area. Specifically, the majority of KGW samples were below the 1:1 line except for three points, indicating that most of the KGW samples were dominated by the dissolution of carbonate rocks. While most of the QGW and PGW samples deviated far from the 1:1 line, the high concentration of Cl^- and $SO_4{}^{2-}$ could be related to agricultural and industrial activities. Excess Cl^- and $SO_4{}^{2-}$ were the evidence of increased pollution. In addition, three KGW samples (C12, C14, C15) from the coal mining runoff area (Zone II_1) were distributed above the 1:1 line and mixed with the QGW and PGW samples, suggesting that the karst groundwater in some runoff areas closely connected with the shallow groundwater and was affected by anthropogenic activities.

The scatter plots of Mg^{2+} versus Ca^{2+} can be used to infer whether the Mg^{2+} and Ca^{2+} in groundwater were derived from the dissolution of dolomite, calcite and gypsum. Theoretically, when the ratio of $\gamma(Ca^{2+})/\gamma(Mg^{2+})$ is 1:1, it indicates that dolomite is dissolved in groundwater. When the ratio of $\gamma(Ca^{2+})/\gamma(Mg^{2+})$ is 2:1, it implies that calcite and dolomite are dissolved at the same time in groundwater. If the ratio of $\gamma(Ca^{2+})/\gamma(Mg^{2+})$ is greater than two, there are other sources of Ca^{2+} in the groundwater, such as gypsum dissolution, silicate dissolution, reverse cation exchange or external input. As shown in Figure 6f, almost all the KGW samples were located between the 2:1 and 3:1 lines. The concentration of Ca^{2+} in the groundwater ranged from 3.4 to 8.5 meq/L with a mean of 4.8 meq/L. The concentration of Mg^{2+} varied from 1.2 to 3.9 meq/L and had an average mean of 2.0 meq/L. It implied that the dissolution of calcite, dolomite and gypsum might occur simultaneously in the carbonate aquifer. In contrast, most of the PGW and QGW samples fell above the 3:1 line. The concentration of Ca^{2+} was much greater than that of Mg^{2+}, which indicated that there might be other factors contributing to excessive Ca^{2+}, such as the dissolution of silicate minerals.

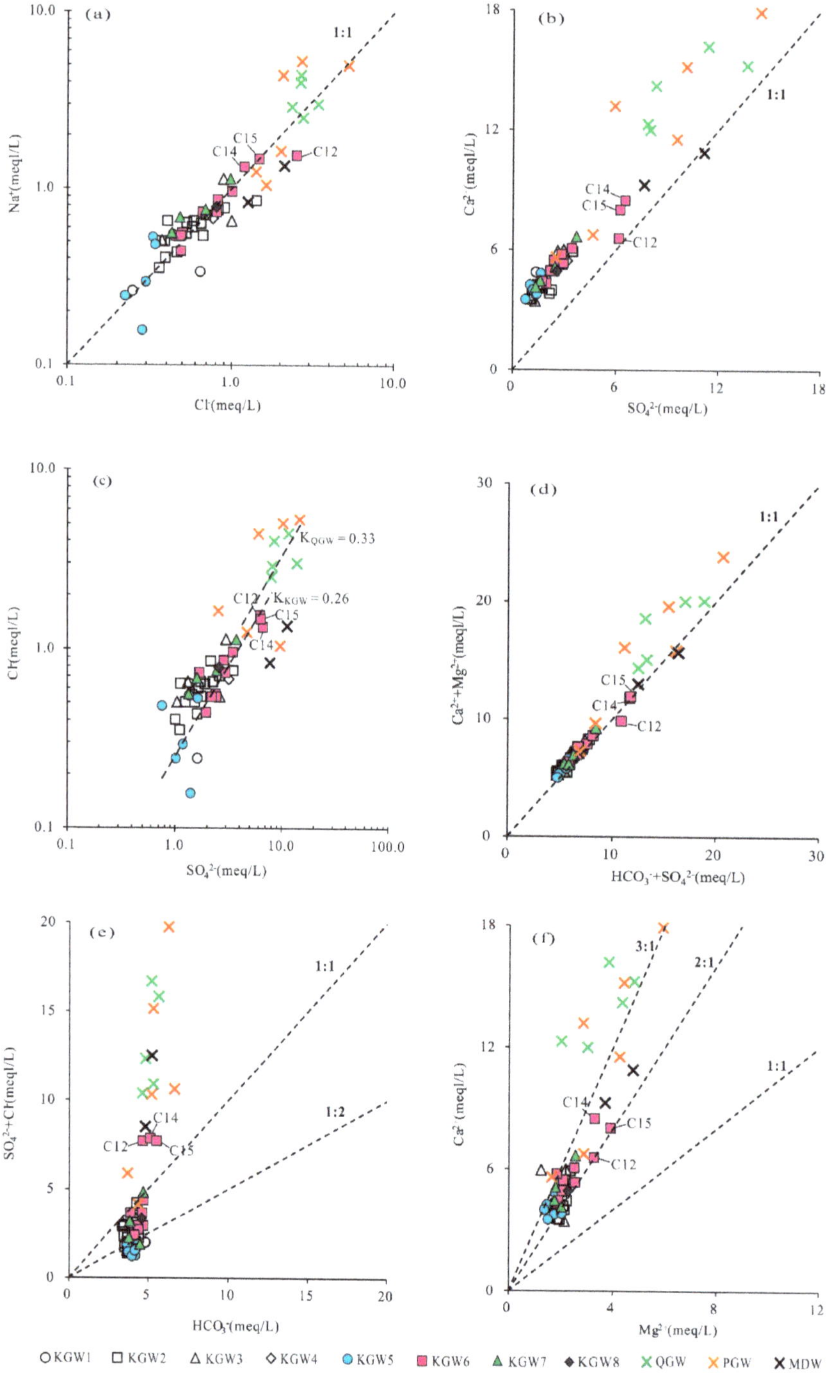

Figure 6. Scatter plots of major ions: (**a**) Cl^- vs. Na^+, (**b**) SO_4^{2-} vs. Ca^{2+}, (**c**) SO_4^{2-} vs. Cl^-, (**d**) (HCO_3^- + SO_4^{2-}) vs. (Ca^{2+} + Mg^{2+}), (**e**) HCO_3^- vs. (Cl^- + SO_4^{2-}), (**f**) Mg^{2+} vs. Ca^{2+}.

5.5. Ion Exchange Processes

Ion exchange is a natural reaction in the hydrochemical evolution of groundwater, which influences the major chemical composition of groundwater. The Schoeller indices (CAI-I and CAI-II) are vital indicators of the ion exchange occurring in the aquifer [32,33]. The CAI (CAI-I and CAI-II) value can be positive or negative; the positive CAI value indicates that the ion exchange takes place between Ca^{2+} and Mg^{2+} from the rocks with Na^+ of the water, and similarly, the negative CAI value means the Na^+ from the rocks was replaced by Ca^{2+} and Mg^{2+} in the water. The ion exchange reactions can be expressed by the following equations (Equations (1) and (2)):

$$\text{CAI-I} = Cl^- - (Na^+ + K^+)/Cl^- \quad (1)$$

$$\text{CAI-II} = Cl^- - (Na^+ + K^+)/(HCO_3^- + SO_4^{2-} + CO_3^{2-} + NO_3^-) \quad (2)$$

Both of the indices (CAI-I and CAI-II) are positive, implying that a reverse ion exchange occurs in the groundwater, whereas the negative values of CAI indicate forward ion exchange in the groundwater.

As shown in Figure 7a,b, there were both positive and negative ion exchanges of all samples in 2013 and 2019, indicating that the forward and reverse ion exchange simultaneously occurred in different types of groundwater.

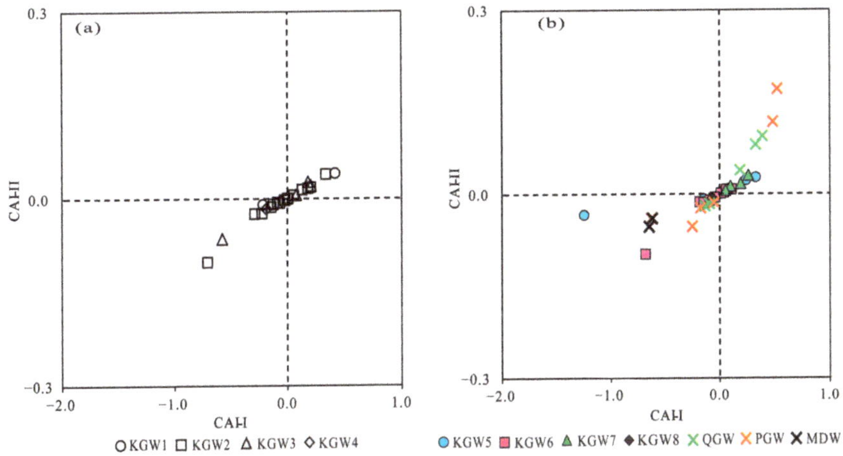

Figure 7. Scatter plot of CAI-I versus CAI-II: (**a**) plot in 2013, (**b**) plot in 2019.

5.6. Saturation Indices

To better understand the equilibrium state between groundwater and minerals, the saturation indices (SI) of calcite, dolomite, gypsum and halite were calculated with PHREEQC [34,35]. Generally, the SI value is less than −0.5, between −0.5 and 0.5 or greater than 0.5, which represents that the mineral is in an unsaturated state (dissolution), saturated state (equilibrium) or oversaturated state (precipitation), respectively [10]. The SI values of calcite and dolomite in the karst groundwater indicated notable differences in 2013 and 2019. As shown in Figure 8a,b, the SI values of calcite and dolomite in the karst groundwater ranged from 0.30 to 1.33 and 0.20 to 2.39, respectively, with an average of 0.86 and 1.41 in 2013. It suggested that the karst groundwater was oversaturated with respect to calcite and dolomite in 2013, and these carbonate mineral phases might have influenced the chemical composition in the HSB. However, the SI values of calcite and dolomite in the karst groundwater varied from 0.16 to 0.66 and −0.08 to 1.13 with a mean of 0.40 and 0.44, respectively, in 2019, indicating that the karst groundwater was saturated with respect to calcite and dolomite in 2019. Especially in the runoff area where coal mines were distributed (Zone II_1), most of the KGW samples changed from the oversaturated to saturated state during 2013–2019.

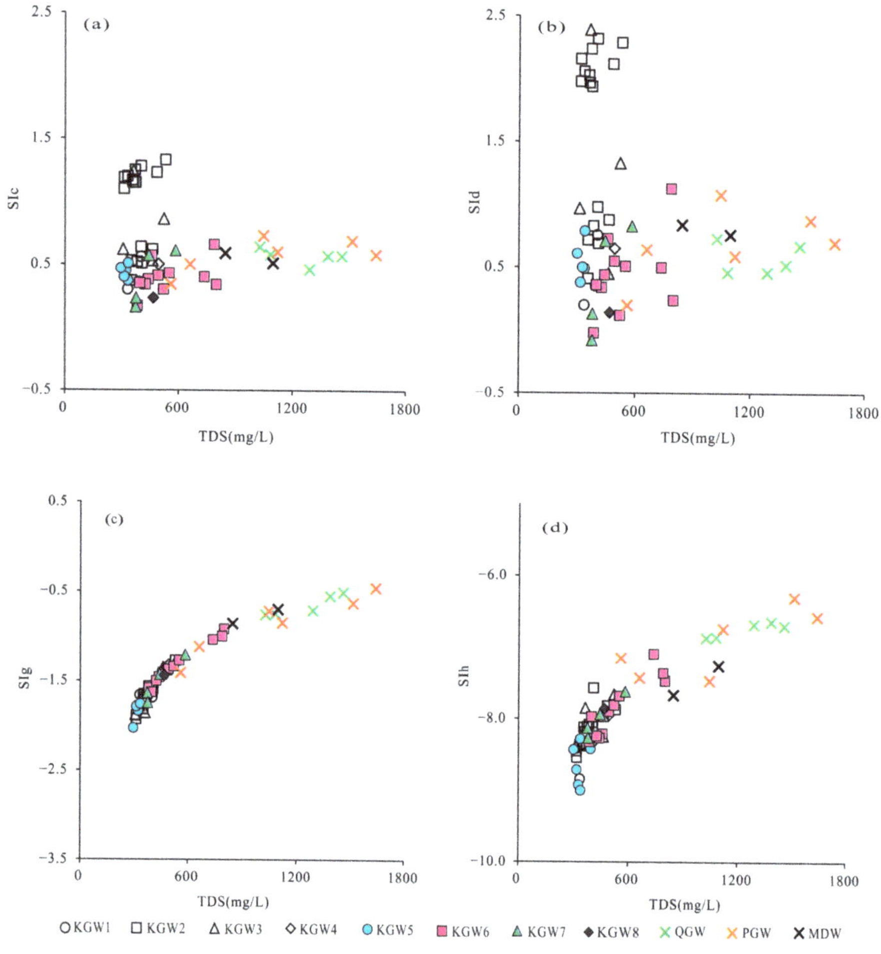

Figure 8. SI values of groundwater samples vs. TDS: (**a**) calcite, (**b**) dolomite, (**c**) gypsum, (**d**) halite.

This might have been related to the overall uplift of the karst groundwater level in the study area. Understanding the temporal and spatial variations in groundwater level is an essential prerequisite for groundwater management [36]. Taking the northern runoff area as an example (Figure 9), only one coal mine (M8) in the northern runoff area was in production (this coal mine was closed in 2016), and the karst groundwater level was between 118 and 123 m in 2012. In recent years, due to the increase in rainfall and the decrease in exploitation, the karst groundwater level in the northern runoff area rose to 124~132 m, with the overall groundwater level rising by 6~8 m and even 15 m in some areas in 2018 [37]. Consequently, the overall groundwater level was relatively low, and the flow was very slowly affected by natural and human factors in 2012–2013, as calcite and dolomite minerals were in an oversaturated state. However, the calcite and dolomite minerals gradually transited to a saturated state with the groundwater level rising during the period of 2018–2019. As time goes on, if the groundwater level continues to rise, the groundwater and minerals will be expected to reach a new equilibrium state again.

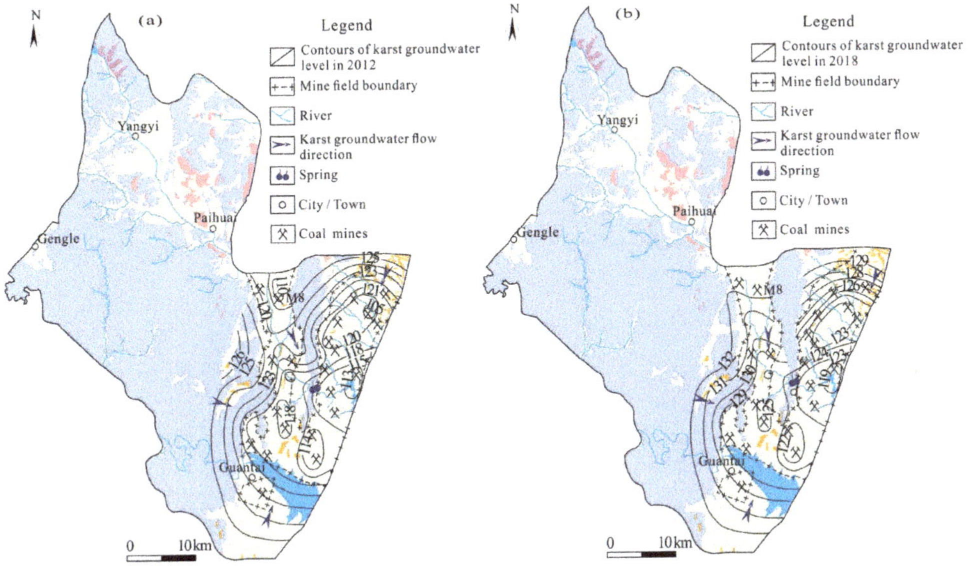

Figure 9. Contour maps of karst groundwater levels (m) in 2012 (**a**) and 2018 (**b**) [37].

The SI values of gypsum and halite in the karst groundwater in 2013 and 2019 are shown in Figures 8c and 8d, respectively. The SI values of gypsum and halite in the karst groundwater ranged from −1.93 to −1.27 and −8.84 to −7.57, respectively, with an average of −1.62 and −8.15 in 2013, respectively, indicating that gypsum and halite minerals were in a dissolved state in 2013. In contrast, the SI values of gypsum and halite in the karst groundwater varied from −2.03 to −0.92 and −9.00 to −7.09 with a mean of −1.49 and −8.09 in 2019, respectively. Compared with 2013, the SI values of gypsum and halite had little change in 2019. In addition, the SI values of gypsum and halite showed an increasing tendency along the TDS, which indicated that the increase in TDS concentration in the karst groundwater might be related to the dissolution of gypsum and halite, whereas there were no significant correlations between the TDS and SI values of calcite and dolomite, which indicated that the dissolution of calcite and dolomite was a very rapid process. Considering the carbonate rocks were widely distributed in the aquifer of the entire region, the dissolution of carbonate rocks had most likely been completed rapidly in the initial stage of the chemical evolution in the karst groundwater. Therefore, it can be considered that carbonate rock dissolution was an important source of HCO_3^-, Ca^{2+} and Mg^{2+} in karst groundwater, which was consistent with the above conclusion.

Furthermore, the SI values of calcite, dolomite, gypsum and halite had little difference in different aquifers. However, the TDS concentrations from Quaternary sediment and most Permian sandstone aquifers were obviously higher than those in carbonate aquifers. It showed that there was a close hydraulic connection between Quaternary sediment groundwater and most Permian sandstone groundwater, which was likely affected by human activities.

5.7. Analyses of the Sources of $SO4^{2-}$ in Groundwater

As previously discussed, the concentration of SO_4^{2-} showed a significant change, especially in the runoff areas impacted by mining activities during 2013–2019. Therefore, it is necessary to identify the source of SO_4^{2-} in the study area.

In general, the main sources of SO_4^{2-} in groundwater include atmospheric precipitation, sulfate mineral dissolution (gypsum, etc.) and sulfide minerals oxidation such as pyrite [20]. The $\delta^{34}S$ value of SO_4^{2-} from atmosphere precipitation is generally around −3~+9‰; the $\delta^{34}S$ content from evaporite and limestone can reach above +20‰, even up to +35‰ from gypsum, and the $\delta^{34}S$ composition of SO_4^{2-} from the oxidation of sulfides varies from-30‰ to +5‰ [20,38].

It can be seen from Table 5 that the concentrations of SO_4^{2-} in the groundwater from the Carboniferous thin-layer limestone aquifer were the highest in the study area followed by the Permian sandstone groundwater, and SO_4^{2-} concentrations in the karst groundwater from the Cambrian and Ordovician aquifers were the lowest. This was likely related to the dissolution of sulfate minerals in different aquifers. The Carboniferous Taiyuan Formation is one of the main coal-bearing strata in Northern China, and the pyrite content in the coal is relatively high. Therefore, the concentrations of SO_4^{2-} in the groundwater from Carboniferous strata were proportionally high.

Table 5. Chemical and sulfur isotopic composition for groundwater taken in the HSB.

	SO_4^{2-} (mg/L)			Cl^- (mg/L)			$\delta^{34}S$ (‰, CDT)		
	Min	Max	Mean	Min	Max	Mean	Min	Max	Mean
PGW	120.5	694.9	481.4	57.8	186.9	140.7	2.7	5.7	4.7
CGW	513.5	2591.1	1262.0	30.9	59.2	43.6	−10.6	−1.0	−5.8
KGW in the recharge area	56.6	67.0	62.7	5.6	19.6	11.9	3.0	14.1	7.1
KGW in the runoff area	93.1	408.9	159.5	15.7	70.4	30.8	2.1	13.0	6.2
KGW in the discharge area	97.4	122.8	110.1	15.8	27.9	21.9	6.3	8.2	7.3

According to Figure 10a, the $\delta^{34}S$ values gradually increased with the decrease in the SO_4^{2-}/Cl^- ratios. The samples from the Carboniferous thin-layer limestone groundwater (CGW) had relatively high SO_4^{2-}/Cl^- ratios and low δ^{34} values, distributing around the end member of sulfide oxidation (pyrite, etc.) (Figure 10b), which implied that the major sources of SO_4^{2-} and the S-isotopic composition were derived from mineral dissolution. While the $\delta^{34}S$ values of the other water samples were close to the end member of precipitation, the concentrations of SO_4^{2-} were much higher than those of precipitation, indicating that the sulfate in the groundwater was probably derived from the combination of gypsum dissolution and sulfide mineral oxidation by meteoric water. These reactions were as follows (Equations (3) and (4)):

$$\text{gypsum dissolution: } CaSO_4{\cdot}2H_2O \rightarrow Ca^{2+} + SO_4^{2-} + 2H_2O \quad (3)$$

$$\text{pyrite oxidation: } 2FeS_2 + 7O_2 + 2H_2O \rightarrow 2Fe^{2+} + 4SO_4^{2-} + 4H^+ \quad (4)$$

There are several relatively impermeable layers among the Permian sandstone, Carboniferous thin-layer limestone and the Cambrian–Ordovician karst aquifer, and they are relatively independent of each other under natural conditions. However, anthropogenic activities such as coal mining and groundwater pumping lead to the hydraulic connection among different aquifers.

As shown in Figure 10b, some samples (E5, E6, D2) from the Permian sandstone groundwater and Cambrian–Ordovician karst groundwater in the runoff areas (Zone II) fell on the mixing line between gypsum dissolution and sulfide oxidation, which lay close to the samples of Carboniferous thin-layer limestone groundwater. The results further showed that there was a certain hydraulic connection between the karst groundwater and other types of groundwater in some parts of the study area.

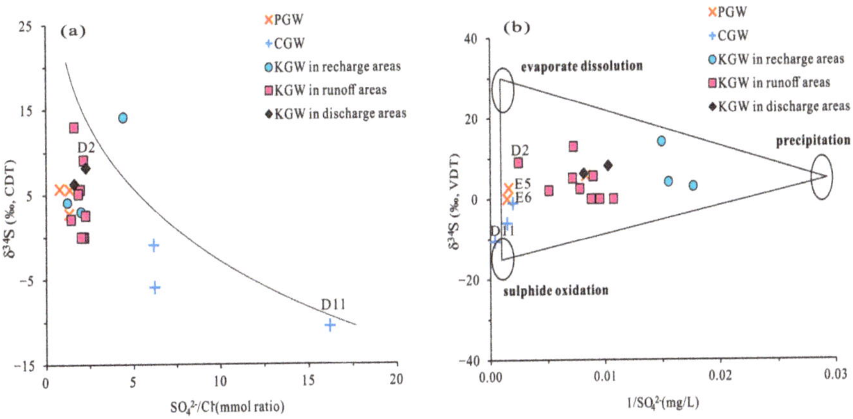

Figure 10. Scatter plot of (**a**) $\delta^{34}S$ versus SO_4^{2-}/Cl^-, (**b**) $\delta^{34}S$ versus $1/SO_4^{2-}$.

5.8. Anthropogenic Factors Affecting the Groundwater Quality

The concentration of NO_3^- ion in groundwater does not exceed 20 mg/L under natural conditions [39]. A total of 56.9% of the groundwater samples exceeded this threshold, indicating that the groundwater was influenced by anthropogenic inputs in the area. Anthropogenic activities such as domestic sewage, manure and agricultural fertilization have been proven to be the main sources of NO_3^- in groundwater. In addition, domestic sewage infiltration into groundwater often occurs in urbanized areas [40]. As shown in Table 4, the concentration of NO_3^- had a good correlation with Na^+, Ca^{2+}, Cl^- and SO_4^{2-}, and showed the best correlation with Cl^- ($r = 0.773$, $p = 0.01$), indicating that NO_3^- and these ions might have had the same anthropogenic origin. In addition to the dissolution of halite, excessive Cl^- was generally related to agricultural activities and domestic pollution.

The variation trend of NO_3^- concentrations in groundwater with well depth is shown in Figure 11a. A total of 80% of the samples from the Quaternary sediment aquifer and 66.7% of the samples from the Permian sandstone aquifer exceeded 50 mg/L, which is the permissible limit for drinking water set by the World Health Organization (WHO) [41]. The results showed that 95.6% of the samples of karst groundwater were within this threshold. In addition, the groundwater samples with NO_3^- exceeding the standard were mainly distributed at the depth of 0~100 m below the surface, and the maximum concentration was up to 198 mg/L. It showed that shallow groundwater was more affected by anthropogenic activities than deep groundwater.

Theoretically, shallow groundwater is greatly affected by evaporation due to its shallow groundwater table. As shown in Figure 11b, compared to the karst groundwater, the shallow Quaternary sediment groundwater and most of the Permian sandstone groundwater were more enriched in heavy isotopes. In addition, the NO_3^- concentrations were significantly higher in several deep karst groundwater samples (B22, C13) in the runoff area, indicating that some karst groundwater was affected by nitrate pollution due to coal mining, agricultural activities and other factors.

Generally, the TDS values of all the groundwater samples increased with the increasing $(Cl^- + SO_4^{2-})/HCO_3^-$ (expressed in molar ratios) (Figure 11c). Similarly, the TDS values of the groundwater samples in the Quaternary sediments aquifer and most Permian sandstone aquifer increased with the increasing of $(NO_3^- + Cl^-)/Na^+$ molar ratio, whereas there were no significant correlations between the TDS values and $(NO_3^- + Cl^-)/Na^+$ in the karst groundwater (Figure 11d).

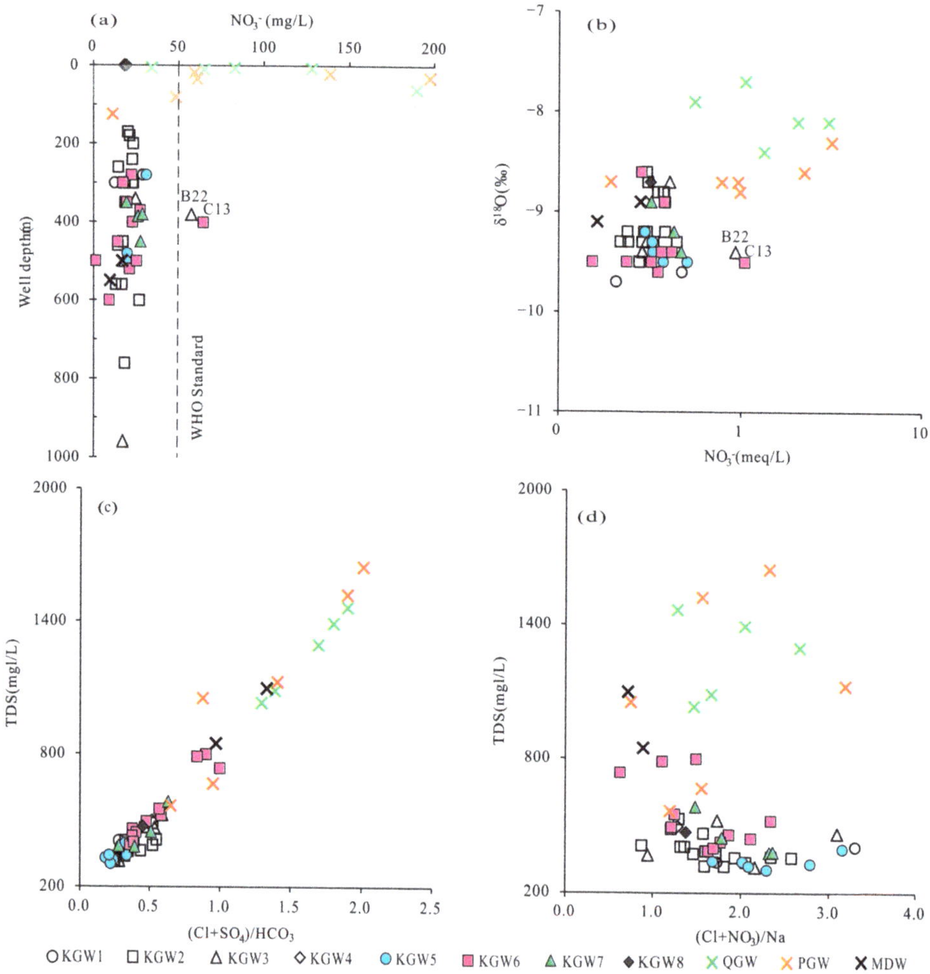

Figure 11. Plots of (**a**) NO_3^- vs. well depth, (**b**) $\delta^{18}O$ vs. NO_3^-, (**c**) TDS vs. $(Cl^- + SO_4^{2-})/HCO_3^-$, (**d**) TDS vs. $(NO_3^- + Cl^-)/Na^+$.

The results showed that the contribution from evaporite dissolution to the karst groundwater exceeded that from anthropogenic influence, while the Quaternary sediment groundwater and most of the Permian sandstone groundwater were affected by agricultural activities and domestic sewage. These were consistent with the fact that the NO_3^- concentrations in the shallow groundwater generally exceeded the WHO standard.

6. Conclusions

This research adopted the comprehensive method of correlation analysis, Piper trilinear diagram, Gibbs plot, ionic ratios and multiple isotopic analyses, which provided an effective method for analyzing the origin of karst groundwater and the hydrochemical process in the HSB. The main conclusions are as follows:

(1) In the past six years (2013–2019), due to the influence of coal mining and other factors, the average concentrations of TDS, Na^+, Cl^-, SO_4^{2-} and other major ions had been increasing in the mining runoff areas, and even exceeded the concentrations in the

discharge area. Especially, the concentration of SO_4^{2-} in the karst groundwater changed the most. A sulfur isotope analysis showed that SO_4^{2-} concentrations in the karst groundwater were potentially derived from the combination of gypsum dissolution and sulfide oxidation by meteoric water.

(2) With the closure of some coal mines, the karst groundwater level in the runoff area began to recover. The SI values of calcite and dolomite in karst groundwater varied greatly during 2013–2019, which reflected the changing runoff conditions in the area, and the karst groundwater transferred gradually from an oversaturated state to saturated state.

(3) Agricultural production and domestic sewage except for mining activities also had a negative impact on the quality of regional groundwater, which caused the increase in the content of NO_3^- and Cl^- in the Quaternary sediment groundwater, Permian bedrock groundwater and a small amount of karst groundwater. It also meant that some samples (C12, C14, C15) from the coal mine runoff area (II_1) had a relatively close hydraulic connection with shallow groundwater.

(4) Given the increasingly serious environmental and geological problems in the HSB, future research work should enhance the hydraulic connection between karst aquifers and coal seams or other aquifers, reduce the pollution of karst groundwater by coal mining and agricultural activities and strengthen the continuous monitoring of groundwater level and hydrochemical abnormal areas, which is important for the management of karst groundwater resources in North China.

Author Contributions: Methodology, M.G.; software, M.G. and C.F.; validation, M.G. and X.L.; formal analysis, X.L. and J.Q.; investigation, X.H., J.M., C.Z. and J.L.; data curation, Z.W.; writing—original draft preparation, M.G.; writing—review and editing, M.G. and C.F.; visualization, M.G.; supervision, X.L. and J.Q.; project administration, Z.W.; funding acquisition, Z.W. All authors have read and agreed to the published version of the manuscript.

Funding: This study was supported by the China Geological Survey's project (Grant No. DD20190252, DD20221812), the National Natural Science Foundation of China (Grant No. 42002264) and the Fundamental Research Funds for the Chinese Academy of Geosciences (No. SK202109, SK202205).

Institutional Review Board Statement: Not applicable.

Informed Consent Statement: Not applicable.

Data Availability Statement: The datasets generated and/or analyzed during the current study are not publicly available.

Acknowledgments: Many thanks to the editors and anonymous reviewers for providing valuable comments that significantly improved this paper.

Conflicts of Interest: The authors declare no conflict of interest.

Abbreviations

The following abbreviations are used in this manuscript:

HSB	Heilongdong Spring Basin
HLD	Heilongdong
YC	Yuecheng (place name)
DWS	Dongwushi (place name)
Z	Sinian
∈	Cambrian
O	Ordovician
C	Carboniferous
P	Permian
T	Triassic

Q	Quaternary sediments
KGW	Karst groundwater in the Cambrian and Ordovician limestone aquifers
SNWTP	South-to-North Water Transfer Project
YJP	Yangjiaopu (place name)
QGW	Groundwater in the Quaternary sediment aquifer
PGW	Groundwater in the Permian sandstone aquifer
CGW	Carboniferous thin-layer limestone groundwater (CGW)
EC	Electrical conductivity
ICP-OES	Inductively Coupled Plasma Optical Emission Spectrometer
CBE	Charge Balance Error
VSMOW	Vienna Standard Mean Ocean Water
TCEA	Thermal conversion elemental analyzer
V-CDT	Vienna Canyon Diablo Troilite
MDW	Mine drainage water
SW	Surface water
KGW1	Karst groundwater in recharge area in 2013
KGW2	Karst groundwater in mining runoff area in 2013
KGW3	Karst groundwater in no mining runoff area in 2013
KGW4	Karst groundwater in discharge area in 2013
KGW5	Karst groundwater in recharge area in 2019
KGW6	Karst groundwater in mining runoff area in 2019
KGW7	Karst groundwater in no mining runoff area in 2019
KGW8	Karst groundwater in discharge area in 2019
LMWL	The local meteoric water line
GMWL	The global meteoric water line
SI	Saturation indices
WHO	The World Health Organization

References

1. Gomez, M.; Perdiguero, J.; Sanz, A. Socioeconomic Factors Affecting Water Access in Rural Areas of Low and Middle Income Countries. *Water* **2019**, *11*, 202. [CrossRef]
2. Santos, S.D.; Adams, E.A.; Neville, G.; Wada, Y.; Sherbinin, A.D.; Bernhardt, E.M.; Adamo, S.B. Urban growth and water access in sub-Saharan Africa: Progress, challenges, and emerging research directions. *Sci. Total Environ.* **2017**, *607–608*, 497–508. [CrossRef]
3. Chiocchini, U.; Castaldi, F.; Barbieri, M.; Eulilli, V. A stratigraphic and geophysical approach to studying the deep-circulating groundwater and thermal springs, and their recharge areas, in Cimini Mountains-Viterbo area, central Italy. *Hydrogeol. J.* **2010**, *18*, 1319–1341. [CrossRef]
4. Carrard, N.; Foster, T.; Willetts, J. Groundwater as a Source of Drinking Water in Southeast Asia and the Pacific: A Multi-Country Review of Current Reliance and Resource Concern. *Water* **2019**, *11*, 1605. [CrossRef]
5. Barbieri, M.; Barberio, M.D.; Banzato, F.; Billi, A.; Boschetti, T.; Franchini, S.; Gori, F.; Petitta, M. Climate change and its effect on groundwater quality. *Environ. Geochem. Health* **2021**, 1–12. [CrossRef] [PubMed]
6. Hao, Z.; Gao, Y.; Green, S.M.; Wen, X.F.; Yang, J.; Xiong, B.L.; Quine, T.A.; He, N.P. Chemical Characteristics of Flow Driven by Rainfall and Associated Impacts on Shallow Groundwater Quality in a Karst Watershed, Southwest China. *Environ. Prog.* **2021**, *8*, 615–636. [CrossRef]
7. Liu, F.; Wang, S.; Wang, L.S.; Shi, L.M.; Song, X.F.; Yeh, T.C.; Zhen, P.N. Coupling hydrochemistry and stable isotopes to identify the major factors affecting groundwater geochemical evolution in the Heilongdong Spring Basin, North China. *J. Geochem. Explor.* **2019**, *205*, 106352. [CrossRef]
8. Stevanović, Z. Global distribution and use of water from karst aquifers. *Geol. Soc. Lond. Spec. Publ.* **2018**, *466*, 217–236. [CrossRef]
9. Li, X.Q.; Zhang, C.C.; Hou, X.W. Characteristics of groundwater circulation and evolution in Jindong large coal base driven by coal mining: An example of Xin'an spring area. *J. China Coal Soc.* **2021**, *46*, 3015–3026.
10. Wang, R.; Li, X.H.; Wei, A.H. Hydrogeochemical characteristics and gradual changes of groundwater in the Baiquan karst spring region, northern China. *Carbonates Evaporites* **2022**, *37*, 47. [CrossRef]
11. Zhang, C.C.; Li, X.Q.; Ma, J.F.; Wang, Z.X.; Hou, X.W. Stable isotope and hydrochemical evolution of shallow groundwater in mining area of the Changzhi Basin, northern China. *Environ. Earth Sci.* **2022**, *81*, 294. [CrossRef]
12. Liang, Y.P.; Gao, X.B.; Zhao, C.H.; Tang, C.L.; Shen, H.Y.; Wang, Z.H.; Wang, Y.X. Review: Characterization, evolution, and environmental issues of karst water systems in Northern China. *Hydrogeol. J.* **2018**, *26*, 1371–1385. [CrossRef]
13. Sun, W.J.; Song, J.X.; Yang, W.K.; Zheng, Y.J.; Li, C.Y.; Kuang, D. Distribution of carbonate rocks and variation analysis of karst water resources in China. *Carbonates Evaporites* **2020**, *35*, 121. [CrossRef]

14. Huang, Q.B.; Qin, X.Q.; Yang, Q.Y.; Liu, P.Y.; Zhang, J.S. Identification of dissolved sulfate sources and the role of sulfuric acid in carbonate weathering using $\delta^{13}C_{DIC}$ and $\delta^{34}S$ in karst area, northern China. *Environ. Earth Sci.* **2016**, *75*, 51.
15. He, K.Q.; Guo, L.; Guo, Y.Y.; Luo, H.L.; Liang, Y.P. Research on the effects of coal mining on the karst hydrogeological environment in Jiaozuo mining area, China. *Environ. Earth Sci.* **2019**, *78*, 434.
16. Liu, P.; Hoth, N.; Drebenstedt, C.; Sun, Y.J.; Xu, Z.M. Hydro-geochemical paths of multi-layer groundwater system in coal mining regions—Using multivariate statistics and geochemical modeling approaches. *Sci. Total Environ.* **2017**, *601–602*, 1–14. [CrossRef]
17. Qian, J.Z.; Peng, Y.X.; Zhao, W.D.; Ma, L.; He, X.R.; Lu, Y.H. Hydrochemical processes and evolution of karst groundwater in the northeastern Huaibei Plain, China. *Hydrogeol. J.* **2018**, *26*, 1721–1729. [CrossRef]
18. Fu, C.C.; Li, X.Q.; Ma, J.F.; Liu, L.X.; Gao, M.; Bai, Z.X. A hydrochemistry and multi-isotopic study of groundwater origin and hydrochemical evolution in the middle reaches of the Kuye River basin. *Appl. Geochem.* **2018**, *98*, 82–93. [CrossRef]
19. Keesari, T.; Roy, A.; Mohokar, H.; Pant, D.; Sinha, U.K. Characterization of mechanisms and processes controlling groundwater recharge and its quality in drought-prone region of central India (Buldhana, Maharashtra) using isotope hydrochemical and end-member mixing modeling. *Nat. Resour. Res.* **2020**, *29*, 1951–1973. [CrossRef]
20. Qu, S.; Wang, G.C.; Shi, Z.M.; Xu, Q.Y.; Guo, Y.Y.; Ma, L.; Sheng, Y.Z. Using stable isotopes (δD, $\delta^{18}O$, $\delta^{34}S$ and $^{87}Sr/^{86}Sr$) to identify sources of water in abandoned mines in the Fengfeng coal mining district, northern China. *Hydrogeol. J.* **2018**, *26*, 1443–1453. [CrossRef]
21. Guo, Y.Y.; Lyu, Z.C.; Wang, G.C.; Ma, L.; Xu, Q.Y.; Huang, X.J.; Gao, S.Z. Hydrogeochemical simulation of groundwater in Eastern Fengfeng mining area. *Coal Geol. Explor.* **2016**, *44*, 101–105.
22. Hao, C.M.; Sun, W.; He, P.Y.; Li, C. The impact of nearly 30 years mining activities on the hydrochemistry characteristic of karst groundwater in Fengfeng coal mining area. *China Min. Mag.* **2015**, *24*, 45–51.
23. Hao, C.M.; Huang, Y.; He, P.Y.; Sun, W. Isotope Drift Characteristics in Ordovician Limestone Karst Water Caused by Coal Mining in Northern China. *Mine Water Environ.* **2019**, *38*, 507–516. [CrossRef]
24. Liu, F.; Zhen, P.N.; Wang, S. Groundwater quality assessment and health risks from nitrate contamination in the Heilongdong Spring Basin, a typical headwater basin of the North China Plain. *Environ. Sci. Pollut. Res.* **2022**, *29*, 17655–17670. [CrossRef]
25. Wang, S. Research on Groundwater Circulation and Hydrochemical Formation Mechanism in the Heilongdong Spring Basin. Ph.D. Thesis, Hebei University of Engineering, Handan, China, 2020.
26. Liu, B.; Guan, Y.Q.; Sun, Y.Z.; Zhang, H.S.; Bian, K. Water inrush type division and water inrush mode in Fengfeng Mining Area. *Saf. Health Coal Mines* **2021**, *52*, 186–194.
27. Wang, Z.X.; Li, X.Q.; Hou, X.W. Hydrogeochemistry of River Water in the Upper Reaches of the Datong River Basin, China: Implications of Anthropogenic Inputs and Chemical Weathering. *Acta Geol. Sin.-Engl. Ed.* **2021**, *95*, 962–975. [CrossRef]
28. Craig, H. Isotopic variations in meteoric waters. *Science* **1961**, *133*, 1702–1703. [CrossRef]
29. Gibbs, R.J. Mechanisms controlling world water chemistry. *Science* **1970**, *170*, 1088–1090. [CrossRef] [PubMed]
30. Wu, X.C.; Li, C.S.; Sun, B.; Geng, F.Q.; Gao, S.; Lv, M.H.; Ma, X.Y.; Li, H.; Xing, L.T. Groundwater hydrogeochemical formation and evolution in a karst aquifer system affected by anthropogenic impacts. *Environ. Geochem. Health* **2020**, *42*, 2609–2626. [CrossRef] [PubMed]
31. Lin, Y.; Ren, H.X.; Wu, Y.Z.; Cao, F.L.; Jia, F.J.; Qu, P.C. The evolution of hydrogeochemical characteristics of a typical piedmont karst groundwater system in a coal-mining area, Northern China. *Environ. Earth Sci.* **2019**, *78*, 557. [CrossRef]
32. Schoeller, H. Qualitative Evaluation of Groundwater Resources. In *Methods and Techniques of Groundwater Investigations and Development*; Unesco: Paris, France, 1965; pp. 54–83.
33. Venkatramanan, S.; Chung, S.Y.; Ramkumar, T.; Gnanachandrasamy, G.; Vasudevan, S.; Lee, S.Y. Application of GIS and hydrogeochemistry of groundwater pollution status of Nagapattinam district of Tamil Nadu, India. *Environ. Earth Sci.* **2015**, *73*, 4429–4442. [CrossRef]
34. Parkhurst, D.L.; Appelo, C.A.J. User's Guide to PHREEQC (Version 2) a Computer Program for Speciation, Batch-Reaction, One-Dimensional Transport, and Inverse Geochemical Calculations: U.S Geological Survey Water-Resources Investigations Report 99-4259. 1999. Available online: https://pubs.er.usgs.gov/publication/wri994259 (accessed on 14 December 2022).
35. Wu, J.H.; Li, P.Y.; Qian, H. Hydrochemical characterization of drinking groundwater with special reference to fluoride in an arid area of China and the control of aquifer leakage on its concentrations. *Environ. Earth Sci.* **2015**, *73*, 8575–8588. [CrossRef]
36. Li, L.P.; Huang, G.X. Groundwater Level Mapping Using Multiple-Point Geo-statistics. *Water* **2016**, *8*, 400. [CrossRef]
37. Ma, Z.J. Study on Karst Water Cycle Evolution Law in Heilongdong Spring Area. Ph.D. Thesis, Hebei University of Engineering, Handan, China, 20 December 2021.
38. Krouse, H.R.; Mayer, B. Sulphur and Oxygen Isotopes in Sulphate. In *Environmental Tracers in Subsurface Hydrology*; Cook, P.G., Herczeg, A.L., Eds.; Springer: Boston, MA, USA, 2000; pp. 195–231.
39. Bahrami, M.; Zarei, A.R.; Rostami, F. Temporal and spatial assessment of groundwater contamination with nitrate by nitrate pollution index (NPI) and GIS (case study: Fasarud Plain, southern Iran). *Environ. Geochem. Health* **2020**, *42*, 3119–3130. [CrossRef] [PubMed]

40. Huang, G.X.; Liu, C.Y.; Sun, J.C.; Zhang, M.; Jing, J.H.; Li, L.P. A regional scale investigation on factors controlling the groundwater chemistry of various aquifers in a rapidly urbanized area: A case study of the Pearl River Delta. *Sci. Total Environ.* **2018**, *625*, 510–518. [CrossRef] [PubMed]
41. WHO. *Guidelines for Drinking-Water Quality*, 4th ed.; WHO: Geneva, Switzerland, 2011; pp. 398–403.

water

MDPI

Article

Key Factors Dominating the Groundwater Chemical Composition in a Grain Production Base: A Case Study of Muling–Xingkai Plain, Northeast China

Chen Su [1,2], Zhuang Li [3], Wenzhong Wang [1,2,*], Zhongshuang Cheng [1,2], Zhaoxian Zheng [1,2] and Zongyu Chen [1,2]

1 Institute of Hydrogeology and Environmental Geology, Chinese Academy of Geological Sciences, Shijiazhuang 050061, China; sc.1219@163.com (C.S.); zshuangcheng@hotmail.com (Z.C.); aslangyeah@126.com (Z.Z.); chenzy@heinfo.net (Z.C.)
2 Key Laboratory of Groundwater Sciences and Engineering, Ministry of Natural Resources, Shijiazhuang 050061, China
3 Shandong Geological Environment Monitoring Station, Jinan 250014, China; 13964000805@163.com
* Correspondence: wangwenzhong@mail.cgs.gov.cn

Abstract: Groundwater quality in the Muling–Xingkai Plain (MXP) is closely related to food security and human health. The chemical composition of groundwater in MXP has attracted great attention. A total of 168 groundwater samples were collected in MXP, and principal component analysis, chemical ion analysis and stable isotopic analysis were used to explore key factors affecting the chemical composition and hydrochemical evolution process of groundwater. Results show sources of chemical ions in groundwater are silicate minerals, carbonate minerals and domestic sewage. Domestic sewage is responsible for groundwater with high levels of Cl^-, SO_4^{2-} and NO_3^-, but a reduction environment can lead to groundwater with a high level of NH_4^+ due to nitrification. Human activity and soil media together influence groundwater chemical composition. Groundwater with a high level of chemical ions is mainly collected from wells near river channels, where coarse-textured soils are overlying aquifers. The black soil far away from river channels can retard the infiltration of wastewater. Agricultural activities do not directly lead to deterioration of groundwater qualities, and agricultural non-point-source pollution does not occur in MXP. Nearly 70% of the population in MXP is living in the southern plain, where the influence of sewage on groundwater chemical composition is obvious. Thus, shallow groundwater far away from river channels is the best choice for irrigation. Some measures should be implemented to control the discharge of domestic sewage for the protection of groundwater. In addition, it is necessary to avoid the transformation of the redox environment of groundwater in the northern plain.

Keywords: groundwater; hydrochemistry; human activities; agricultural area; Muling–Xingkai Plain

Citation: Su, C.; Li, Z.; Wang, W.; Cheng, Z.; Zheng, Z.; Chen, Z. Key Factors Dominating the Groundwater Chemical Composition in a Grain Production Base: A Case Study of Muling–Xingkai Plain, Northeast China. *Water* **2022**, *14*, 2222. https://doi.org/10.3390/w14142222

Academic Editor: Aldo Fiori

Received: 30 May 2022
Accepted: 12 July 2022
Published: 14 July 2022

1. Introduction

Groundwater is one of the most valuable natural resources in the world, and it can support nearly all kinds of human activities [1–5]. Generally, the development and utilization of groundwater resources are closely related to the chemical composition and qualities of groundwater. The chemical composition of groundwater can be influenced by natural processes (such as hydrogeological conditions, redox conditions and interaction of groundwater with minerals) and human activities (e.g., exploitation, sewage discharge and fertilizer application). Many important physicochemical and ecological processes are disrupted by changes in groundwater qualities, and even the rational utilization of groundwater resources is influenced by high levels of chemical ions [6–8]. It is necessary to ensure the safety of groundwater quality. The issues of groundwater chemical characteristics and their dominating factors have attracted more attention in the world. Relevant studies have been conducted in coastal areas, arid or semi-arid areas, karst areas and even rapidly urbanized areas [9–12]. At present, agricultural activities have become a critical cause and a source of groundwater pollution. Irrigation return flow water with chemical pollutants can lead to the groundwater and surface water being unsuitable or less valuable for other water uses [13–16]. Therefore, groundwater chemical characteristics under the influence of human activities in agricultural areas have become a research focus [17–21]. It is of great significance to identify the key factors influencing the chemical composition and hydrochemical evolution process of groundwater for rational utilization of water resources and food security in agricultural areas [22–24].

Muling–Xingkai Plain (MXP) is one of the biggest and most important grain production bases in NE China, where plants and crops grow in rich, dark soil. A great deal of pesticides and fertilizer is frequently applied, which could lead to deterioration of water quality. However, black soils with weak infiltration capacity can retard the infiltration of irrigation water and other surface water. Human activities coupled with special soil conditions result in a complex chemical evolution process of groundwater in MXP. Previous studies showed that agricultural activities have influenced groundwater qualities in local areas [25]. Recent studies show domestic sewage in residential zones is the source of contamination of groundwater [26,27]. Food security and human health in MXP are closely related to groundwater qualities. It is necessary to perform a thorough analysis to understand the chemical characteristics of groundwater and to distinguish the main factors affecting the chemical composition of groundwater in MXP.

Chemical ion analysis and multivariate statistical analysis have been widely applied to assess hydrochemical processes and geochemical evolution in complex systems [28–30]. These methods coupled with isotopic analysis were effectively applied to delineate the hydrochemical evolution process and to distinguish the source of chemical ions in groundwater [31–34]. Generally, correlation among chemical variables is the basis of chemical ion analysis [35]. Multivariate analysis is mainly applied to classify samples and to further identify the principal components. These methods are effective ways to distinguish the key factors influencing groundwater chemical composition [36,37].

The chemical evolution process of groundwater and dominant factors affecting chemical composition under the anthropogenic influence and special soil conditions are investigated using chemical ion analysis and multivariate statistical analysis in MXP. The aims are to (1) show the chemical composition of groundwater, (2) identify the chemical evolution process of groundwater, (3) distinguish various factors affecting the chemical composition of groundwater and (4) provide some implications for the protection of groundwater. The results will be beneficial for groundwater resource management in agricultural areas with similar geological or hydrogeological conditions.

2. Description of Study Area

2.1. Geographical Conditions

MXP occupies an area of about 10,000 km^2 in NE China, extending between longitudes of 131°30′–133°40′ E and latitudes of 45°05′–46°17′ N. Wandashan Mountain is found along the north side and west side of the study area. The east and south boundaries of the study area are Wusuli River and Xingkai Lake, respectively. Abuqing River, Qihulin River and Muling River are three major rivers (Figure 1), and rivers flow through the plain from west to east. The annual average air temperature is 1.9 degrees centigrade, and the temperature is highest in July (21 °C) and lowest in January (−21 °C). Mean annual rainfall is 540–680 mm, with June to September accounting for about 70% of it.

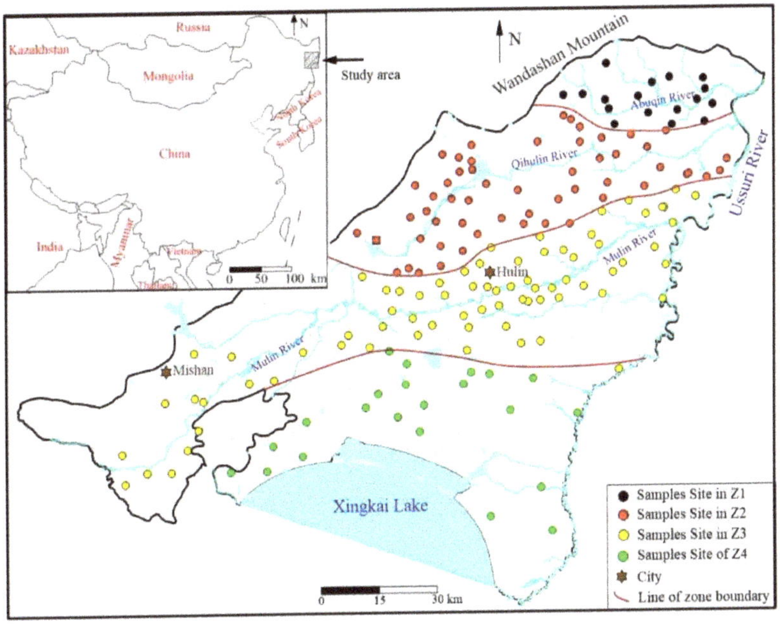

Figure 1. Study area and sampled sites.

2.2. Hydrogeological Setting

Four hydrogeologic subdivisions are divided based on rivers and lakes in the plain. Abuqin zone (Z1) and Qihulin zone (Z2) are located in the northern plain, and Muling zone (Z3) and Xingkai zone (Z4) are distributed in the middle and southern part (Figure 1).

Aquifers principally consist of Quaternary alluvial and lacustrine deposits and Neogene sediments. The thickness of the aquifer is 50–200 m, and the depth to groundwater is 2–20 m. Phreatic aquifers are widely overlain by thick black clay layers with small infiltration coefficients, especially the aquifer in Z4. Sandy sediments are mainly distributed along river channels. Groundwater flows from west to east, similar to the flow direction of rivers in MXP.

Recharge sources of groundwater are atmospheric precipitation, surface water and irrigation water. Groundwater is principally discharged by evaporation, artificial extraction and lateral outflow into river channels.

2.3. Land Uses and Human Activities

Large areas of natural grasslands and wetlands in the MXP have been exploited and turned into croplands since 1950s. In addition, large areas of gardens and woodland were turned into croplands, and many dry farmlands distributed in piedmont zones were turned into rice fields recently. The farmlands are widely distributed in MXP.

Groundwater is the main irrigation water in Z1, Z2 and Z3. Rice fields in Z4 are mainly irrigated by surface water abstracted from Xingkai Lake. A large amount of agricultural effluents has seeped underground, which may lead to a change in groundwater quality. Meanwhile, the quality and yield of crops may be influenced by groundwater quality. Mishan city and Hulin city are located in Z3, and villages are distributed widely in MXP. Domestic sewage in the rural area is discharged into leak channels, and it poses a threat to groundwater quality.

3. Materials and Methods

This study was conducted based on 164 groundwater samples collected in 2016–2018, and the numbers of samples in Z1, Z2, Z3 and Z4 are 17, 54, 71 and 22, respectively (Figure 1). Samples were filtrated by using 0.45 μm filter membranes and then stored under a temperature of 4 °C. Field parameters pH, temperature (T), electrical conductivity (EC) and redox potential (Eh) were measured on-site.

All analyses were carried out at the Institute of Hydrogeology and Environmental Geology, Chinese Academy of Geological Sciences. Analyses for total concentrations of four major cations (K^+, Na^+, Ca^{2+}, Mg^{2+}) were measured by inductively coupled plasma mass spectrometry. NO_3^-, Cl^-, Br^- and SO_4^{2-} analyses of water samples were measured by spectrophotometry. HCO_3^- was measured by acid–base titration. Total dissolved solids (TDS) were measured gravimetrically. The relative error was less than 3% for all analyzed elements.

In this study, chemical ion analysis was carried out based on the molar ratio or milliequivalent ratio among chemical ions. It can suggest the source of the chemical ions, water–rock interaction and even the influence of human activities. Principal component analysis (PCA) was applied to analyze possible processes influencing the chemical composition of groundwater using the IBM SPSS program (version 19, 2010, SPSS Inc. Chicago, IL, USA). Parameters Eh and EC were eliminated for samples in Z1, Z2 and Z3 due to more than 50% of data values being missing (Table 1). In addition, nearly half of the samples were below the detection limit value for NH_4^+, and the values were replaced by 0.5 times the detection limit. Rotation of principal components was carried out in PCA, and the expressions "strong", "moderate", and "weak" were applied to describe the factor loading values of >0.75, 0.75–0.50 and 0.50–0.30, respectively [38].

Table 1. Statistics of groundwater chemistry data in four zones of Muling–Xingkai Plain (MXP).

Parameter	Unit	Z1 (n = 17)					Z2 (n = 54)					Z3 (n = 75)					Z4 (n = 22)				
		Min.	Max.	Mean	SD	CV (%)	Min.	Max.	Mean	SD	CV (%)	Min.	Max.	Mean	SD	CV (%)	Min.	Max.	Mean	SD	CV (%)
PH	Standard	5.87	7.80	6.63	0.52	8	6.09	7.60	6.74	0.38	6	5.92	7.46	6.74	0.31	5	6.38	7.15	6.77	0.23	3
TDS	$mg{\cdot}L^{-1}$	64.70	442.40	195.48	93.35	48	84.59	746.10	243.40	122.94	51	90.05	1179.00	248.20	154.13	62	106.00	1154.00	278.58	208.11	75
K^+	$mg{\cdot}L^{-1}$	0.52	2.61	1.45	0.60	41	0.22	49.69	3.09	7.50	243	0.33	22.92	2.87	3.96	138	0.48	42.57	4.29	8.54	199
Na^+	$mg{\cdot}L^{-1}$	5.40	52.17	17.37	11.46	66	8.42	63.79	20.27	12.36	61	3.75	96.68	20.07	13.32	66	6.21	79.05	27.50	16.17	59
Ca^{2+}	$mg{\cdot}L^{-1}$	6.01	79.08	29.87	17.57	59	10.07	120.40	38.87	25.36	65	8.79	167.20	36.17	22.15	61	12.65	191.40	38.09	36.05	95
Mg^{2+}	$mg{\cdot}L^{-1}$	2.78	29.10	11.84	7.34	62	4.37	44.26	12.81	7.89	62	4.27	55.84	12.44	8.14	65	3.72	81.35	13.90	15.36	111
Cl^-	$mg{\cdot}L^{-1}$	–	64.93	7.76	15.41	198	0.35	59.72	11.51	14.91	129	–	184.30	20.03	29.04	145	0.35	308.90	37.09	66.53	179
SO_4^{2-}	$mg{\cdot}L^{-1}$	1.61	28.21	6.90	6.24	90	1.62	74.18	22.91	21.07	92	1.70	181.30	24.24	31.56	130	1.64	125.80	23.59	33.24	141
HCO_3^-	$mg{\cdot}L^{-1}$	18.23	504.40	174.93	125.42	72	33.63	642.00	183.49	115.47	63	24.46	354.60	133.93	72.41	54	66.03	550.30	153.16	104.06	68
NH_4^+	$mg{\cdot}L^{-1}$	0.02	7.70	1.53	2.52	165	0.01	7.00	1.00	1.63	163	0.02	5.30	0.66	1.31	198	0.02	3.90	0.69	1.27	185
NO_3^-	$mg{\cdot}L^{-1}$	1.75	104.30	9.58	23.83	249	0.20	147.70	9.74	24.17	248	1.75	399.00	32.61	68.66	211	1.78	95.68	21.24	29.95	141
EC	$ms{\cdot}cm^{-1}$																0.13	2.05	0.50	0.40	80
Eh	mv																−164.00	135.00	10.58	80.16	758

Note: SD: standard deviation; CV: coefficient of variation; –: below detectable limit.

4. Results and Discussion

4.1. Chemical Characteristics

4.1.1. General Chemistry

The descriptive statistics of groundwater chemistry data in four zones are presented in Table 1 and Figure 2. pH is predominantly slightly acidic to near neutral with TDS 64.7–1179 mg/L, and ionic composition is dominated by Ca^{2+} (6.01–191.4 mg·L^{-1}), Na^{+} (3.75–96.68 mg·L^{-1}) and HCO_3^- (18.23–642 mg·L^{-1}). The order of abundance of cations is $Ca^{2+} > Na^{+} > Mg^{2+} > K^{+} > NH_4^{+}$, and anions follow the order of $HCO_3^- > SO_4^{2-} > Cl^- > NO_3^-$. The differences in chemical ion concentrations in the four zones are obvious. Samples in Z1 have the lowest levels of Ca^{2+}, Mg^{2+}, Na^{+}, K^{+}, Cl^- and SO_4^{2-} but the highest level of HCO_3^-. The ion concentrations of samples in Z4 are highest, except NH_4^+. Concentrations of chemical ions in Z2 are nearly identical to those in Z3, except NO_3^- and NH_4^+. Nearly all of the chemical parameters in the four zones are characterized by a high coefficient of variation. The groundwater chemical composition is influenced by multiple factors.

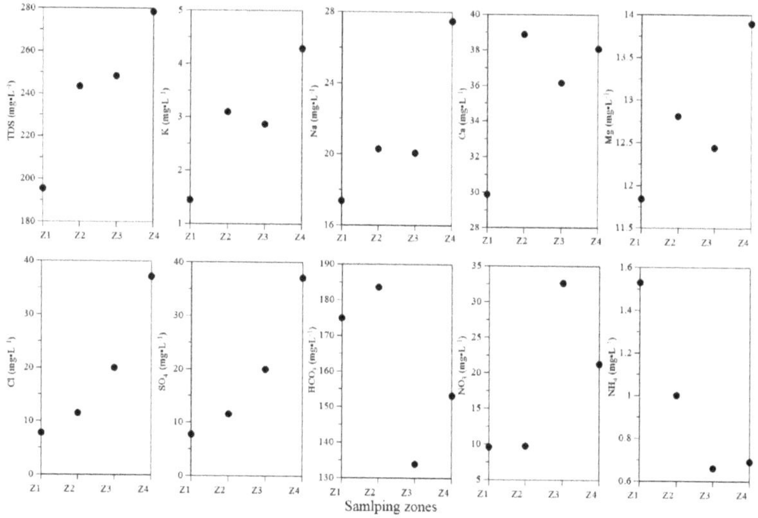

Figure 2. Concentration of chemical parameters in four zones.

4.1.2. Groundwater Types

With respect to cations, most samples are scattered in zone B in the lower-left triangle, indicating a mixed type. Only a few samples in Z3 and Z4 are Ca-type (Figure 3). With respect to anions, most groundwater samples are plotted in zone E in the lower-right triangle. Most groundwater samples are HCO_3-type.

As shown in the central diamond plot in Figure 3, more than half of the samples are scattered in zone 1, which indicates that alkaline earth elements and bicarbonate are the most common in groundwater chemistry. A large number of samples in Z1 and Z2 are scattered in zone 1, but more than half of the samples in Z3 and Z4 are scattered in zone 4. Zone 4 means samples are mixed chemical types. The chemical composition of samples in the southern part of MXP is significantly influenced by multiple factors.

The water type of groundwater samples in MXP is predominantly Ca-HCO_3. However, samples with water types of Ca·Mg·Na-HCO_3, Ca·Mg-HCO_3, Ca·Mg-HCO_3·SO_4 and Ca·Na-HCO_3 are widely distributed in the study area. At present, some samples are characterized by the water type of SO_4, even NO_3, which hardly occurred before 1963 [27]. The chemical composition of groundwater in MXP is influenced by human activities.

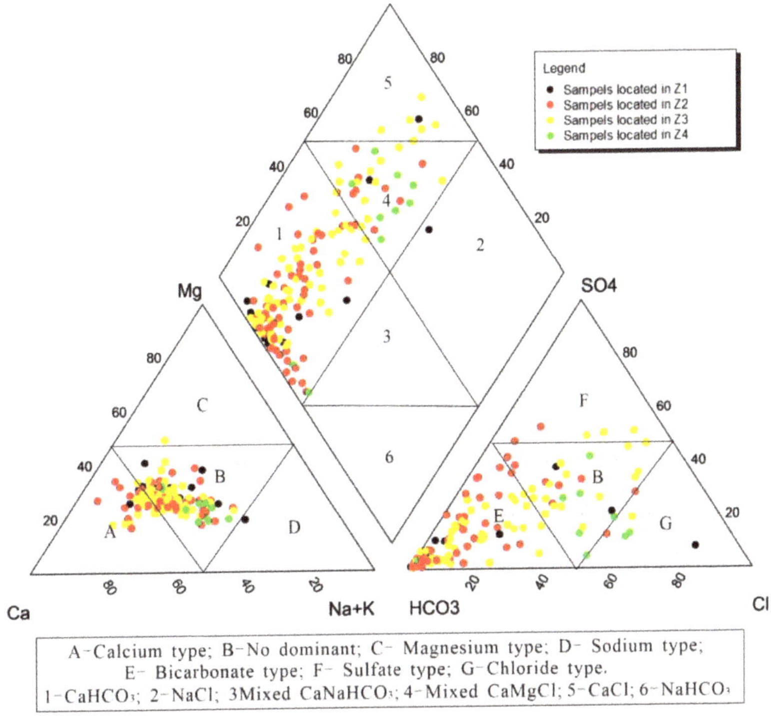

Figure 3. Piper third-line diagram of groundwater samples in MXP.

4.2. Hydrochemical Evolution Process

4.2.1. Chemical Ion Analysis

(1) Gibbs Plot

Gibbs diagram can be used to detect the primary formation mechanism of the water chemistry [39]. The weight ratios of $Na^+/(Na^++Ca^{2+})$ and $Cl^-/(Cl^-+HCO_3^-)$ of most samples are less than 0.5, with TDS 200–500 mg/L (Figure 4). Rock dissolution is the predominant mechanism controlling the groundwater chemistry [40]. However, many groundwater samples do not fall in the banana-shaped area in the Gibbs diagram (Figure 4). The chemical composition is influenced by other factors, such as human activities and cation exchange reactions. Samples in and out of the banana-shaped area are defined as the "general samples" and the "special samples" in this paper, respectively. Only a few special samples were collected from wells in Z1 and Z2 (Table 2), suggesting the slight effect of external factors on the chemical compositions of groundwater in the northern plain.

Table 2. The number of the special samples in four zones.

Zones	Total Samples	Number of Special Samples	C (%)
Z1	17	3	17.65
Z2	54	9	16.67
Z3	71	16	22.53
Z4	22	10	45.45

Note: C: ratio of the number of special samples to total samples.

chemical ion concentration and the stable isotopic value under the effect of evaporation [45]. When some chemical ions increase sharply without dissolution of related minerals, but stable isotopic values of groundwater stay rather constant, the wastewater from human activities may be the principal cause of the high level of related chemical ions.

Figure 7 illustrates the relationships between the concentration of chemical variables (Cl^-, SO_4^{2-}, NO_3^-, Na^+, Mg^{2+} and Ca^{2+}) and the $\delta^{18}O$ value of groundwater. The elevation of chemical ions for most general samples results from evaporation and mineral dissolution. However, concentrations of chemical ions for special samples increase sharply with little modifications in stable isotopic values, especially for Cl^-, SO_4^{2-} and NO_3^-. Major anthropogenic components in the groundwater include Cl^- and NO_3^-, as well as SO_4^{2-}. The chemical composition of special samples is influenced severely by human activities. The result of stable isotope analysis is consistent with the results discussed above.

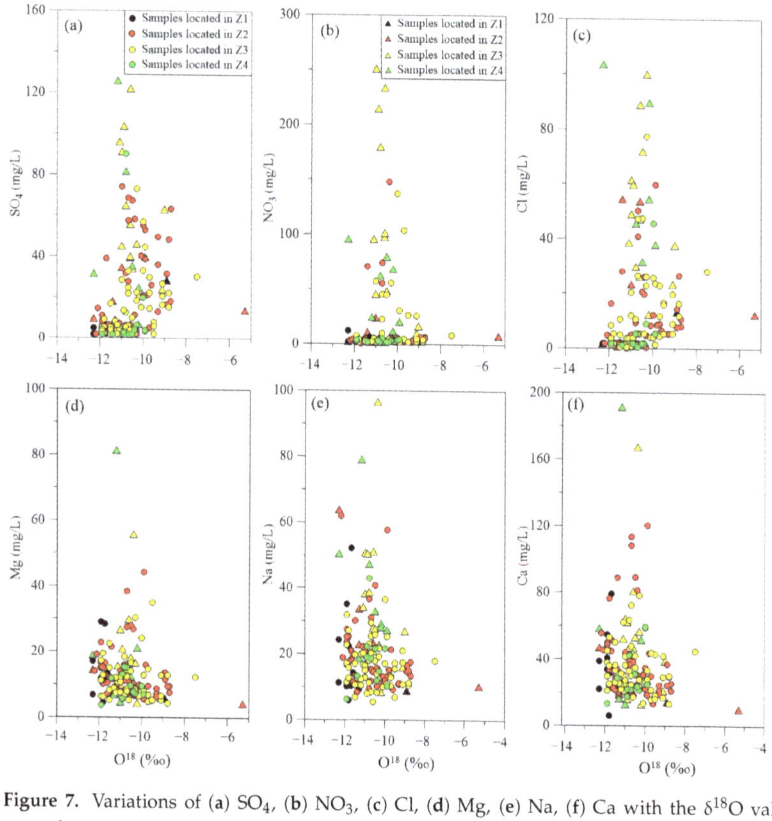

Figure 7. Variations of (**a**) SO_4, (**b**) NO_3, (**c**) Cl, (**d**) Mg, (**e**) Na, (**f**) Ca with the $\delta^{18}O$ value of groundwater.

4.2.3. PCA

PCA is used to further discuss the key factors dominating the chemical composition of groundwater in four zones, and the results are shown in Table 3.

(1) Z1

The chemical composition of groundwater samples in Z1 is controlled by a three-factor model, and the cumulative variance of the three PCs is 87.81%. PC1 comprises strong loading of Ca^{2+}, TDS, Mg^{2+}, HCO_3^-, Na^+ and PH, and TDS values increase with the concentrations of HCO_3^-, Ca^{2+}, Na^+ and Mg^{2+}. Water–rock interactions are responsible for their occurrence. Based on the chemical analysis mentioned above, the dissolution of silicate and carbonate minerals is responsible for the chemical composition of groundwater.

Potassic fertilizers and urea fertilizers are widely applied in MXP. When the topsoil with high permeability occurs in farmlands, a clear correlation between K^+ and NH_4^+ will be indicated in groundwater. Thus, PC3 is the indication of agricultural activities.

Potential sources of Cl^- in groundwater include natural sources (dissolution of minerals), agricultural chemicals, animal waste and septic effluent. In PC2, no clear correlation is observed between Cl^- and K^+. Thus, fertilizer is not the main source of Cl^-. In addition, a high level of Cl^- cannot result from the dissolution of minerals as discussed above. Cl^- and NO_3^- are the major components of domestic sewage [43]. Previous studies showed that the discharge of domestic sewage in MXP had resulted in a high level of NO_3^-. Chemical parameters Cl^- and NO_3^- associated with factor 2 imply the influence process of domestic sewage.

(2) Z2

A three-factor model can be used to explain the chemical composition of groundwater samples in Z2, and the cumulative variance of three PCs is 79.83%. As in the analysis of Z1, PC1 is the indication of water–rock interaction. PC2 with high positive loading of Cl^- and SO_4^{2-} and moderate loading of NO_3^- indicates that domestic sewage is responsible for groundwater chemistry. PC3 with moderate positive loading of NH_4^+ and K^+ suggests groundwater quality is affected by agricultural production.

(3) Z3

Groundwater chemistry in Z3 is controlled by a three-factor model, and the cumulative variance of the three PCs is 88.74%.

PC1 comprises strong loading of TDS, Na^+, Ca^{2+}, Mg^{2+}, SO_4^{2+}, Cl^- and NO_3^- and moderate loading of K^+. The strong correlations among TDS, Na^+, Ca^{2+}, Mg^{2+} and K^+ show that water–rock interaction is responsible for chemical composition. However, strong correlations among NO_3^-, Cl^- and SO_4^{2-} are mainly caused by human activities. The TDS value is increased with concentrations of SO_4^{2-}, Cl^- and NO_3^-, which suggests that infiltration of domestic sewage is responsible for its high concentrations in groundwater. Thus, the chemical compositions of groundwater are dominated by water–rock interaction and domestic sewage.

PC2, which comprises strong loading of PH and HCO_3^- and positive loading of Ca^{2+} and Mg^{2+}, indicates the slight dissolution process of carbonate minerals [7].

PC3, which only comprises strong loading of NH_4^+ and weak loading of NO_3^-, indicates reduction conditions.

(4) Z4

Groundwater chemical composition in Z4 is controlled by a two-factor model, and the cumulative variance of the two PCs is 83.08%.

PC1, which comprises strong loading of TDS, K^+, Na^+, Ca^{2+}, Mg^{2+}, SO_4^{2+}, Cl^-, HCO_3^- and EC, implies water–rock interactions are responsible for chemical composition.

Halite and gypsum minerals are not the main constituents of rock and stratum. Chlorine-bearing fertilizers are not applied frequently. Thus, the high concentration of Cl^- and SO_4^{2-} cannot be attributed to the dissolution of minerals. Domestic wastewater is the main source of Cl^- and SO_4. PC1 indicates that water–rock interaction and domestic sewage control groundwater chemical composition.

PC2 comprises strong positive loading of PH and NH_4^+ and negative loading of NO_3^- and EH. Redox conditions may dominate the content of NH_4^+ and NO_3^-. Other ions do not show a significant correlation with NH_4^+ and NO_3^-, which implies that organic matter decomposition is dominating the concentration of nitrogen species. PC2 can be regarded as "natural reduction conditions". Large areas of natural grasslands and wetlands were exploited and turned into croplands in the last few decades in MXP. Abundant herbaceous plants were buried underground. The organic matter decomposition has produced plenty of NH_3-N. Thus, redox conditions can influence the content of nitrogen species.

Table 3. Principal component loading for groundwater samples in four zones.

Chemical Parameter	Z1			Z2			Z3			Z4	
	PC1	PC2	PC3	PC1	PC2	PC3	PC1	PC2	PC3	PC1	PC2
PH	**0.820**	−0.390	−0.037	**0.860**	−0.272	−0.301	−0.115	**0.957**	−0.084	0.228	**0.775**
TDS	**0.965**	0.140	0.131	**0.819**	0.559	0.036	**0.995**	−0.011	0.029	**0.996**	−0.040
K^+	−0.210	−0.108	**0.913**	0.482	0.448	**0.589**	**0.684**	−0.211	0.016	**0.943**	0.213
Na^+	**0.845**	0.136	−0.271	**0.771**	−0.069	0.215	**0.935**	−0.069	−0.044	**0.881**	−0.253
Ca^{2+}	**0.969**	0.000	0.108	**0.777**	0.563	−0.111	**0.943**	0.185	0.096	**0.982**	0.049
Mg^{2+}	**0.952**	−0.033	−0.003	**0.779**	0.576	−0.046	**0.922**	0.228	0.068	**0.978**	0.063
HCO_3^-	**0.920**	−0.350	0.136	**0.969**	0.114	0.180	−0.015	**0.760**	0.420	**0.783**	0.587
SO_4^{2-}	−0.428	0.128	−0.624	−0.075	**0.826**	0.026	**0.866**	−0.331	−0.273	**0.823**	−0.140
Cl^-	−0.117	**0.953**	−0.234	0.129	**0.804**	0.018	**0.861**	−0.212	−0.232	**0.944**	−0.235
NH_4^+	0.016	−0.091	**0.892**	0.037	0.019	**0.700**	−0.068	−0.003	**0.979**	−0.012	**0.669**
NO_3^-	0.019	**0.976**	−0.060	0.177	**0.566**	−0.042	**0.896**	−0.253	−0.476	0.262	**−0.835**
EC										**0.981**	−0.014
Eh										0.127	**−0.812**
Eigenvalue	5.25	2.21	2.20	4.44	2.90	1.44	6.39	1.85	1.53	7.86	2.94
Explained variance (%)	47.74	20.09	19.98	40.38	26.40	13.05	58.06	16.80	13.88	60.45	22.63
Cumulative % of variance	47.74	67.83	87.81	40.38	66.77	79.83	58.06	74.86	88.74	60.45	83.08

4.3. Factors Dominating the Groundwater Chemical Characteristics

As the discussion above mentioned, water–rock interaction, infiltration of domestic sewage, agricultural activities and redox conditions are the main factors dominating the chemical composition of groundwater in MXP (Figure 8).

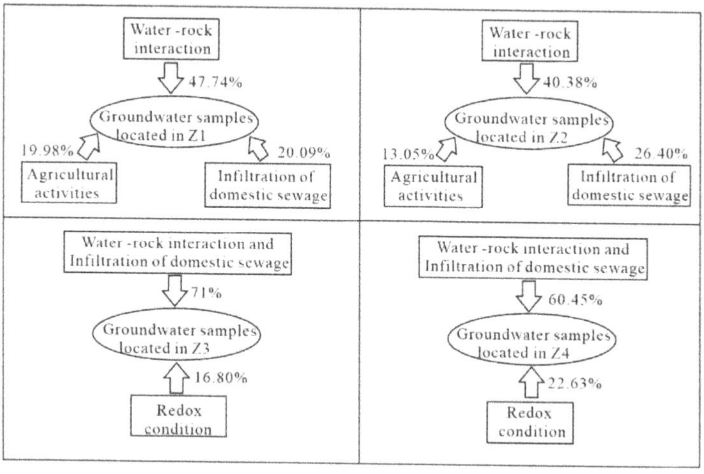

Figure 8. Factors dominating the groundwater chemical characteristics in different zones in MXP.

Agricultural activities only slightly influence the chemical composition of groundwater in the northern plain (Z1 and Z2). The redox environment of groundwater in the southern plain (Z3 and Z4) is a key factor influencing the concentration of NO_3^- and NH_4^+. The thick silt layer overlaying aquifers can lead to a reduction condition. Organic matter decomposition can produce plenty of NH_3-N. A high level of NO_3^- is closely related to nitrification.

Nearly all the special samples were collected from wells near river channels, where coarse-textured sediments are overlying aquifers (Figure 9). Coarse-textured sediments mainly consist of sand and gravel, and they can allow water to infiltrate downwards easily. Most villages in MXP lack sewage disposal systems, and domestic sewage is discharged into a seepage pit directly. Thus, the discharge of domestic sewage in the rural areas near the rivers often leads to abnormal groundwater types.

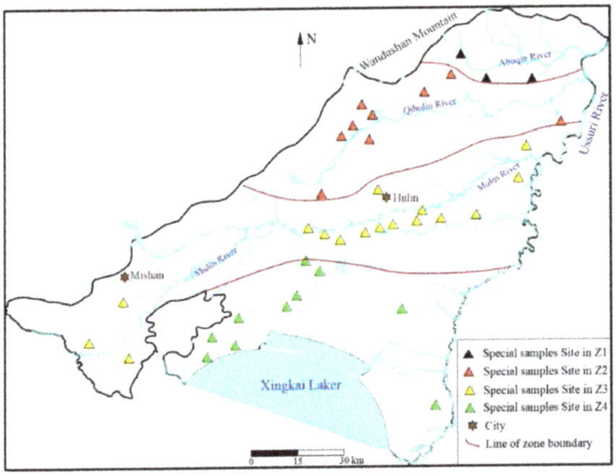

Figure 9. The distribution of the special samples in four zones.

Most of the plants grow in rich, black soil in MXP. Black soils can deter or retard the downward infiltration of water. Thus, agricultural activities do not influence the chemical composition of groundwater significantly due to the weak permeability of black soil. The dissolution of silicate minerals and carbonate minerals is the original source of groundwater. It is the most important factor dominating the chemical composition of groundwater in MXP. The saturation index (SI) data show that calcite, dolomite, gypsum and halite are in an unsaturated state (Table 4). The dissolution of minerals is the critical factor dominating the chemical composition of groundwater.

Table 4. Saturation index of groundwater samples in four zones.

Saturation Index	Z1			Z2			Z3			Z4		
	Max	Min	Mean	Max	Min	Mean	Max	Min	Mean	Max	Min	Mean
SI (Calcite)	0.37	−3.34	−1.95	0.26	−2.49	−1.25	0.08	−2.79	−1.56	0.44	−1.78	−1.45
SI (Dolomite)	0.58	−5.52	−2.24	0.29	−5.2	−2.64	−0.15	−4.52	−3.02	0.75	−3.84	−2.56
SI (Gypsum)	2.61	−3.75	−3.23	−0.61	−3.65	−2.74	−1.91	−3.56	−2.75	−1.31	−3.85	−2.85
SI (Halite)	−7.25	−9.62	−8.35	−7.29	−9.68	−8.89	−6.91	−9.54	−8.51	−6.23	−9.7	−8.88

Domestic sewage from rural areas near rivers is a significant factor affecting the chemical composition of groundwater in local areas. Nearly 70% of the population in MXP is living in the southern plain. That is why most of the special samples are located in Z3 and Z4.

Though MXP is an important grain production base in China, agricultural activities do not lead to the deterioration of groundwater quality. Thick black soil layers deter the infiltration of wastewater. Soil media are an important natural factor influencing the chemical composition.

5. Suggestions for Groundwater Management

Agricultural activities do not directly lead to deterioration of groundwater qualities, and agricultural non-point-source pollution does not occur in MXP. However, the discharge of domestic sewage has led to the abnormity of groundwater quality near river channels. Therefore, shallow groundwater far away from river channels and residential zones can be used as irrigation water based on food safety and human health.

The sampling depth of special samples usually ranges from 4 to 30 m. Chemical compositions of deep groundwater with a depth of more than 80 m are relatively stable (Figure 10), and major chemical ion concentrations hardly exceed the drinking water standard. Deep groundwater is the best choice for water supply.

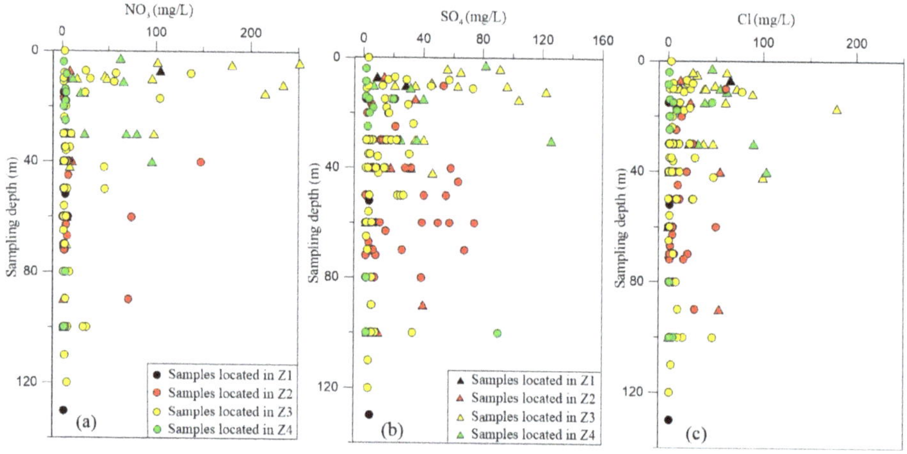

Figure 10. Variations of (**a**) NO_3, (**b**) SO_4^{2-}, (**c**) Cl with sampling depth (well depth).

Nitrification has led to the transformation of NH_4^+ into NO_3^-. Thus, samples in Z3 and Z4 are characterized by a low level of NH_4^+. However, samples in Z1 and Z2 are characterized by a high level of NH_4^+. Human activities, such as digging many wells and overpumping, can increase dissolved oxygen in groundwater and accelerate nitrification. Such activities may increase the concentration of NO_3^- in groundwater.

To avoid the deterioration of groundwater quality and protect groundwater resources, local governmental should (1) construct sewage disposal systems immediately in rural areas, (2) supervise the indiscriminate discharge of various wastewaters, (3) impose control on the construction of wells and groundwater pumping and (4) protect the thick black soil layer to retard the infiltration of wastewater.

6. Conclusions

(1) Dissolution of silicate minerals and carbonate minerals is the most important factor dominating the chemical composition of groundwater in MXP. Groundwater in MXP is predominantly Ca-HCO_3 in composition. Human activities have significantly influenced the chemical composition of groundwater in the residential zone near rivers. Thus, the NO_3 type and SO_4 type exist.

(2) Agricultural activities only slightly influence the chemical composition of groundwater in the northern plain, and human activities have significantly influenced the chemical composition of groundwater in the southern area. Groundwater in the southern plain is characterized by a high level of NH_4^+, which is related to the decomposition of organic matters in a reduction condition. The samples in the northern plain are characterized by a high level of NH_4^+, and it is necessary to avoid the transformation of a reduction condition into an oxidation environment.

(3) Due to the widespread distribution of thick black soils in MXP, agricultural non-point-source pollution does not occur. The discharge of domestic sewage mainly influenced the chemical composition of shallow groundwater. So, deep groundwater with a depth of more than 80 m is the best choice for water supply. In addition, shallow groundwater far away from river channels should be used as irrigation water.

(4) Government agencies should adopt some strategies to protect groundwater resources. These strategies include regulating the amount and type of fertilizers applied to

farmland, constructing sewage disposal systems in rural areas, and strengthening the supervision of the discharge of wastewater and construction of deep wells.

Author Contributions: Methodology, C.S.; software, C.S. and Z.C. (Zhongshuang Cheng); validation, C.S. and Z.L.; formal analysis, W.W. and C.S.; investigation, C.S., Z.Z. and Z.L.; data curation, Z.C. (Zhongshuang Cheng); writing—original draft preparation, C.S. and W.W.; writing—review and editing, C.S. and W.W.; visualization, Z.L. and C.S.; supervision, Z.C. (Zongyu Chen); project administration, C.S. and Z.C. (Zhongshuang Cheng); funding acquisition, C.S. All authors have read and agreed to the published version of the manuscript.

Funding: This study was financially supported by the National Natural Science Foundation of China (Grant No. 41602268) and China Geological Survey's project (Grant No. DD20160311).

Institutional Review Board Statement: Not applicable.

Informed Consent Statement: Not applicable.

Data Availability Statement: The data presented in this study are available on request from the corresponding author.

Conflicts of Interest: The authors declare no conflict of interest.

References

1. Foster, S.; Garduno, H.; Evans, R. Quaternary Aquifer of the North China Plain? Assessing and achieving groundwater resource sustainability. *Hydrogeol. J.* **2004**, *12*, 81–93. [CrossRef]
2. Turhan, A.; Ozmen, N. Influence of Chloride on Growth, Fruit Yield and Quality Parameters of Processing Pepper. *J. Agric. Nat.* **2021**, *24*, 1139–1144. [CrossRef]
3. Wang, Y.X.; Zheng, C.M.; Ma, R. Review: Safe and sustainable groundwater supply in China. *Hydrogeol. J.* **2018**, *26*, 1301–1324. [CrossRef]
4. Zhang, Y.; Wu, J.; Xu, B. Human health risk assessment of groundwater nitrogen pollution in Jinghui canal irrigation area of the loess region, northwest China. *Environ. Earth Sci.* **2018**, *77*, 273. [CrossRef]
5. Deng, L.; Xu, B.; Yang, X.T.; Hu, A.Y. Water quality and health risk assessment based on hydrochemical characteristics of tap and large-size bottled water from the main cities and towns in Guanzhong Basin, China. *Environ. Earth Sci.* **2021**, *80*, 139. [CrossRef]
6. Li, P.; Tian, R.; Xue, C.; Wu, J. Progress, opportunities, and key fields for groundwater quality research under the impacts of human activities in China with a special focus on western China. *Environ. Sci. Pollut. R.* **2017**, *24*, 13224–13234. [CrossRef] [PubMed]
7. Huang, G.X.; Liu, C.Y.; Sun, J.C.; Zhang, M.; Jing, J.H.; Li, L.P. A regional scale investigation on factors controlling the groundwater chemistry of various aquifers in a rapidly urbanized area: A case study of the Pearl River Delta. *Sci. Total Environ.* **2018**, *625*, 510–518. [CrossRef]
8. Yin, Z.Y.; Luo, Q.K.; Wu, J.F.; Xu, S.H.; Wu, J.C. Identification of the long-term variations of groundwater and their governing factors based on hydrochemical and isotopic data in a river basin. *J. Hydrol.* **2021**, *597*, 125604. [CrossRef]
9. Shi, X.; Wang, Y.; Jiao, J.J.; Zhong, J.; Wen, H.; Dong, R. Assessing major factors affecting shallow groundwater geochemical evolution in a highly urbanized coastal area of Shenzhen City, China. *J. Geochem. Explor.* **2018**, *184*, 17–27. [CrossRef]
10. El Alfy, M.; Alharbi, T.; Mansour, B. Integrating geochemical investigations and geospatial assessment to understand the evolutionary process of hydrochemistry and groundwater quality in arid areas. *Environ. Monit. Assess.* **2018**, *190*, 277. [CrossRef]
11. Hassen, I.; Hamzaoui-Azaza, F.; Bouhlila, R. Application of multivariate statistical analysis and hydrochemical and isotopic investigations for evaluation of groundwater quality and its suitability for drinking and agriculture purposes: Case of Oum Ali-Thelepte aquifer, central Tunisia. *Environ. Monit. Assess.* **2016**, *188*, 135. [CrossRef] [PubMed]
12. Liu, C.Q.; Li, S.L.; Lang, Y.C.; Xiao, H.Y. Using δ^{15}N- and δ^{18}O-values to identify nitrate sources in Karst ground water, Guiyang, Southwest China. *Environ. Sci. Technol.* **2006**, *40*, 6928–6933. [CrossRef] [PubMed]
13. Wang, H.; He, P.; Shen, C.; Wu, Z. Effect of irrigation amount and fertilization on agriculture non-point source pollution in the paddy field. *Environ. Sci. Pollut. R.* **2019**, *26*, 10363–10373. [CrossRef]
14. Mohammed, A.M.; Refaee, A.; El-Din, G.K.; Harb, S. Hydrochemical characteristics and quality assessment of shallow groundwater under intensive agriculture practices in arid region, Qena, Egypt. *Appl. Water Sci.* **2022**, *12*, 92. [CrossRef]
15. Lwimbo, Z.D.; Komakech, H.C.; Muzuka, A.N. Impacts of emerging agricultural practices on groundwater quality in kahe catchment, Tanzania. *Water* **2019**, *11*, 2263. [CrossRef]
16. Pulido-Bosch, A.; Rigol-Sanchez, J.P.; Vallejos, A.; Andreu, J.M.; Ceron, J.C.; Molina-Sanchez, L. Impacts of agricultural irrigation on groundwater salinity. *Environ. Earth Sci.* **2018**, *77*, 197. [CrossRef]
17. Ledesma-Ruiz, R.; Pastén-Zapata, E.; Parra, R.; Harter, T.; Mahlknecht, J. Investigation of the geochemical evolution of groundwater under agricultural land: A case study in northeastern Mexico. *J. Hydrol.* **2015**, *521*, 410–423. [CrossRef]

18. Van Geldern, R.; Schulte, P.; Mader, M.; Baier, A.; Barth, J.A.C.; Juhlke, T.R. Insights into agricultural influences and weathering processes from major ion patterns. *Hydrol. Process.* **2018**, *32*, 891–903. [CrossRef]
19. Merz, C.; Lischeid, G. Multivariate analysis to assess the impact of irrigation on groundwater quality. *Environ. Earth Sci.* **2019**, *78*, 274. [CrossRef]
20. Li, C.X.; Li, G.Z. Impact of China's water pollution on agricultural economic growth: An empirical analysis based on a dynamic spatial panel lag model. *Environ. Sci. Pollut. R.* **2020**, *28*, 6956–6965. [CrossRef]
21. Feng, W.; Lu, H.W.; Yao, T.C.; Guan, Y.L.; Xue, Y.X.; Yu, Q. Water environmental pressure assessment in agricultural systems in Central Asia based on an Integrated Excess Nitrogen Load Model. *Sci. Total Environ.* **2022**, *803*, 149912. [CrossRef] [PubMed]
22. Ayotte, J.D.; Szabo, Z.; Focazio, M.J.; Eberts, S.M. Effects of human-induced alteration of groundwater flow on concentrations of naturally-occurring trace elements at water-supply wells. *Appl. Geochem.* **2011**, *26*, 747–762. [CrossRef]
23. Wang, L.H.; Dong, Y.H.; Xie, Y.Q.; Song, F.; Wei, Y.Q.; Zhang, J.Y. Distinct groundwater recharge sources and geochemical evolution of two adjacent sub-basins in the lower Shule River Basin, northwest China. *Hydrogeol. J.* **2016**, *24*, 1967–1979. [CrossRef]
24. Xiao, Y.; Hao, Q.C.; Zhang, Y.H.; Zhu, Y.C.; Yin, S.Y.; Qin, L.M.; Li, X.H. Investigating sources, driving forces and potential health risks of nitrate and fluoride in groundwater of a typical alluvial fan plain. *Sci. Total Environ.* **2022**, *802*, 149909. [CrossRef] [PubMed]
25. Cao, Y.J.; Tang, C.Y.; Song, X.F.; Liu, C.M.; Zhang, Y.H. Characteristics of nitrate in major rivers and aquifers of the Sanjiang Plain. *J. Environ. Monit.* **2012**, *14*, 2624. [CrossRef]
26. Cui, X.S.; Zheng, Z.X.; Cheng, Z.S.; Su, C. Hydrochemical distribution characteristic and formation mechanism of shallow groundwater in the north of Muling-Xingkai Plain. *South-toNorth Water Transf. Water Sci. Technol.* **2018**, *16*, 146–153. (In Chinese)
27. Su, C.; Zhang, F.E.; Cui, X.S.; Cheng, Z.Y.; Zheng, Z.X. Source characterization of nitrate in groundwater using hydrogeochemical and multivariate statistical analysis in the muling-xingkai plain, northeast china. *Environ. Monit. Assess.* **2020**, *192*, 456. [CrossRef]
28. Xue, Y.Q.; Wu, J.C.; Ye, S.J.; Zhang, Y.X. Hydrogeological and hydrogeochemical studies for salt water intrusion the south coast of Laizhou Bay, China. *Groundwater* **2000**, *38*, 38–45. [CrossRef]
29. Beal, L.K.; Wong, C.I.; Bautista, K.K.; Jenson, J.W.; Banner, J.L.; Lander, M.A.; Gingerich, S.B.; Partin, J.W.; Hardt, B.; van Oort, N.H. Isotopic and geochemical assessment of the sensitivity of groundwater resources of Guam, Mariana Islands, to intra- and inter-annual variations in hydroclimate. *J. Hydrol.* **2019**, *568*, 174–183. [CrossRef]
30. Shuai, G.Y.; Shao, J.L.; Cui, Y.L.; Zhang, Q.L.; Guo, Y.T. Hydrochemical Characteristics and Quality Assessment of Shallow Groundwater in the Xinzhou Basin, Shanxi, North China. *Water* **2021**, *13*, 1993. [CrossRef]
31. Newman, B.D.; Havenor, K.C.; Longmire, P. Identification of hydrochemical facies in the Roswell Artesian Basin, New Mexico (USA), using graphical and statistical methods. *Hydrogeol. J.* **2016**, *24*, 819–839. [CrossRef]
32. Liu, J.T.; Hao, Y.J.; Gao, Z.J.; Wang, M.; Liu, M.X.; Wang, Z.Y.; Wang, S. Determining the factors controlling the chemical composition of groundwater using multivariate statistics and geochemical methods in the Xiqu coal mine, North China. *Environ. Earth Sci.* **2019**, *78*, 36. [CrossRef]
33. Blarasin, M.; Matiatos, I.; Cabrera, A.; Lutri, V.; Giacobone, D.; Becher Quinodoz, F.; Matteoda, E.; Eric, C.; Felizzia, J.; Giuliano Albo, J. Characterization of groundwater dynamics and contamination in an unconfined aquifer using isotope techniques to evaluate domestic supply in an urban area. *J. S. Am. Earth. Sci.* **2021**, *110*, 103360. [CrossRef]
34. Hakimi, Y.; Orban, P.; Deschamps, P.; Brouyere, S. Hydrochemical and isotopic characteristics of groundwater in the Continental Intercalaire aquifer system: Insights from Mzab Ridge and surrounding regions, North of the Algerian Sahara. *J. Hydrol. Reg. Stud.* **2021**, *34*, 100791. [CrossRef]
35. Yang, Q.C.; Li, Z.J.; Ma, H.Y.; Wang, L.C.; Martín, J.D. Identification of the hydrogeochemical processes and assessment of groundwater quality using classic integrated geochemical methods in the Southeastern part of Ordos basin, China. *Environ. Pollut.* **2016**, *218*, 879–888. [CrossRef] [PubMed]
36. Güler, C.; Kurt, M.A.; Alpaslan, M.; Akbulut, C. Assessment of the impact of anthropogenic activities on the groundwater hydrology and chemistry in Tarsus coastal plain (Mersin, SE Turkey) using fuzzy clustering, multivariate statistics and GIS techniques. *J. Hydrol.* **2012**, *414–415*, 435–451. [CrossRef]
37. Liu, S.; Tang, Z.H.; Gao, M.S.; Hou, G.H. Evolutionary process of saline-water intrusion in Holocene and late Pleistocene groundwater in southern Laizhou Bay. *Sci. Total Environ.* **2017**, *607–608*, 586–599. [CrossRef]
38. Liu, C.W.; Lin, K.H.; Kuo, Y.M. Application of factor analysis in the assessment of groundwater quality in the blackfoot disease area in Taiwan. *Sci. Total Environ.* **2003**, *313*, 77–89. [CrossRef]
39. Peng, L.; Shi, Q.D.; Wan, Y.B.; Shi, H.B.; Kahaer, Y.J.; Abudu, A. Impact of Flooding on Shallow Groundwater Chemistry in the Taklamakan Desert Hinterland: Remote Sensing Inversion and Geochemical Methods. *Water* **2022**, *14*, 483. [CrossRef]
40. Marandi, A.; Shand, P. Groundwater chemistry and the Gibbs Diagram. *Appl. Geochem.* **2018**, *97*, 209–212. [CrossRef]
41. Gaillardet, J.; Dupré, B.; Louvat, P.; Allègre, C.J. Global silicate weathering and CO_2 consumption rates deduced from the chemistry of large rivers. *Chem. Geol.* **1999**, *159*, 3–30. [CrossRef]
42. Rehman Qaisar, F.U.; Zhang, F.; Pant, R.R.; Wang, G.; Khan, S.; Zeng, C. Spatial variation, source identification, and quality assessment of surface water geochemical composition in the Indus River Basin, Pakistan. *Environ. Sci. Pollut. R.* **2018**, *25*, 12749–12763. [CrossRef] [PubMed]

43. Lang, Y.C.; Liu, C.Q.; Zhao, Z.Q.; Li, S.L.; Han, G.L. Geochemistry of surface and ground water in Guiyang city, China: Water/rock interaction and pollution in a karst hydrological system. *Appl. Geochem.* **2006**, *21*, 887–903. [CrossRef]
44. Ahmed, A.; Clark, I. Groundwater flow and geochemical evolution in the Central Flinders Ranges, South Australia. *Sci. Total Environ.* **2016**, *572*, 837–851. [CrossRef] [PubMed]
45. Carrasco-Cantos, F. Application of stable isotopes ($\delta^{34}S$-SO_4, $\delta^{18}O$-SO_4, $\delta^{15}N$-NO_3, $\delta^{18}O$-NO_3) to determine natural background and contamination sources in the Guadalhorce River Basin (southern Spain). *Sci. Total Environ.* **2015**, *506–507*, 46–57. [CrossRef]

water

MDPI

Article

Evaluation of Water Quality of Groundwater of Sanghar District, Sindh, Pakistan: Chemical and Multivariate Analysis

Abdul Qayoom Landar [1], Taj Muhammad Jahangir [1,2,3,*], Muhammad Yar Khuhawar [1,2,3], Muhammad Farooque Lanjwani [3] and Faheem Yar Khuhawar [4]

1 Institute of Advanced Research Studies in Chemical Sciences, University of Sindh, Jamshoro 76080, Pakistan; aqayoom1978@gmail.com (A.Q.L.); mykhuhawar@usindh.edu.pk (M.Y.K.)
2 High Tech Central Resource Laboratory, University of Sindh, Jamshoro 76080, Pakistan
3 Dr. M. A. Kazi Institute of Chemistry, University of Sindh, Jamshoro 76080, Pakistan; mfarooquechemist@yahoo.com
4 Department of Telecommunication, Mehran University of Engineering and Technology, Jamshoro 76062, Pakistan; faheem.khuhawar@faculty.muet.edu.pk
* Correspondence: tmj.iarscs@usindh.edu.pk; Tel.: +92-22-921-3213

Abstract: Sanghar District is located in the central part of Sindh Province and shares a boarder with India to the east. This work examines the water quality of the groundwater of three subdistricts, Sanghar, Khipro, and Jan Nawaz Ali, mostly used for human consumption, cattle farming, and irrigation. A total of 74 representative samples were collected and analyzed for 26 different parameters, including anions, cations, trace, and toxic elements. The total dissolved salts (TDSs) contained in 41 samples (55.4%), the major cations, K, Na, Mg, and Ca, in 44.6–93.2% of samples, and the major anions, Cl, HCO_3, and SO_4, in 68.6–81.0% of samples were within the permissible guidelines of the World Health Organization (WHO). The samples containing elements Cr (24.0%), Pb (29.7%), Ni (39.2%), Cd (40.5%), As (10.8%), and F (39.18%) were above the permissible limits of the WHO. The groundwater samples were examined for water quality index (WQI), contamination index (C_d), chronic daily intake indices (CDIs), hazard quotient indices (HQ), principal component analysis (PCA), piper diagrams, Gibbs diagrams, and cluster analysis to ascertain nature of the groundwater present in the study area. The samples were also examined for suitability for irrigation by sodium percentage (Na%), sodium adsorption ratio (SAR), Kelly's index (KI), permeability index (PI), and Wilcox diagrams. Samples in the range of 0 to 58.33% were inappropriate for irrigation.

Keywords: water quality; chemical assessment; groundwater; water quality index; health hazard

Citation: Landar, A.Q.; Jahangir, T.M.; Khuhawar, M.Y.; Lanjwani, M.F.; Khuhawar, F.Y. Evaluation of Water Quality of Groundwater of Sanghar District, Sindh, Pakistan: Chemical and Multivariate Analysis. *Water* **2024**, *16*, 856. https://doi.org/10.3390/w16060856

Academic Editors: Liangping Li and Guanxing Huang

Received: 5 December 2023
Revised: 11 January 2024
Accepted: 18 January 2024
Published: 16 March 2024

1. Introduction

Water is the most abundant chemical substance and is a basic source for the lives of people, plants, and animals on Earth. Water has a central role to play for every cell and organ in the human body. The main sources of water for human consumption are surface and groundwater [1]. Water may become hazardous if it does not meet the required standards for drinking, irrigation, and industries [2]. Groundwater is broadly used in rural areas as a source of drinking water, mostly in third-world countries; thus, the quality of groundwater needs to be tested for possible health effects [3]. A number of studies have been conducted to examine the quality of groundwater from many parts of the world, including Sindh Province of Pakistan. One study analyzed 218 samples from Sindh Province and evaluated them for irrigation and drinking purposes [4]. The results of the physicochemical analysis indicated that 17.8–62.8% of samples were not suitable for drinking. Higher values were reported from Southern Sindh due to the effects of sea intrusion. Lanjwani et al. [5] analyzed 21 samples from the district of Qambar Shahdadkot and reported water quality for human consumption and irrigation. The TDSs in 81% of samples, major cations in 50% of samples, and Cd, Pb, and Ni in 28.5–57.1% of samples exceeded the permissible limits of the WHO. Lanjwani et al. [6] analyzed 25 samples from Ratodero subdistrict, Larkana district, Sindh, for their physicochemical parameters. The results of the analysis based on the water quality index (WQI) indicated that 35% of samples were of the poor water category. Another work assessed [7,8] 425 samples for the water quality of the primary schools of selected districts of Sindh. Groundwater samples made up 62% of the total samples used for drinking in schools, and the results of the WQI indicated that 26% of schools were supplied poor to very poor drinking water. Rind et al. [9] analyzed 30 groundwater samples for 26 different parameters from Hala subdistrict, Matiari district, Sindh. Fluoride and arsenic were reported to be 40% and 43% above the WHO limits, and the results of the WQI suggested that 57.7% of samples were poor to very poor for drinking water. Another study [10] reported the contamination of arsenic in the roundwater of Indus Valley, Pakistan. Nearly 1200 samples were analyzed for a number of physicochemical parameters. The mobilization of arsenic in groundwater was related to elevated pH dissolution. Higher arsenic contents were reported near the Indus river and its tributaries. Khuhawar et al. [11] assessed the quality of the groundwater of the Thar Desert of Sindh Province, where 2193 samples were taken from dug wells. The TDSs of 57.5% samples were higher than 3000 mg/L, 14% were within 1500–3000 mg/L, and 27% were below 1500 mg/L. Ullah et al. [12] analyzed 61 samples from different sites in Sanghar District and analyzed for 12 different parameters, including arsenic. Arsenic was reported to be in the range of 5–25 μg/L. Bashed et al. [13] analyzed 39 groundwater samples for 11 different parameters from the subdistrict of Khipro, Sanghar District. The results of the analysis indicated that the majority of the samples were suitable for cultivation, except for a few samples. Another study [14] examined the groundwater quality of Hangu district, Pakistan, for the purposes of drinking and irrigation. The water quality was suitable for cultivation, except for 5% of the sampling sites. Another study [15] evaluated the groundwater quality of Northern Algeria using the WQI and GIS. The results indicated that nitrate ions exceeded the WHO limits in all water samples, and calcium, chloride, and sulphate occurred at alarming concentrations. Kaur et al. [16] analyzed 24 groundwater samples in Malawi, Southwestern Punjab, India, for major cations, anions, and other physicochemical parameters. The results indicated a major number of parameters above the Indian standards. Adimalla et al. [17] analyzed 105 groundwater samples from Central Telangana, India, for drinking and irrigation. The samples were suitable for drinking, but 51% and 71% exceeded the permissible limits for fluoride and nitrate, respectively. Li et al. [18] analyzed 74 groundwater samples for irrigation purposes from Pengyang, China. The quality of the groundwater for irrigation was reported to be excellent to good. Fifty groundwater samples were assessed for drinking water quality from Qorveh and Dehgolan, Kurdistan, Iran. The results based on the WQI indicated that 36% of the samples were excellent and 64% were good for drinking [19].

The present work critically examines the groundwater quality by analyzing 74 samples collected from 3 subdistricts of Sanghar District for drinking, domestic, and irrigation purposes. A total of 26 different physicochemical parameters were analyzed, and results were compared with WHO permissible limits. The results were analyzed on multivariate statistical procedures, including WQI, C_d, CDI, HQ, PCA, Gibbs diagrams, and coefficient of correlation (r), to ascertain the nature of groundwater available in the study area. The work also reports the quality of groundwater for irrigation.

Geography and Geology

Sanghar District lies between 25°58′13″ N latitude and 69°24′4″ E longitude. The area of the district is 10,608 km^2 with a total population of about 2.049 million [20]. It is located at the Centre of Sindh and has a common border with India in the east. The Sanghar town is the district headquarter and is at a distance of about 56 km from Benazirabad (Nawabshah) and Mirpurkhas cities. The main industry of the district is agriculture. The district is administratively divided into six subdistricts: Jam Nawaz Ali, Khipro, Sanghar, Shahdadpur, Singhro, and Tando Adam Khan. The district is also divided into 70 union councils. The areas of Sanghar District mainly consists of semiarid land, a part of the great Thar desert (mostly in the subdistrict of Khipro), and cropped areas, irrigated by the Mithrao and Nara canal system, connected with the river Indus. The main crops cultivated are rice, wheat, and cotton. The climatic conditions of district Sanghar are subtropical. The average monthly temperature varies within 21.2–39.11 °C with an average yearly temperature of 32.3 °C. It receives about 15.87 millimeters of precipitation annually.

Sanghar District is located in the lower Indus basin; the stratigraphic sequence drilled in the southern part of the Indus basin indicates from Jurassic to recent and the sediments source is the Indian craton [21]. The lower Goru formation of the southern Indus basin was deposited in a shallow to deep marine environment, with seven distinct lithological units composed of various sand and shale intervals [22]. The study area has composition of silts, sand, and clay of tertiary rocks [23]. The significance of soil differs from one place to another by sand and clay, with approximately 250 to 450 feet thickness of sandy layers under a shallow aquifer [24]. They are satisfactory to intermediate micaceous sands having well-sorted bands and lenses of clay and silt [25]. The water tables have a mean depth 3.93 m and range between 1.39–12.76 m [26]. Shakir et al. examined Cretaceous sands of Khipro, Sanghar District area, by integrated seismic structural analysis, and studied Naimat Basal 01 and Siraj South 01 wells using petrophysical interpretation [27].

2. Materials and Methods

2.1. Sample Collection

The 74 samples were collected from three subdistricts of the districts Sanghar (32), Khipro (30), and Jam Nawaz Ali (12), from villages and populated towns randomly from most of the union councils to cover the study area, where the groundwater for human needs is used for drinking and agriculture (Table 1) (Figure 1). The water samples were gathered from hand pumps, motor pumps, and tube wells. Two clean plastic bottles (1.5 L) were filled from each sampling station, after allowing water from the source to drain for 5 min. A bottle was used for physicochemical analysis, and another was used for metal analysis [28,29]. The bottles for metal analysis were acidified with 1.5 mL hydrochloric acid or nitric acid. Hand pumps and motor pumps were bored at 75–150 feet and tube wells at 150–350 feet depth [30].

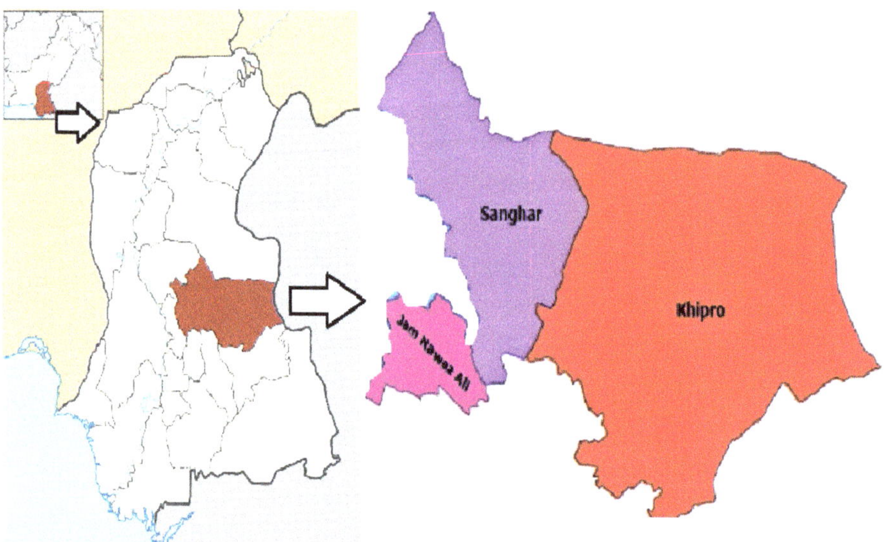

Figure 1. Pakistan, Sindh, and Sanghar subdistricts.

Table 1. Water analysis of Sanghar with descriptive statistics (n = 74).

Parameters	Minimum	Maximum	Mean	Std. Deviation	Std. Error
pH	6.74	8.65	7.7284	0.34037	0.03957
Conductivity µs/cm	346	5730	1675	1231	143
Salinity g/L	0.2	3.1	0.858	0.6697	0.0779
TDS mg/L	221	3667	1072	787	91
NO_2 µg/L	0.09	400	42	95	11
NO_3 mg/L	0.65	6.39	2.8199	1.26219	0.14673
T.PO_4 mg/L	0.26	7.76	1.4619	1.40125	0.16289
O.PO_4 mg/L	0.14	3.07	0.7942	0.75109	0.08731
T.H mg/L	1	800	244	170	19
Na mg/L	20	1349	185	254	29
K mg/L	5	103	20	21	2.5
Ca mg/L	44	272	98	45	5
Mg mg/L	10	149	36.39	27	3.212
Chloride mg/L	28	1230	211	227	26
Alkalinity mg/L	104	820	265	127	14
SO_4 mg/L	27	1319	216	256	29
Mn µg/L	3	282	36	40	4
As µg/L	0	10	4	7.8	0.9
Fe µg/L	1.2	256	27	41	4
Co µg/L	9.8	48	26	9	1
Cu µg/L	0	113	24	38	4
Ni µg/L	0	155	45	38	4
Pb µg/L	0	66	14	15	2
Cd µg/L	0	107	19	27	3
F mg/L	0.1	26	4	5.4	0.6
Cr µg/L	0	96	29	31	3.6
KI meq/L	0.2	76	24	22	2.6
SAR meq/L	0.4	4.9	1.3	0.84	0.0984
PI meq/L	67	112	94	8.6	1.0068
Na% meq/L	27	82	52	11.5	1.3386

2.2. Sample Analysis

The samples were analyzed for 26 different parameters (Table S1) following standard analytical procedures. The analyses were carried out at least in triplicate, and average values are reported. The electrical conductivity (EC), salinity, total dissolved salts (TDS), and pH were measured utilizing an Orion 115 conductivity meter (Orion Pvt Ltd., Boston, MA, USA). Chloride, alkalinity, and total hardness were determined by titrimetric methods [31]. Nitrate, nitrite, total phosphate, orthophosphate, sulphate, and fluoride were measured using spectrophotometric procedures utilizing a Hitachi 220 double beam spectrophotometer (Hitachi (Pvt) Ltd., Tokyo, Japan). Na, K, Ca, Mg, Cu, Ni, Mn, Co, Fe, Cd, Pb, and Cr were determined by flame atomic absorption spectrometry (Perkin Elmer, AA 800, Singapore) at the conditions recommended by the manufacturer. The equipment was controlled by the computer with Winlab 36 software. The analyses were carried out in triplicate (n = 3) with integration time of 4 s and time of delay of 4 s. For the analysis of Ca, Mg, K, and Na, the samples were diluted 10–25 times with distilled water. For determination of trace elements, the samples were preconcentrated by the factor of 10 by evaporation of water at 80–90 °C on an electrical hot plate. The solutions were filtered if required before analysis. The arsenic was estimated using the E. Merck kit method.

The standard deviation (SD), relative standard deviation (RSD), mean, scatter diagrams, Gibbs diagrams, and Wilcox diagrams were drawn on the Excel 2013 program. Coefficient of correlation®, principal component analysis (PCA), and cluster analysis were carried out on SPSS 20 (SPSS Inc., Chicago, IL, USA) program. The piper diagrams were drawn on Aquachem software 11.

2.3. Water Quality Index (WQI)

WQI was calculated using the reported procedure [32]. The parameters for water quality were ascribed the weight (wi) corresponding to their hazardous effects. The weight of 5 was given to As, Pb, and Cd owing to their expressing effects on water quality [28,29]. The K, Na, and Ca were allowed a weight (wi) of 3. The relative weight (Wi) is accounted from the following relation, where n designates the number of parameters [33,34].

$$Wi = wi/\Sigma^{n}{}_{i=1}\ wi. \quad (1)$$

Quality rating (Qi) is determined by dividing the concentration of each sample (Ci) with the WHO standard (Si) of the corresponding parameter for drinking water, then multiplying the equation by 100.

$$Qi = (Ci/Si) \times 100. \quad (2)$$

Then,

$$WQI = \Sigma\ Wi \times Qi. \quad (3)$$

2.4. Gibbs Diagrams

The Gibbs diagrams are used to calculate the sources of anions (Cl^-, HCO_3^-) and cations (Na^+, Ca^{2+}, K^+) in groundwater by plotting against TDS in the Origin program. The values are evaluated with the aid of the following formula:

$$Anions = Cl^-/(Cl^- + HCO_3^-). \quad (4)$$

$$Cations = (Na^+ + K^+)/(Na^+ + K^+ + Ca^{2+}). \quad (5)$$

All the concentrations are in milliequivalent (m.eq/L).

2.5. Contamination Index (C_d)

Contamination index (C_d) was calculated by Backman et al.'s [35] method to evaluate the harmful effect of the parameters on the human health and environment, by comparing the observed values of the parameters with WHO permissible limits. The C_d is estimated by using the following Formula (6) [5]:

$$Cfi = (CAi/CNi) - 1. \quad (6)$$

where Cfi = Contamination factor of the ith parameters; CAi = observed value of ith parameter; CNi = higher permissible level of ith parameter by the WHO for human consumption.

2.6. Chronic Daily Intake Indices (CDIs)

Metal ions enter the human body by different ways: food, drinking, and inhalation. However, by the drinking water is considered more important. The CDIs of Sanghar District were estimated by using the following relation (7) [36]:

$$CDI = C \times DI/BW. \quad (7)$$

where C stands for the values of the parameters, DI represents daily intake (2 L/day), and BW is the average body weight (72 kg).

2.7. Hazard Quotient Indices (HQ)

The HQ is calculated for noncarcinogenic effects of metal ions on the human body. HQ was estimated from the following relation (8):

$$HQ = CDI/RfD. \quad (8)$$

CDI stands for chronic daily intake indices and RfD for reference doze of oral harmfulness. The values described by US EPA are for Cd 5.0×10^{-4}, Ni 2.0×10^{-2}, Cu 2.7×10^{-2}, Pb 3.5×10^{-3}, Cr 3.0×10^{-3}, Zn 3.0×10^{-1}, As 3.0×10^{-4}, Mn 2.4×10^{-2}, and Fe 3.0×10^{-1} mg/kg-day [37]. The value of HQ less than 1 is considered safe, and an HQ value above 1 is considered unsafe [34].

2.8. Suitability of the Groundwater for Irrigation

The parameters sodium percent (% Na), sodium adsorption ratio (SAR), Kelly's index (KI), and permeability index (PI) were calculated by using the following Equations (9)–(12). All the concentrations were in m.eq/L.

$$Na\% = [(Na^+ + K^+)/(Ca^{2+} + Mg^{2+} + Na^+ + K^+)] \times 100. \quad (9)$$

$$SAR = [Na^+]/[(Ca^{2+}\ Mg^{2+})/2]^{1/2}. \quad (10)$$

$$KI = Na^+/Ca^{2+} + Mg^{2+} \quad (11)$$

$$PI = [(Na^+ + HCO_3^-)/(Na^+ + Ca^{2+} + Mg^{2+})] \times 100. \quad (12)$$

3. Results and Discussion

The results describe the physicochemical analysis of 74 groundwater samples from the Sanghar (32), Khipro (30), and Tando Jam Nawaz Ali (12) subdistricts of Sanghar District, Sindh.

3.1. Sanghar Subdistrict

The samples collected were from hand pumps, 20 (62.5%), motor pumps, 8 (25%), and tube wells, 4 (12.5%). The water from hand pumps and motor pumps was mostly used for the drinking of humans as well as cattle, but the water from tube wells was consumed for irrigation. The results of analysis are summarized in Table 1 and the results with mean, minimum–maximum, and standard deviation are summarized in Table 1. All the results obtained were compared with the permissible limits of WHO [36]. The pH of the Sanghar subdistrict ranged from 6.74–8.65, and two samples (6.2%) crossed the upper limit of 8.5, which may be due to the presence of carbonate salts in the groundwater. More samples were slightly on the alkaline side. EC and TDS were observed within 399–5730 µS/cm and 255.4–3667.2 mg/L. EC of 16 (50%) and TDS of 13 (40.6%) samples crossed the WHO limits (EC 1562.5 µS/cm and TDS 1000 mg/L). Higher amount of TDS gives an unpleasant taste to the water. Total hardness was between 120–788 mg/L, and only one sample crossed the limit 500 mg/L. The high concentration of total hardness is objectionable for aesthetic sense and may cause gastrointestinal problems. The alkalinity was indicated within 120–362 mg/L, and 8 (25%) samples crossed the permissible limit of 300 mg/L. High amounts of alkalinity may cause dry skin by dissolving skin oil. The chloride is highly soluble in water and is distributed in groundwater as salts of sodium, potassium, and calcium. The sources of sulphate may be associated with agriculture activities, including the use of the fertilizers. The chloride and sulphate were observed between 28–750 mg/L and 27–762 mg/L, and 13 (40.6%) for chloride and 5 (15.6%) for sulphate were above the limit for both 250 mg/L. Based on average values, $Cl^- > SO_4^{2-} > HCO_3^-$. The high concentration of nitrate-N may cause a disease commonly known as blue baby in children. High concentration of nitrite-N may cause cancer due to the formation of N-nitroso and nitrosamine compounds [38]. The application of fertilizers for irrigation and the use of detergents for washing purposes may be responsible for the higher concentration of phosphate-P in groundwater. The results of nitrate, nitrite, total phosphate, and orthophosphate for all samples were observed within WHO permissible limits (Table S1) [39]. Sodium is required for the muscles and nerve functioning, but the higher concentration of sodium may be responsible for high blood pressure and kidney damage in humans. High concentration of potassium may be due to mineral dissolution, agricultural activities, and plant material decomposition. Sodium and potassium were observed within 22–390 mg/L and 6–117 mg/L. The sodium in 13 (40.6%) and potassium in 21 (65.6%) samples were above the permissible limits of 200 mg/L and 12 mg/L, respectively. Calcium and magnesium have similar functions in the human body. High concentration of magnesium may cause vomiting or diarrhea. The results of calcium and magnesium were within 44–272 mg/L and 10–130 mg/L, and calcium in 01 (3.1%) and magnesium in 5 (15.6%) samples crossed the limits of 150 mg/L and 75 mg/L, respectively. Based on the average values, $Na^+ > Ca^{2+} > Mg^{2+} > K^+$. Iron and copper are essential elements for humans, but higher intake of copper may be responsible for neurological or liver complications. Manganese is mostly present in groundwater together with iron, but high amounts of manganese may account for lungs, eyes blindness, and nerve damage. The concentrations of Fe, Cu, Mn, and Co for all samples were within the limits of WHO. Chromium is present in a number of oxidation states, but chromium (VI) is more present in natural waters. It can enter cell tissues and may prove hazardous. Lead is a main pollutant and is considered as toxic. Lead is harmful for multiple body systems and can particularly affect children [40]. Cadmium is also toxic and can act as an agent to cause cancer. Nickel affects the absorption of iron in the human body, and higher concentrations of nickel in the human body may be responsible for causing vomiting and headaches. The concentrations of Cr for 15 (46.8%), Pb for 10 (31.2%), Cd for 15 (46.8%), and Ni for 13 (40.6%) samples crossed WHO guidelines (Table S2). The arsenic in inorganic form is widely present on Earth. The arsenic contamination of groundwater is reported from different countries, including Pakistan [41]. The initial poisonous effects of arsenic are skin and nail abnormalities like hyperpigmentation. The arsenic concentration in 3 (9.4%) crossed the limit of 10 µg/L. Fluoride at low levels is beneficial for dental caries, but fluoride concentration above 1.5 mg/L in water causes dental fluorosis. The fluoride concentrations above 4 mg/L in water causes bones fluorosis, where the skeleton is affected. The concentration of fluoride was found to be between 0.21–10.7 mg/L and the fluoride in 16 (50%) samples was above the limit of 1.5 mg/L. The values of standard deviation (SD) for a number of parameters present are high (Table 1), indicating that a wide variation in the parameter may be due to the changes in the geological settings within the study area.

Suitability of Groundwater for Irrigation

The groundwater of Sanghar subdistrict was evaluated for irrigation based on Na%, SAR, KI, and PI. The sodium% (Na%) was calculated within 21.48–60.46%. Samples 7 (21.8%) were of good quality for irrigation with Na% within 20–40%. Samples 24 (74%) were permissible with Na% within 40–60%. Sample 01 (3.1%) was in the uncertain category with Na% 60–80 (Table S3). The results of sodium adsorption ratio (SAR) for 32 groundwater samples were observed within 0.47–2.84 and the indicated values were below the required sodium absorption ratio value of 6. All the samples based on SAR values were suitable for irrigation. The Kelly's index (KI) of samples were calculated within 0.24–1.42, and 75% of samples indicated KI values less than 01 and were considered suitable for irrigation. However, 25% samples showed KI values more than 01 and were not considered suitable for irrigation. The permeability index (PI) was calculated within 23.83–142.09. A total of 62.5% of the samples were observed with PI values above 75% and were considered as class I, good for irrigation. A total of 21.9% samples were within PI values between 75–50% and were considered as class II, suitable for irrigation. However, 15.6% of the samples indicated PI values below than 50% and were classified as class III, as inappropriate for agriculture.

3.2. *Khipro Subdistrict*

The groundwater (30) samples were collected from Khipro subdistrict, comprising 10 samples from motor pumps, 18 from hand pumps, and 2 from dug wells. The samples were gathered from villages and towns used for drinking. The pH of all samples was within WHO guidelines of 6.5–8.5 [40]. The EC and TDS were observed between 385–5710 μS/cm and 246–3654 mg/L. EC and TDS of 12 (40%) samples were observed above WHO limits (EC 1500 μg/cm and TDS 1000 mg/L). The total hardness was detected within 140–800 mg/L, and 3 (10%) samples were found above the limit of 500 mg/L. The alkalinity was indicated between 138–820 mg/L, and 7 (23.3%) samples crossed the limit (300 mg/L). Chloride and sulphate were observed within 40–1230 mg/L and 32–1319 mg/L. The chloride for 9 (30%) and sulphate for 6 (20%) samples crossed the limit of 250 mg/L. Based on the average values, $SO_4^{2+} > Cl^- > HCO_3^-$ Table S1). Nitrate in all the samples was within permissible limits, but nitrite in 01 sample, orthophosphate in 02 samples, and total phosphates in 04 samples crossed the limits of WHO. The sodium was detected within 22–1349 mg/L and 9 (30%) samples were above the limit of 200 mg/L. Potassium was found between 5–103 mg/L and 15 (50%) samples crossed the limit of 12 mg/L. The calcium was observed within 46–260 mg/L and 4 (13.3%) samples were beyond the limit of 150 mg/L. The magnesium was within 10–149 mg/L and 3 (10%) samples were above the limit of 75 mg/L. Based on the average values, $Na^+ > Ca^{2+} > Mg^{2+} > K^+$ (Table S2). The concentrations of Fe, Cu, Mn, and Co for all the (30) samples were within the permissible limits, but 9 (30%) samples for Cr and Pb and 11 (36.6%) samples for Ni and Cd crossed the limits of WHO for drinking water. The arsenic contents in 3 (10%) samples crossed the permissible limit of 10 μg/L. The contents of fluoride in groundwater samples varied between 0.11–18.0 mg/L, and 10 (33.3%) samples were above the acceptable limit of 1.5 mg/L. The area is devoid of industrial activities and the livelihood of most of the inhabitants depends on the agriculture. Thus, high concentrations of salts and fluoride may be due to geological reasons.

Quality of the Water for Irrigation

The Na% was calculated within 18.3–76.55%, and 9 (30%) samples were within the range of 20 to 40%, indicating good quality. Eighteen samples (60%) were in the range of 40 to 60% and were acceptable for irrigation. Three (10%) samples were found within 60–80% and were indicated as doubtful. The SAR values were observed within 0.39–6.26. A total of 29 (96.6%) samples were within acceptable limits of SAR (6) for irrigation, and only one sample crossed the limit. The KI values were calculated within 0.19–3.3, and 76.7% of samples indicated KI values less than 1, and were good for irrigation. A total of 23.3% samples had KI values more than 01 and were not suitable. PI of the samples were indicated within 19.75–312.79, and 43.3% of samples were in class I, with PI values greater than 75%. A total of 36.7% of samples were within class II, with PI values of 50–75% (Table S3). However, 20% samples were in class III, with PI values less than 50%, and were inappropriate for irrigation.

3.3. *Jam Nawaz Ali Subdistrict*

Twelve samples were collected from Jam Nawaz Ali subdistrict: 3 from motor pumps, 01 from dug well, and 8 from hand pumps. Two samples were gathered from Jam Nawaz Ali town and 10 samples from different villages, mostly used for drinking purposes. The pH of all the samples was observed within the limits 6.5–8.5. EC and TDS of the samples were observed within 346–2798 μS/cm and 221–1791 mg/L. Seven (58.3%) samples for EC and eight (66.6%) samples for TDS were within WHO limits (Table S1). The total hardness and alkalinity were noted within 126–400 mg/L and 104–304 mg/L, respectively. All the samples for total hardness and 11 (91.2%) samples for alkalinity were within limits of drinking water. The chloride and sulphate were indicated between 35–257 mg/L and 43–414 mg/L, respectively. Eleven (91.2%) samples for chloride and nine (75%) samples for sulphate were within permissible limits of 250 mg/L for both. The nitrate and nitrite for all samples and orthophosphate and total phosphate for 10 (83.3%) samples were within limits. Sodium and potassium were observed within 20–272 mg/L and 5–32 mg/L, respectively. Na in 9 (75%)

and K in 7 (58.3%) samples were within acceptable limits. Similarly, calcium and magnesium were indicated within 45–140 mg/L and 13–50 mg/L. The Ca and Mg contents in all the samples were within permissible limits for drinking water. The concentrations of Fe, Cu, Mn, and Co in all (12) samples were in acceptable limits, but concentrations of Cr and Ni in 7 (58.3%), Ni in 9 (75%), and Cd in 8 (66.6%) samples were within agreeable limits for drinking. The arsenic contents in 10 samples (83.3%) were within 10 µg/L, and fluoride in 9 (75%) samples were within permissible limits of 1.5 mg/L (Table S2).

Water Quality for Irrigation

The Na% for Jam Nawaz Ali subdistrict varied from 18.42–47.04. Seven (58.3%) samples were within Na% 20–40 and were of good quality. Five (41.66) samples were calculated within Na% 40–60 and were considered of acceptable quality for irrigation (Table S3). The SAR values varied from 0.52–2.21 and were within the acceptable limit of 6 for irrigation. The KI values were calculated between 0.88–3.8. 4 (33.3%) and samples indicated that KI values were less than 1 and were acceptable for irrigation. Eight (66.66%) samples showed KI values more than 1 and were doubtful for irrigation. PI values for the samples were observed within 84–106 and were above 75%. All the samples were considered good for irrigation.

3.4. *Water Quality Index (WQI)*

WQI provides a numerical value that indicates the quality of drinking water. The calculated values of WQI are categorized into five groups for drinking water: excellent < 50, good 50–100, poor 100–200, very poor 200–300, and unfit > 300 [42]. The values of WQI for the collected samples (n = 74) from Sanghar, Khipro, and Jam Nawaz Ali subdistricts of district Sanghar were within 22.47–374.25 (Table 2). The samples were divided into categories: excellent, 26 (35.14%), good, 31 (41.89), poor, 12 (16.22%), very poor, 2 (2.70%), and unfit, 3 (4.05%). The results indicated that more samples were present in excellent to good water categories for drinking.

Table 2. Water Quality Index of Sanghar District.

Sanghar Sub District			Khipro Sub District			Jam Nawaz Ali Sub District		
Sample ID	**WQI**	**State**	**Sample ID**	**WQI**	**State**	**Sample ID**	**WQI**	**State**
S1	75	Poor	K1	49	Good	J1	43	Good
S2	48	Good	K2	42	Good	J2	112	Unfit
S3	212	Unfit	K3	65	Poor	J3	74	Poor
S4	46	Good	K4	49	Good	J4	83	Very Poor
S5	22	Excellent	K5	105	Unfit	J5	108	Unfit
S6	128	Unfit	K6	374	Unfit	J6	35	Good
S7	99	Very Poor	K7	345	Unfit	J7	19	Excellent
S8	91	Very Poor	K8	323	Unfit	J8	81	Very Poor
S9	119	Unfit	K9	96	Very Poor	J9	42	Good
S10	65	Poor	K10	65	Poor	J10	64	Poor
S11	37	Good	K11	96	Very Poor	J11	30	Good
S12	29	Good	K12	67	Poor	J12	34	Good
S13	117	Unfit	K13	51	Poor			
S14	167	Unfit	K14	70	Poor			
S15	43	Good	K15	98	Very Poor			

Table 2. *Cont.*

Sanghar Sub District			Khipro Sub District			Jam Nawaz Ali Sub District		
Sample ID	WQI	State	Sample ID	WQI	State	Sample ID	WQI	State
S16	55	Poor	K16	30	Good			
S17	73	Poor	K17	64	Poor			
S18	38	Good	K18	135	Unfit			
S19	69	Poor	K19	28	Good			
S20	73	Poor	K20	53	Poor			
S21	57	Poor	K21	57	Poor			
S22	28	Good	K22	34	Good			
S23	81	Very Poor	K23	43	Good			
S24	80	Very Poor	K24	134	Unfit			
S25	107	Unfit	K25	64	Poor			
S26	37	Good	K26	44	Good			
S27	39	Good	K27	50	Good			
S28	33	Good	K28	52	Poor			
S29	73	Poor	K29	247	Unfit			
S30	51	Poor	K30	128	Unfit			
S31	42	Poor						
S32	102	Unfit						

3.5. Contamination Index (C_d)

The individual parameters above the permissible limit of WHO are added up to obtain the contamination index (C_d). The values of parameters indicate whether the sample is hazardous for drinking purposes. The C_d values calculated for all 74 samples varied within 0–42.8 (Table 3). The 59 (79.7%) samples had C_d values less than 1, and were considered suitable for drinking. Seven samples (9.4%) had C_d values within 3 and were considered moderately acceptable for drinking. Eight (10.8%) samples were above the value of 3 and were not considered suitable for drinking.

Table 3. Water analysis of Sanghar for contamination index.

Sanghar Sub District		Khipro Sub District		Jam Nawaz Ali Sub District	
Sample ID	Contamination Index	Sample ID	Contamination Index	Sample ID	Contamination Index
S1	−1	K1	−8	J1	−5
S2	−7	K2	−10	J2	−0.7
S3	19	K3	−7	J3	−2
S4	−7	K4	−9	J4	−1
S5	−9	K5	−2	J5	0.7
S6	8.5	K6	42	J6	−7.3
S7	−1	K7	38	J7	−9.2
S8	0.2	K8	37	J8	0.2
S9	2	K9	−0.101	J9	−7
S10	−2	K10	−0.5	J10	−3
S11	−7	K11	−2	J11	−8

Table 3. *Cont.*

Sanghar Sub District		Khipro Sub District		Jam Nawaz Ali Sub District	
Sample ID	**Contamination Index**	**Sample ID**	**Contamination Index**	**Sample ID**	**Contamination Index**
S12	2	K12	−7	J12	−9
S13	0.85	K13	−8		
S14	11	K14	−4		
S15	−7	K15	−2		
S16	−5	K16	−13		
S17	−0.2	K17	−5		
S18	−10	K18	1		
S19	−3	K19	−10		
S20	−2	K20	−8		
S21	−8	K21	−8		
S22	−11	K22	−6		
S23	3	K23	−8		
S24	−1	K24	2		
S25	2	K25	−3		
S26	−3	K26	−10		
S27	−8	K27	−8		
S28	−9	K28	−8		
S29	−0.4	K29	16		
S30	−5	K30	1		
S31	−9				
S32	2				

3.6. Gibbs Diagrams

The Gibbs diagrams [43] are used to assess the sources of cations and anions (m.eq/L) against TDS (mg/L). The Gibbs diagrams are mostly used to examine the relationships between water quality and aquifer characteristics. The allotment of cations or anions versus TDS are used to evaluate the dominance of rock, evaporation, or precipitation in Gibbs diagrams (Figure 2). The cations and anions were predominantly within rock–water dominance and a few samples (samples 3, 4.0%) were within evaporation dominance. The samples within precipitation dominance were not observed. The rock–water dominance showed synergy between the chemistry of the rocks and leaching of the water under the subsurface.

3.7. Chronic Daily Intake Indices (CDIs)

Human health risk assessment is determined from the magnitude of negative effects on human health due to the exposure to heavy metals, mainly through the path of drinking water. The CDI values for the drinking water of three subdistricts of Sanghar District for iron varied from 3.3×10^{-5} to 7.1×10^{-3} with an average value of 7.7×10^{-4}, copper 0–3.1×10^{-3} with an average value of 6.7×10^{-4}, cadmium 0–2.97×10^{-3} with an average value of 5.3×10^{-4}, chromium 0–2.66×10^{-3} with an average value of 8.2×10^{-4}, manganese 8.3×10^{-5} to 7.83×10^{-3} with an average value of 1.03×10^{-3}, nickel 0–4.3×10^{-3} with an average value of 1.27×10^{-3}, lead 0–1.87 with an average value of 0.40, and arsenic 0–6.94×10^{-4} with an average value of 1.16×10^{-4} μg/kg-day. The CDI values of the elements decreased in the following order: Pb > Ni > Mn > Cr > Fe > Cu > Cd > As. The mean values of CDI of all the elements were below 1, but the CDI values for Pb at some locations were higher than 1 (Table 4). The area is mostly based on agricultural lands, and high value of CDI for Pb at some locations may be due to geological reasons owing to rock–water interactions.

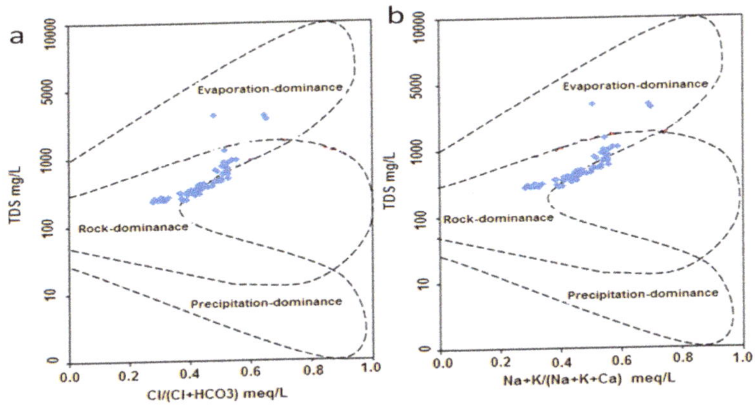

Figure 2. Gibbs diagram: (**a**) $Cl/(Cl + HCO_3)$ vs. TDS, (**b**) $Na + K/(Na + K + Ca)$ vs. TDS.

Table 4. Water analysis of Sanghar for chronic daily index (CDI).

Samples No.	Fe	Cu	Cd	Cr	Mn	Ni	Pb	As
S1	5.80×10^{-4}	1×10^{-3}	1×10^{-4}	1×10^{-3}	1.3×10^{-3}	$4. \times 10^{-4}$	2.14×10^{-1}	1×10^{-4}
S2	5.30×10^{-4}	1×10^{-4}	0	1.9×10^{-3}	1.9×10^{-3}	3×10^{-4}	0	0
S3	3.22×10^{-3}	0	7×10^{-4}	0	2.2×10^{-3}	2.7×10^{-3}	1.92×10^{-1}	0
S4	7.12×10^{-3}	5×10^{-4}	0	1.1×10^{-3}	6×10^{-4}	1×10^{-3}	2.44×10^{-1}	0
S5	1.74×10^{-3}	2.50×10^{-3}	0	1.3×10^{-3}	1.3×10^{-3}	0	1.97×10^{-1}	7×10^{-4}
S6	3.70×10^{-4}	1×10^{-4}	7×10^{-4}	1.5×10^{-3}	1.4×10^{-3}	2.1×10^{-3}	12.2×10^{1}	0
S7	8.0×10^{-4}	2×10^{-4}	3×10^{-4}	0	4×10^{-4}	2.4×10^{-3}	3×10^{-1}	0
S8	1.57×10^{-3}	2.5×10^{-3}	2×10^{-4}	1.8×10^{-3}	1.9×10^{-3}	1.5×10^{-3}	2.47×10^{-1}	0
S9	2.51×10^{-3}	0	0	1.5×10^{-3}	2.2×10^{-3}	3.1×10^{-3}	0	0
S10	0	2×10^{-4}	0	1.1×10^{-3}	6×10^{-4}	2.6×10^{-3}	9.64×10^{-1}	1×10^{-4}
S11	0	0	0	2×10^{-3}	1.3×10^{-3}	4×10^{-4}	3.44×10^{-1}	3×10^{-4}
S12	9.20×10^{-4}	1×10^{-4}	0	2.7×10^{-3}	1.4×10^{-3}	5×10^{-4}	2.03×10^{-1}	1×10^{-4}
S13	1.41×10^{-3}	1×10^{-4}	3×10^{-4}	0	4×10^{-4}	1.5×10^{-3}	1.78×10^{-1}	0
S14	3.50×10^{-4}	1×10^{-4}	5×10^{-4}	1.2×10^{-3}	3×10^{-4}	2.6×10^{-3}	9.58×10^{-1}	0
S15	7.30×10^{-4}	2×10^{-4}	0	1.3×10^{-3}	5×10^{-4}	3×10^{-4}	2.17×10^{-1}	1×10^{-4}
S16	2.70×10^{-4}	3.1×10^{-3}	0	9×10^{-4}	2×10^{-4}	1.5×10^{-3}	2.33×10^{-1}	1×10^{-4}
S17	1.20×10^{-4}	2.7×10^{-3}	3×10^{-4}	1.4×10^{-3}	7×10^{-4}	1.3×10^{-3}	8.19×10^{-1}	1×10^{-4}
S18	8.70×10^{-4}	1×10^{-4}	0	0	8×10^{-4}	0	1.19×10^{-1}	0
S19	3.30×10^{-4}	1×10^{-4}	0	0	2.9×10^{-3}	4.3×10^{-3}	7.75×10^{-1}	7×10^{-4}
S20	5.60×10^{-4}	2×10^{-4}	1×10^{-4}	1.9×10^{-3}	1.8×10^{-3}	4×10^{-4}	2.06×10^{-1}	3×10^{-4}
S21	2.40×10^{-4}	1×10^{-4}	1×10^{-4}	0	1.3×10^{-3}	4×10^{-4}	1.64×10^{-1}	0
S22	2.70×10^{-4}	2×10^{-4}	3×10^{-5}	1.2×10^{-3}	2.1×10^{-3}	1×10^{-3}	2.14×10^{-1}	1×10^{-4}
S23	5.60×10^{-4}	0	3×10^{-4}	1×10^{-3}	1×10^{-3}	4×10^{-4}	17.6×10^{1}	0
S24	3.70×10^{-4}	1.6×10^{-3}	2×10^{-4}	2.3×10^{-3}	6×10^{-4}	3×10^{-4}	1.58×10^{-1}	0

Table 4. *Cont.*

Samples No.	Fe	Cu	Cd	Cr	Mn	Ni	Pb	As
S25	4.25×10^{-3}	2.8×10^{-3}	2×10^{-4}	1.8×10^{-3}	3×10^{-4}	1.6×10^{-3}	0	1×10^{-4}
S26	2.50×10^{-4}	$1. \times 10^{-4}$	0	2×10^{-3}	1.8×10^{-3}	1.7×10^{-3}	12.3×10^{1}	7×10^{-4}
S27	1.20×10^{-4}	0	0	1.6×10^{-3}	4×10^{-4}	7×10^{-4}	2.17×10^{-1}	0
S28	3.40×10^{-4}	3×10^{-3}	0	1.3×10^{-3}	3×10^{-4}	4×10^{-4}	2.42×10^{-1}	0
S29	1.90×10^{-4}	1×10^{-4}	2×10^{-4}	2.4×10^{-3}	6×10^{-4}	3×10^{-4}	4.28×10^{-1}	1×10^{-4}
S30	1.60×10^{-4}	0	0	1.4×10^{-3}	3×10^{-4}	8×10^{-4}	2.42×10^{-1}	3×10^{-4}
S31	2.88×10^{-3}	2.9×10^{-3}	0	1.2×10^{-3}	1.8×10^{-3}	1.8×10^{-3}	0	0
S32	5.80×10^{-4}	2×10^{-4}	3×10^{-4}	1.8×10^{-3}	2.2×10^{-3}	1×10^{-3}	2.42×10^{-1}	1×10^{-4}
K1	6.67×10^{-5}	0	0	0	5×10^{-4}	5×10^{-4}	2.03×10^{-1}	1×10^{-4}
K2	1.80×10^{-4}	1×10^{-4}	2.50×10^{-3}	0	4×10^{-4}	9×10^{-4}	2.44×10^{-1}	0
K3	3.30×10^{-4}	2×10^{-4}	0	0	3×10^{-4}	2.5×10^{-3}	8.61×10^{-2}	0
K4	3.60×10^{-4}	2×10^{-4}	$4. \times 10^{-4}$	0	7.8×10^{-3}	9×10^{-4}	2.28×10^{-1}	0
K5	1.10×10^{-4}	3×10^{-4}	1.5×10^{-3}	1×10^{-4}	2.1×10^{-3}	2.5×10^{-3}	0	0
K6	3.33×10^{-5}	2×10^{-4}	3×10^{-4}	5×10^{-4}	8×10^{-4}	2.6×10^{-3}	10.8×10^{1}	0
K7	1.06×10^{-3}	2.0×10^{-4}	1.3×10^{-3}	5×10^{-4}	2.4×10^{-3}	9×10^{-4}	2.19×10^{-1}	0
K8	5.10×10^{-4}	2.0×10^{-4}	7×10^{-4}	4×10^{-4}	0	2.6×10^{-3}	9.33×10^{-1}	0
K9	1.60×10^{-4}	2.2×10^{-3}	2.8×10^{-3}	0	4×10^{-4}	2.1×10^{-3}	9.22×10^{-1}	0
K10	5.0×10^{-4}	1×10^{-4}	2.2×10^{-3}	0	9×10^{-4}	4×10^{-4}	18.4×10^{1}	0
K11	3.40×10^{-4}	0	0	0	2×10^{-4}	4×10^{-4}	1.75×10^{-1}	0
K12	5.28×10^{-5}	3×10^{-4}	2.1×10^{-3}	0	4×10^{-4}	3.8×10^{-3}	2.31×10^{-1}	0
K13	1.30×10^{-3}	3.1×10^{-3}	0	0	2.4×10^{-3}	1.4×10^{-3}	2.47×10^{-1}	0
K14	2.80×10^{-4}	2×10^{-4}	0	0	4×10^{-4}	2.5×10^{-3}	9.0×10^{-1}	0
K15	1.80×10^{-4}	1×10^{-4}	2.6×10^{-3}	1×10^{-4}	1.1×10^{-3}	4×10^{-4}	2.19×10^{-1}	0
K16	6.94×10^{-5}	0	3×10^{-3}	0	8×10^{-4}	6×10^{-4}	1.97×10^{-1}	0
K17	2.80×10^{-4}	1×10^{-4}	9×10^{-4}	0	3×10^{-4}	4×10^{-3}	2.11×10^{-1}	7×10^{-4}
K18	4.80×10^{-4}	2×10^{-4}	1.4×10^{-3}	1×10^{-4}	4×10^{-4}	1×10^{-3}	2.42×10^{-1}	0
K19	5.60×10^{-4}	2×10^{-4}	8×10^{-4}	0	2.2×10^{-3}	4×10^{-4}	2.06×10^{-1}	7×10^{-4}
K20	2.40×10^{-4}	1×10^{-4}	1.1×10^{-3}	1×10^{-5}	6×10^{-4}	4×10^{-4}	1.64×10^{-1}	0
K21	2.70×10^{-4}	2×10^{-4}	3×10^{-4}	0	3×10^{-4}	1×10^{-3}	2.14×10^{-1}	0
K22	5.60×10^{-4}	0	1.1×10^{-3}	0	2×10^{-4}	4×10^{-4}	17.6×10^{1}	1×10^{-4}
K23	3.70×10^{-4}	1.6×10^{-3}	1×10^{-3}	0	9×10^{-4}	3×10^{-4}	1.58×10^{-1}	7×10^{-4}
K24	4.25×10^{-3}	2.8×10^{-3}	1.2×10^{-3}	2×10^{-4}	1.1×10^{-3}	1.6×10^{-3}	0	0
K25	2.50×10^{-4}	1×10^{-4}	3×10^{-4}	0	1.3×10^{-3}	1.7×10^{-3}	12.3×10^{1}	1×10^{-4}
K26	1.20×10^{-4}	0	6×10^{-4}	0	0	7×10^{-4}	2.17×10^{-1}	0
K27	3.40×10^{-4}	3×10^{-3}	1.7×10^{-3}	0	4×10^{-4}	4×10^{-4}	2.42×10^{-1}	0
K28	1.90×10^{-4}	1×10^{-4}	1.1×10^{-3}	0	8×10^{-4}	3×10^{-4}	4.28×10^{-1}	1×10^{-4}
K29	4.80×10^{-4}	2×10^{-4}	1.9×10^{-3}	3×10^{-4}	3×10^{-4}	1×10^{-3}	2.42×10^{-1}	0
K30	5.28×10^{-5}	0	0	2×10^{-4}	2.9×10^{-3}	4×10^{-4}	3.42×10^{-1}	0
J1	9.20×10^{-4}	1×10^{-4}	0	2.6×10^{-3}	2.1×10^{-3}	5×10^{-4}	2.03×10^{-1}	0
J2	1.41×10^{-3}	1×10^{-4}	2×10^{-4}	0	0	1.5×10^{-3}	1.78×10^{-1}	0
J3	3.50×10^{-4}	1×10^{-4}	2×10^{-4}	0	6×10^{-4}	2.6×10^{-3}	9.58×10^{-1}	0
J4	7.30×10^{-4}	2×10^{-4}	2×10^{-4}	1.7×10^{-3}	2.1×10^{-3}	3×10^{-4}	2.17×10^{-1}	0
J5	2.70×10^{-4}	3.10×10^{-3}	2×10^{-4}	3×10^{-4}	2×10^{-4}	1.5×10^{-3}	2.33×10^{-1}	0
J6	1.20×10^{-4}	2.70×10^{-3}	0	1×10^{-3}	2×10^{-4}	1.3×10^{-3}	8.19×10^{-1}	1×10^{-4}

Table 4. Cont.

Samples No.	Fe	Cu	Cd	Cr	Mn	Ni	Pb	As
J7	8.70×10^{-4}	1×10^{-4}	0	2.4×10^{-3}	4×10^{-4}	0	1.19×10^{-1}	7×10^{-4}
J8	3.30×10^{-4}	1×10^{-4}	0	2.2×10^{-3}	2×10^{-4}	4.3×10^{-3}	7.75×10^{-1}	1×10^{-4}
J9	1.60×10^{-4}	0	0	2.1×10^{-3}	4×10^{-4}	8×10^{-4}	2.42×10^{-1}	0
J10	2.88×10^{-3}	2.9×10^{-3}	0	2.4×10^{-3}	2×10^{-4}	1.8×10^{-3}	0	0
J11	5.80×10^{-4}	2×10^{-4}	0	1×10^{-3}	2×10^{-4}	1×10^{-3}	2.42×10^{-1}	7×10^{-4}
J12	0	0	0	1.9×10^{-3}	3×10^{-4}	5×10^{-4}	2.03×10^{-1}	0
Min	0	0	0	0	0	0	0	0
Max	7.12×10^{-3}	3.1×10^{-3}	3×10^{-3}	2.7×10^{-3}	7.8×10^{-3}	4.3×10^{-3}	18.4×10^{1}	7×10^{-4}
Mean	7.70×10^{-4}	7×10^{-4}	5×10^{-4}	8×10^{-4}	1×10^{-3}	1.3×10^{-3}	4.07×10^{-1}	1×10^{-4}

3.8. Hazard Quotient Indices (HQ)

HQ for noncarcinogenic effect is related to dose received at risk point as compared to the reference dose (RfD) and is calculated by dividing chronic daily intake (CDI) by RfD [7]. HQ calculated for the groundwater samples were in the range for iron of 4.3×10^{-5} to 1.01×10^{-2} with average value of 1.20×10^{-3}, copper 0–0, cadmium 0–5.94 with average value of 1.11, chromium 0–1.8×10^{-3} with average value of 5.59×10^{-4}, manganese 5.95×10^{-4} to 5.59×10^{-2} with average value of 7.95×10^{-3}, nickel 0–2.15×10^{-1} with average value of 6.47×10^{-2}, lead 0–118 with average value of 15.97, and As 0–16.2 with average value of 2.85×10^{-1} mg/kg-day (Table 5). The average values of cadmium and lead crossed the safe limit of 1 and may be considered hazardous for human health [44]. The average values of HQ were observed in the following deceasing order: Pb > Cd > As > Ni > Mn > Fe > Cr.

Table 5. HQ results of heavy metals.

Samples No.	Fe	Cd	Cr	Mn	Ni	Pb	As
S1	8.20×10^{-4}	2.61×10^{-1}	7×10^{-4}	9.30×10^{-3}	1.79×10^{-2}	1.18×10^{2}	3.23×10^{-1}
S2	7.60×10^{-4}	1.61×10^{-1}	1.20×10^{-3}	1.39×10^{-2}	1.51×10^{-2}	5.43×10^{1}	0
S3	4.60×10^{-3}	13.7×10^{1}	0	1.57×10^{-2}	1.35×10^{-1}	5.72×10^{1}	0
S4	1.02×10^{-2}	1.17×10^{-1}	7×10^{-4}	4.60×10^{-3}	5.19×10^{-2}	5.65×10^{1}	0
S5	2.49×10^{-3}	6.11×10^{-2}	8×10^{-4}	8.90×10^{-3}	0	5.25×10^{1}	16.2×10^{1}
S6	5.30×10^{-4}	13.5×10^{1}	1×10^{-3}	9.70×10^{-3}	1.07×10^{-1}	0	0
S7	1.15×10^{-3}	6.61×10^{-1}	0	2.80×10^{-3}	1.21×10^{-1}	0	0
S8	2.24×10^{-3}	4.94×10^{-1}	1.20×10^{-3}	1.39×10^{-2}	7.65×10^{-2}	0	0
S9	3.58×10^{-3}	1.28×10^{-1}	1×10^{-3}	1.57×10^{-2}	1.56×10^{-1}	0	0
S10	0	1.44×10^{-1}	8×10^{-4}	4.60×10^{-3}	1.28×10^{-1}	59.4×10^{1}	3.23×10^{-1}
S11	0	1×10^{-1}	1.30×10^{-3}	8.90×10^{-3}	2.13×10^{-2}	0	6.46×10^{-1}
S12	1.31×10^{-3}	6.67×10^{-2}	1.80×10^{-3}	9.70×10^{-3}	2.74×10^{-2}	53.2×10^{1}	3.23×10^{-1}
S13	2.02×10^{-3}	6.33×10^{-1}	0	2.80×10^{-3}	7.74×10^{-2}	67.9×10^{1}	0
S14	5×10^{-4}	10.5×10^{1}	8×10^{-4}	2.20×10^{-3}	1.32×10^{-1}	54.8×10^{1}	0
S15	1.05×10^{-3}	1.39×10^{-1}	9×10^{-4}	3.40×10^{-3}	1.61×10^{-2}	3.40×10^{1}	3.23×10^{-1}
S16	3.80×10^{-4}	1.28×10^{-1}	6×10^{-4}	1.20×10^{-3}	7.42×10^{-2}	83.3×10^{1}	3.23×10^{-1}
S20	7.90×10^{-4}	2.94×10^{-1}	1.30×10^{-3}	1.29×10^{-2}	2.22×10^{-2}	95.7×10^{1}	6.46×10^{-1}
S21	3.40×10^{-4}	2.33×10^{-1}	0	8.90×10^{-3}	1.75×10^{-2}	56.3×10^{1}	0

Table 5. *Cont.*

Samples No.	Fe	Cd	Cr	Mn	Ni	Pb	As
S17	1.70×10^{-4}	5.44×10^{-1}	1×10^{-3}	5×10^{-3}	6.35×10^{-2}	68.7×10^{1}	3.23×10^{-1}
S18	1.24×10^{-3}	1.28×10^{-1}	0	6×10^{-3}	0	0	0
S19	4.70×10^{-4}	6.11×10^{-2}	0	2.04×10^{-2}	2.15×10^{-1}	2.68×10^{1}	16.2×10^{1}
S22	3.90×10^{-4}	5×10^{-2}	8.00×10^{-4}	1.51×10^{-2}	4.82×10^{-2}	49.4×10^{1}	3.23×10^{-1}
S23	8.00×10^{-4}	6.94×10^{-1}	6.00×10^{-4}	6.90×10^{-3}	2.04×10^{-2}	2.66×10^{1}	0
S24	5.20×10^{-4}	4.78×10^{-1}	1.50×10^{-3}	4.60×10^{-3}	1.65×10^{-2}	60.2×10^{1}	0
S25	6.07×10^{-3}	4.06×10^{-1}	1.20×10^{-3}	2.20×10^{-3}	7.83×10^{-2}	64.8×10^{1}	3.23×10^{-1}
S26	3.50×10^{-4}	6.67×10^{-2}	1.30×10^{-3}	1.29×10^{-2}	8.50×10^{-2}	2.28×10^{1}	16.2×10^{1}
S27	1.70×10^{-4}	7.22×10^{-2}	1.10×10^{-3}	3.20×10^{-3}	3.44×10^{-2}	33.2×10^{1}	0
S28	4.90×10^{-4}	1.06×10^{-1}	9.00×10^{-4}	2.20×10^{-3}	2.03×10^{-2}	2.15×10^{1}	0
S29	2.70×10^{-4}	3.78×10^{-1}	1.60×10^{-3}	4.2×10^{-3}	1.72×10^{-2}	57.1×10^{1}	3.23×10^{-1}
S30	2.30×10^{-4}	1.50×10^{-1}	1.00×10^{-3}	2×10^{-3}	3.83×10^{-2}	45.5×10^{1}	6.46×10^{-1}
S31	4.11×10^{-3}	6.11×10^{-2}	8.00×10^{-4}	1.29×10^{-2}	8.78×10^{-2}	59.4×10^{1}	0
S32	8.20×10^{-4}	5.50×10^{-1}	1.20×10^{-3}	1.59×10^{-2}	5.06×10^{-2}	4.89×10^{1}	3.23×10^{-1}
K1	0	0	0	3.80×10^{-3}	2.26×10^{-2}	44.0×10^{1}	3.23×10^{-1}
K2	2.60×10^{-4}	49.4×10^{1}	0	3×10^{-3}	4.61×10^{-2}	0	0
K3	4.70×10^{-4}	0	0	1.80×10^{-3}	1.24×10^{-1}	3.42×10^{1}	0
K4	5.10×10^{-4}	8.56×10^{-1}	0	5.60×10^{-2}	4.65×10^{-2}	60.2×10^{1}	0
K5	1.50×10^{-4}	29.4×10^{1}	0	1.53×10^{-2}	1.24×10^{-1}	67.1×10^{1}	0
K6	0	5×10^{-1}	3×10^{-4}	5.40×10^{-3}	1.29×10^{-1}	1.19×10^{1}	0
K7	1.51×10^{-3}	25.6×10^{1}	3×10^{-4}	1.73×10^{-2}	4.69×10^{-2}	67.1×10^{1}	0
K8	7.30×10^{-4}	13.9×10^{1}	3×10^{-4}	6×10^{-4}	1.29×10^{-1}	0	0
K9	2.20×10^{-4}	5.67×10^{1}	0	2.80×10^{-3}	1.07×10^{-1}	67.1×10^{1}	0
K10	7.20×10^{-4}	44.8×10^{1}	0	6.30×10^{-3}	1.92×10^{-2}	56.3×10^{1}	0
K11	4.90×10^{-4}	0	0	1.60×10^{-3}	2.19×10^{-2}	67.9×10^{1}	0
K12	0	42.8×10^{1}	0	2.80×10^{-3}	1.88×10^{-1}	23.9×10^{1}	0
K13	1.86×10^{-3}	0	0	1.75×10^{-2}	7.14×10^{-2}	63.3×10^{1}	0
K14	4×10^{-4}	0	0	3×10^{-3}	1.24×10^{-1}	0	0
K15	2.50×10^{-4}	52.8×10^{1}	0	7.90×10^{-3}	1.88×10^{-2}	3×10^{1}	0
K16	0	59.4×10^{1}	0	5.80×10^{-3}	3.11×10^{-2}	61×10^{1}	0
K17	3.90×10^{-4}	17.2×10^{1}	0	2×10^{-3}	2.01×10^{-1}	2.59×10^{1}	16.2×10^{1}
K18	6.80×10^{-4}	27.2×10^{1}	0	3×10^{-3}	4.79×10^{-2}	2.56×10^{1}	0
K19	7.90×10^{-4}	15×10^{1}	0	1.57×10^{-2}	2.22×10^{-2}	5.11×10^{1}	16.2×10^{1}
K20	3.40×10^{-4}	2.22×10^{0}	0	4.60×10^{-3}	1.75×10^{-2}	48.6×10^{1}	0
K21	3.90×10^{-4}	5.56×10^{-1}	0	2×10^{-3}	4.82×10^{-2}	64.0×10^{1}	0
K22	8×10^{-4}	21.7×10^{1}	0	1.20×10^{-3}	2.04×10^{-2}	68.7×10^{1}	3.23×10^{-1}
K23	5.20×10^{-4}	19.4×10^{1}	0	6.70×10^{-3}	1.65×10^{-2}	2.50×10^{1}	16.2×10^{1}
K24	6.07×10^{-3}	24.4×10^{1}	2×10^{-4}	7.90×10^{-3}	7.83×10^{-2}	61×10^{1}	0
K25	3.50×10^{-4}	6.11×10^{-1}	0	8.90×10^{-3}	8.50×10^{-2}	54.8×10^{1}	3.23×10^{-1}

Table 5. *Cont.*

Samples No.	Fe	Cd	Cr	Mn	Ni	Pb	As
K26	1.70×10^{-4}	11.7×10^{1}	0	6×10^{-4}	3.44×10^{-2}	58.6×10^{1}	0
K27	4.90×10^{-4}	34.4×10^{1}	0	3×10^{-3}	2.03×10^{-2}	67.1×10^{1}	0
K28	2.70×10^{-4}	21.1×10^{1}	0	6×10^{-3}	1.72×10^{-2}	57.1×10^{1}	3.23×10^{-1}
K29	6.80×10^{-4}	37.2×10^{1}	2×10^{-4}	2.20×10^{-3}	4.79×10^{-2}	45.5×10^{1}	0
K30	0	0	1×10^{-4}	2.04×10^{-2}	2.13×10^{-2}	59.4×10^{1}	0
J1	1.31×10^{-3}	1.22×10^{-1}	1.70×10^{-3}	1.51×10^{-2}	2.74×10^{-2}	4.89×10^{1}	0
J2	2.02×10^{-3}	4.50×10^{-1}	0	6×10^{-4}	7.74×10^{-2}	44×10^{1}	0
J3	5×10^{-4}	4.28×10^{-1}	0	4.60×10^{-3}	1.32×10^{-1}	0	0
J4	1.05×10^{-3}	4×10^{-1}	1.20×10^{-3}	1.51×10^{-2}	1.61×10^{-2}	3.42×10^{1}	0
J5	3.80×10^{-4}	4.39×10^{-1}	2×10^{-4}	1.60×10^{-3}	7.42×10^{-2}	62×10^{1}	0
J6	1.70×10^{-4}	3.94×10^{-2}	7×10^{-4}	1.40×10^{-3}	6.35×10^{-2}	6.71×10^{1}	3.23×10^{-1}
J7	1.24×10^{-3}	3.33×10^{-2}	1.60×10^{-3}	3×10^{-3}	0	1.19×10^{1}	16.2×10^{1}
J8	4.70×10^{-4}	9.44×10^{-2}	1.40×10^{-3}	1.40×10^{-3}	2.15×10^{-1}	67.1×10^{1}	3.23×10^{-1}
J9	2.30×10^{-4}	1.06×10^{-1}	1.40×10^{-3}	3.20×10^{-3}	3.83×10^{-2}	94.9×10^{1}	0
J10	4.11×10^{-3}	1.28×10^{-1}	1.60×10^{-3}	1.40×10^{-3}	8.78×10^{-2}	56.3×10^{1}	0
J11	8.20×10^{-4}	4.50×10^{-2}	7×10^{-4}	1.40×10^{-3}	5.06×10^{-2}	49.4×10^{1}	16.2×10^{1}
J12	0	1.17×10^{-1}	1.20×10^{-3}	2.20×10^{-3}	2.26×10^{-2}	2.66×10^{1}	0
Min	0	0	0	6×10^{-4}	0	0	0
Max	1.02×10^{-2}	59.4×10^{1}	1.80×10^{-3}	5.60×10^{-2}	2.15×10^{-1}	1.18×10^{2}	16.2×10^{1}
Mean	1.20×10^{-3}	11.1×10^{1}	6×10^{-4}	8×10^{-3}	6.47×10^{-2}	1.60×10^{1}	2.85×10^{-1}

3.9. Scatter Diagrams for Weathering Processes

The scatter diagrams were drawn using Microsoft Excel 2013 using the concentrations of cations and anions in meq/L. The graph of Na against HCO_3 indicated that Na and HCO_3 were balanced with each other and showed a linear relationship along the trend line. The results of HCO_3 increased with the increase in Na, with $r^2 = 0.9466$ (Figure 3a). The sample dots above the trend line show carbonate weathering, and dots below the trend line indicate silicate weathering [45]. The scatter diagram of HCO_3 + SO_4 against Ca + Mg (Figure 3b) indicates that points fell on both sides of the equiline. The points falling above the equiline indicate weathering of anions (HCO_3 and SO_4), mainly HCO_3, and points falling along and below the equiline are due to the silicate weathering [45,46]. More sample dots fell along and below the equiline and the silicate weathering was considered to be responsible for the calcium in groundwater. The scatter diagram of SO_4 against Ca (Figure 3c) indicates sample dots on both the sides of the trend line, suggesting weathering of both gypsum and silicate. However, more sample dots fell below the trend line due to silicate weathering. The scatter diagram of Na versus Ca + Mg (Figure 3d) indicates sample dots on both the sides along the trend line. The dots below the trend line may be designated as direct cation exchange, and those above as reverse cation exchange. The scatter diagram of Na against Cl (Figure 3e) indicates that most of the sample dots fell along the trend line, but a few sample dots fell away from the trendline on both the sides. The coefficient of determination (r^2) of Na versus Cl was observed to be 0.9444, indicating dissolution of halite [47]. The diagram of EC against Na/Cl (Figure 3f) did not indicate a straight line, but the trend line was inclined upward, suggesting that evaporation may not be the major process involved in the hydrochemistry of groundwater. The sample dots were scattered on both sides of the trend line, but more sample dots were present above the trend line, indicating that silicate weathering or forward ion exchange process was dominant over sodium reduction process or reverse ion exchange represented by the sample dots below trend line.

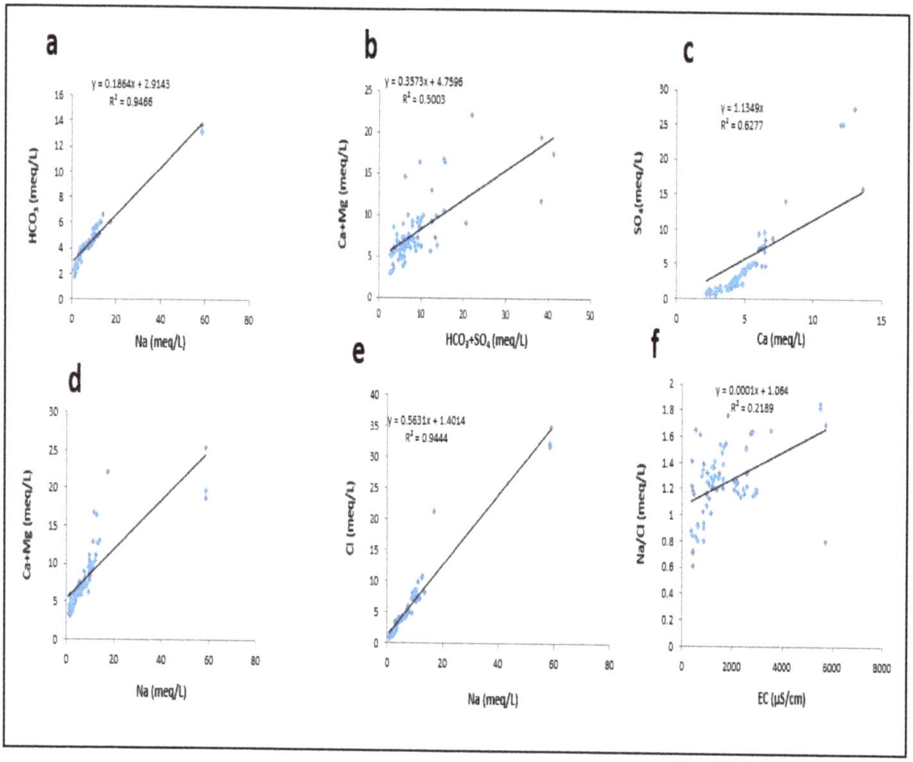

Figure 3. Scatter diagrams: (**a**) Na vs. HCO_3, (**b**) Ca + Mg vs. HCO_3 + SO_4, (**c**) Ca vs. SO_4, (**d**) Na vs. Ca, (**e**) Na vs. Cl, (**f**) EC vs. Na/Cl. All concentrations are in meq/L.

3.10. Cluster Analysis (CA)

The CA indicates similarities and dissimilarities among the samples based on the linkage distance method. Similar results for the samples are grouped in the same cluster. The dendrogram (Figure 4) indicated three clusters. Cluster 1 is again divided into two (clusters 1A and 1B). Cluster 1A comprised 35 and 1B 14 samples. Cluster 2 was based on 21 samples, whereas Cluster 3 consisted of 4 samples. Clusters 1 and 2 are connected by a common line and these have some similarities. Cluster 3 had higher values than clusters 1 and 2, and Cluster 2 had a higher value than Cluster 1. Similarly, Cluster 1A had the lowest values for water quality. The majority of the samples indicated as excellent for water quality in WQI were present in Cluster 1A.

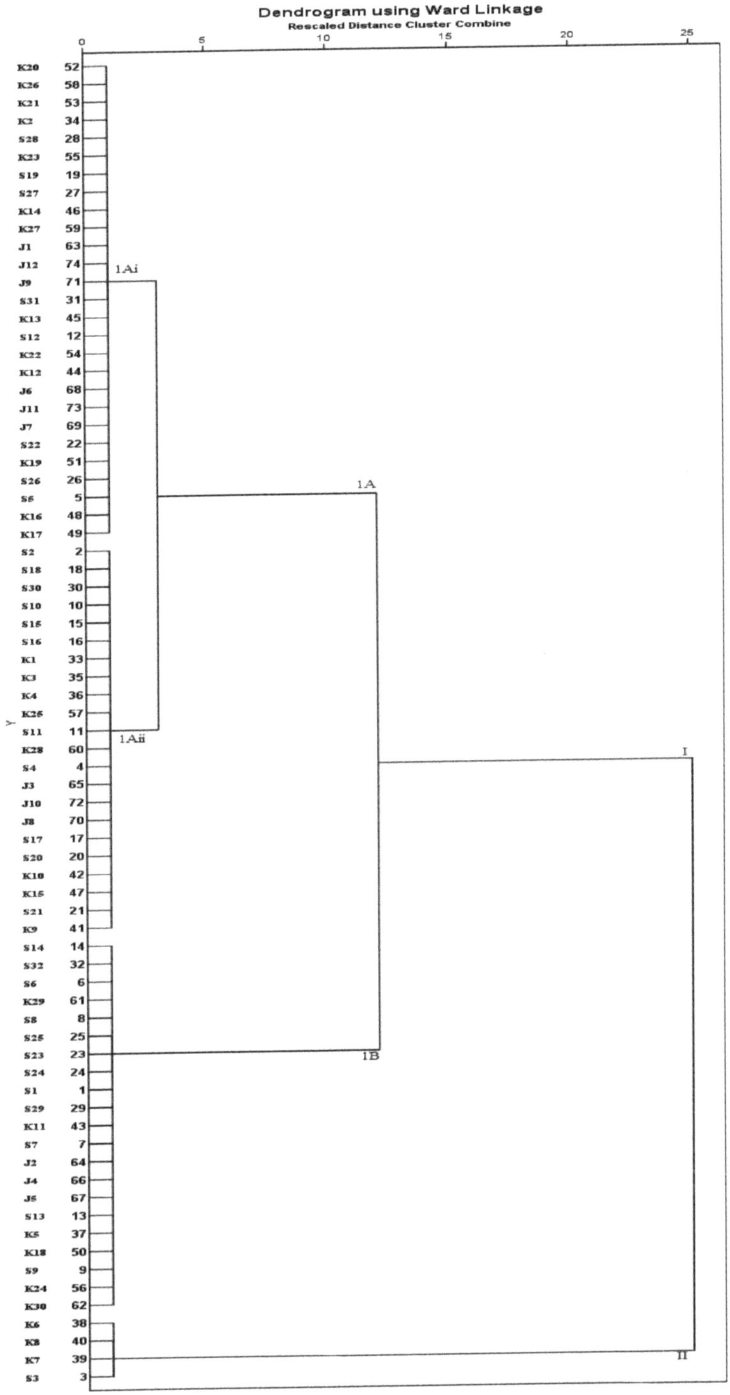

Figure 4. Cluster analysis of sampling stations.

3.11. Piper Diagrams

The Piper diagrams were drawn by using anions (HCO_3^-, Cl^-, SO_4^{2-}) and cations (Na^+, K^+, Ca^2, Mg^{2+}) values to obtain a triangle on the right side for anions and the left side for cations. The plots in the triangles show the composition of cations and anions in water samples [48]. The symbols indicated blue, black and red were from Jam Nawaz Ali, Sanghar and Kot Ghulam Muhammad. The Piper diagrams indicate the geochemistry of the area. Some samples in the cations triangle are gathered at the middle and are of mixed type (Figure 5). More samples are towards $Na^+ + K^+$ and others are towards Ca^{2+}. The right-side triangle for anions also indicates some samples in the middle and these are due to mixed type, but more are towards HCO_3^- and a few towards Cl^-. The quality of groundwater is controlled by rock–water interactions and anthropogenic activities due to agriculture. The diamond shape indicates some samples within mixed-type water, and others in the NaCl zone. Some samples are also in the $Ca(HCO_3)_2$ area.

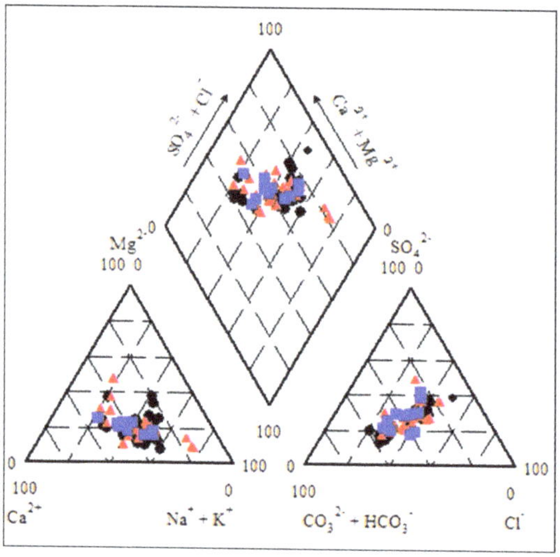

Figure 5. Piper diagrams of water analysis of Sanghar District.

3.12. Principal Component Analysis (PCA)

PCA was calculated based on 26 different parameters, and seven components were obtained with eigenvalues greater than 1 with total cumulative% of 80.07. Component 1 had an eigenvalue of 10.964 with variance% and cumulative% 42.168 (Figure 6, Table 6). Salinity, chloride, conductivity, TDS, sulphate, calcium, alkalinity, sodium, potassium, fluoride, total hardness, and magnesium indicated strong correlation (0.847–0.979) and supported the results of the coefficient of correlation. Other parameters indicated low to negative correlation. Component 2 had an eigenvalue of 2.811 with variance% 10.812 and cumulative% 52.979. Orthophosphate, nitrite, and total phosphate indicated strong to medium correlation (0.696–0.945) and showed anthropogenic activity due to the use of fertilizers in agricultural lands. Component 3 had an eigenvalue of 1.998 with variance% 7.683 and cumulative% 60.663. Cadmium and nitrate indicated medium correction (0.520–0.530). Component 4 was observed with eigenvalue of 1.319 with variance% 5.074 and cumulative% 66.396. Cobalt and arsenic indicated medium correlation (0.531–0.553). Component 5 had an eigenvalue of 1.319 with variance% 5.074 and cumulative% 71.470. Lead indicated medium correlation (0.602). Component 6 was observed with eigenvalue 1.128 with variance% 4.338 and cumulative% 75.809. Nickel indicated medium correlation (0.619). Component 7 showed an eigenvalue of 1.261 with variance% 4.261 and cumulative% 80.070. pH indicated medium correlation (0.670). The maximum loading was obtained with maximum positive correlations in PCA1, and loading decreased continuously from PCA1 to PCA 7.

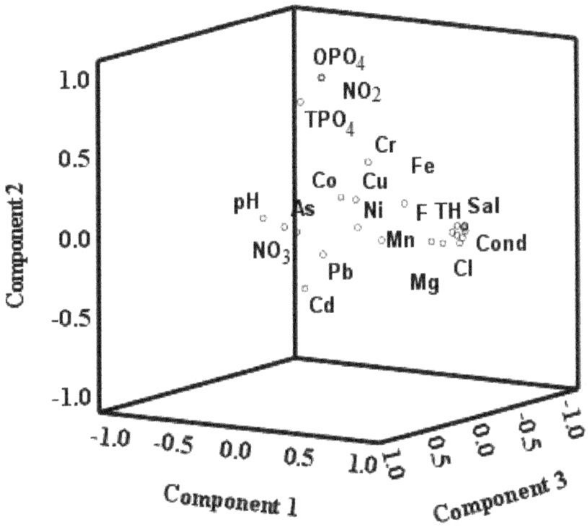

Figure 6. Dendrogram with three components.

3.13. Correlation Coefficient

The correlation coefficient is used to examine nearness among the parameters [48]. The results of correlation coefficient were obtained by plotting the average values of 26 parameters for 74 samples (Table 7). Conductivity and TDS indicated strong correlation with TH, Ca^{2+}, Mg^{2+}, Na^{+}, K^{+}, Cl^{-}, alkalinity, SO_4^{2-}, and F^{-} (0.799–0.981), indicating that these have a similar origin, mostly due to geology. Nitrate and total phosphate had strong correlation (0.864), but these had low to negative correlations with other parameters, indicating the effect of fertilizers on the agricultural lands. Na^{+} and Cl^{-} were strongly correlated (0.972), indicating possible dissolution of halite in an underground aquifer. Calcium indicated strong correlation with alkalinity (0.90), supporting the dissolution of calcite and dolomite. The pH, trace metals, and arsenic had low to negative correlation with most of the parameters.

Table 6. Water analysis for PCA of Sanghar District for 3 subdistricts for total variance.

Component	Initial Eigenvalues			Extraction Sums of Squared Loadings			Rotation Sums of Squared Loadings		
	Total	% of Variance	Cumulative%	Total	% of Variance	Cumulative%	Total	% of Variance	Cumulative%
1	10.96	42.17	42.17	10.96	42.17	42.17	10.67	41.06	41.06
2	2.81	10.81	52.98	2.81	10.81	52.98	2.78	10.71	51.77
3	1.998	7.68	60.66	1.998	7.68	60.66	1.80	6.91	58.67
4	1.49	5.73	66.40	1.49	5.73	66.40	1.55	5.95	64.63
5	1.32	5.07	71.47	1.32	5.07	71.47	1.44	5.55	70.17
6	1.13	4.34	75.81	1.13	4.34	75.81	1.37	5.27	75.44
7	1.11	4.26	80.07	1.11	4.26	80.07	1.204	4.630	80.070
8	0.898	3.45	83.52						
9	0.83	3.18	86.70						
10	0.68	2.60	89.31						
11	0.59	2.27	91.58						
12	0.51	1.98	93.56						
13	0.43	1.66	95.21						
14	0.36	1.40	96.61						
15	0.30	1.14	97.76						
16	0.20	0.78	98.54						
17	0.17	0.64	99.18						
18	0.09	0.34	99.53						
19	0.06	0.25	99.78						
20	0.02	0.09	99.86						
21	0.020	0.08	99.94						
22	0.01	0.05	99.99						
23	0.002	0.007	99.995						
24	0.001	0.004	99.999						
25	0.000	0.001	100.000						
26	2.202×10^{-6}	8.468×10^{-6}	100.000						

Table 7. Water analysis of Sanghar for correlation matrix.

Parameters	pH	Cond	NO_3	$T.PO_4$	TH	Na	K	Ca	Mg	Cl	Alk	SO_4	Mn	As	Fe	Cu	Ni	Pb	Cd	F	Cr
pH	1.00																				
Conduct	−0.309	1.00																			
TDS	−0.309	1.00																			
NO_3	−0.011	0.022	1.00																		
$T.PO_4$	0.082	−0.106	0.105	1.00																	
T.H	−0.298	0.799	0.000	−0.417	1.00																
Na	−0.249	0.864	0.034	−0.105	0.724	1.00															
K	−0.255	0.854	−0.076	−0.161	0.780	0.864	1.00														
Ca	−0.325	0.981	0.030	−0.112	0.806	0.863	0.876	1.00													
Mg	−0.325	0.830	0.052	−0.176	0.737	0.702	0.689	0.783	1.00												
Cl	−0.291	0.933	0.019	−0.148	0.804	0.972	0.921	0.935	0.796	1.00											
Alk	−0.320	0.906	0.042	−0.132	0.769	0.973	0.847	0.900	0.763	0.965	1.00										
SO_4	−0.265	0.947	0.068	−0.052	0.765	0.954	0.874	0.934	0.813	0.970	0.955	1.00									
Mn	−0.309	0.040	−0.045	−0.091	0.132	0.012	0.110	0.050	0.055	0.044	0.021	0.026	1.00								
As	0.361	−0.384	0.071	0.122	−0.284	−0.249	−0.259	−0.417	−0.283	−0.293	−0.349	−0.284	0.191	1.00							
Fe	−0.166	0.147	−0.203	0.006	0.121	0.025	0.101	0.154	0.045	0.083	0.042	0.068	0.152	−0.090	1.00						
Cu	0.045	−0.108	−0.224	0.047	−0.147	−0.098	−0.116	−0.116	−0.171	−0.132	−0.111	−0.104	0.054	−0.034	0.293	1.00					
Ni	−0.032	0.207	0.018	−0.096	0.129	0.172	0.167	0.182	0.259	0.199	0.200	0.196	0.013	0.021	0.073	−0.009	1.00				
Pb	0.033	0.074	−0.052	−0.059	0.105	0.152	0.113	0.065	0.166	0.126	0.126	0.134	−0.028	.017	−0.257	−0.164	0.196	1.00			
Cd	0.068	0.069	0.118	−0.026	0.189	0.064	0.054	0.068	0.118	0.073	0.085	0.105	−0.132	−0.176	−0.135	−0.031	0.007	0.107	1.00		
F	−0.186	0.848	0.189	−0.008	0.733	0.841	0.724	0.829	0.719	0.839	0.865	0.902	−0.025	−0.293	0.027	−0.104	0.146	0.103	0.250	1.00	
Cr	−0.011	−0.097	−0.225	−0.062	−0.229	−0.089	−0.059	−0.135	−0.042	−0.095	−0.109	−0.103	0.107	0.120	0.129	0.083	−0.164	−0.100	−0.464	−0.216	1.00

3.14. Wilcox Diagrams for Quality of Groundwater

The quality of groundwater is tested on the basis of the Wilcox diagram for irrigation. Two parameters are used to draw the Wilcox diagram: (1) salinity, represented as C by conductivity (μS/cm); and (2) sodium adsorption ratio (SAR), represented by (S) (Figure 7). Ten samples (13.5%) were present in C_1S_1 (low salinity and low SAR values), which were suitable for all purposes. A total of 36 (48.6%) samples were in C_2S_1 with medium salinity and low SAR (Figure 7). These samples can be used for irrigation, where moderate percolation. A total of 24 samples (32.4%) were located in C_3S_1 (high salinity and low SAR) and required salinity control and drainage facilities for irrigation. One sample (1.4%) was in C_4S_2 (very high salinity and moderate SAR), and three samples (4%) were present in C_4S_3 (very high salinity and high SAR). These samples were unsuitable for irrigation. The results of the Wilcox diagram indicated that 5.4% samples were unsuitable for irrigation, and more samples can be used for irrigation with some drainage facilities. Another Wilcox diagram was also drawn based on two parameters: (1) conductivity (μS/cm) and (2) Na% (Figure 8). The results indicated that 14 (18.9%) samples were in the excellent category and can be used for all the irrigational purposes. A total of 26 samples (35%) were in the good category and can also be used for irrigation. A total of 20 samples (27%) were in the doubtful category and could be used with some drainage facilities. A total of 13 samples (17.5%) were in the doubtful to unfit category and 1 sample (1.4%) was unfit for irrigation. The results indicated that 54% samples were present in the excellent to good category for irrigation.

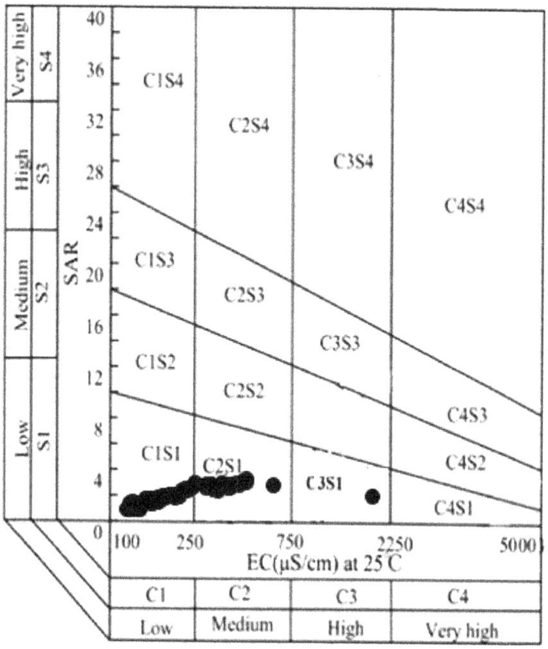

Figure 7. Wilcox diagram EC vs. SAR.

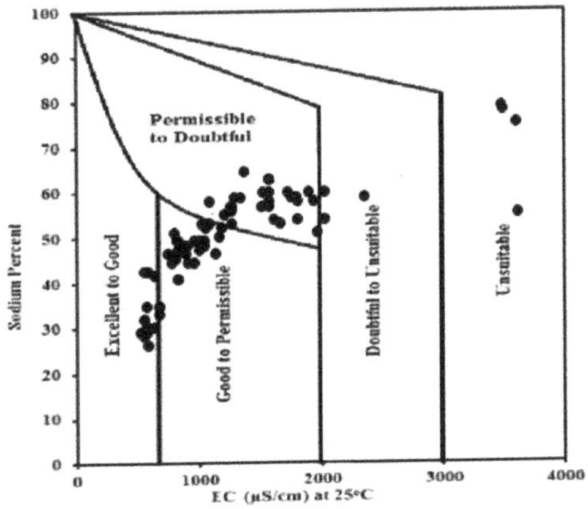

Figure 8. Wilcox diagram EC vs. Na percent.

4. Conclusions

Three subdistricts (Sanghar, Khipro, and Jam Nawaz Ali) of Sanghar District were examined for water quality for drinking and irrigation. The results of water quality were compared with WHO permissible limits. A number of parameters at different sampling stations crossed the permissible limits and were not considered suitable for human consumption. A total of 77% of the samples based on WQI were in excellent to good water quality. The results of contamination index (C_d) also indicated that 89% of the samples had C_d less than 3 and were considered suitable for drinking. The sources of contamination are mostly geological, except at the towns where some anthropogenic contamination can be expected. Gibbs diagrams indicated that the hydrochemistry of the groundwater was mostly based on rock–water interactions. The results of coefficient of correlation indicated that major cations and anions had good correlation among each other and supported that samples had a similar geochemical setting. The results of CDI and HQ indicated that the presence of lead in groundwater was a concern for human health. The water quality for irrigation indicated 54–100% suitability for irrigation.

Recommendations:

(1) Awareness programs may be initiated to inform the inhabitants about the locations of the contaminated groundwater and their possible health effects.
(2) Alternate sources of drinking water may be used for drinking, particularly surface water from the river Indus.
(3) If a dependable source of drinking water is not available, then the groundwater may be joined to suitable filters to reduce the concentrations of contaminants.

Supplementary Materials: The following supporting information can be downloaded at: https://www.mdpi.com/article/10.3390/w16060856/s1.

Author Contributions: Conceptualization, T.M.J. and M.Y.K.; methodology, M.Y.K.; software, F.Y.K., T.M.J., A.Q.L., T.M.J. and M.Y.K.; data curation, T.M.J. and M.Y.K.; writing—original draft preparation, T.M.J.; writing—review and editing, T.M.J.; visualization, M.F.L.; supervision, T.M.J. and M.Y.K.; funding acquisition, M.Y.K. All authors have read and agreed to the published version of the manuscript.

Funding: This research received no external funding.

Data Availability Statement: Data is contained within the article or Supplementary Materials.

Acknowledgments: The work acknowledges faculty and staff along with equipment, chemicals, and counselling facilities by High Tech Central Resource Laboratory HTCRL/IARSCS, University of Sindh, Jamshoro.

Conflicts of Interest: The authors declare no conflicts of interest.

References

1. Brohi, R.O.Z.; Khuhawar, M.Y.; Mahar, R.B.; Ibrahim, M.A. Novel bimetallic nano particles for sorption of mercury (II) from drinking water: Adsorption experiment and computational studies. *J. Water Process Eng.* **2021**, *39*, 101727. [CrossRef]
2. Singh, A.; Sharma, A.; Verma, R.K.; Chopade, R.L.; Pandit, P.P.; Nagar, V.; Aseri, V.; Choudhary, S.K.; Awasthi, G.; Awasthi, K.K.; et al. Heavy Metal Contamination of Water and Their Toxic Effect on Living Organisms. In *The Toxicity of Environmental Pollutants*; IntechOpen: Rijeka, Croatia, 2022.
3. Daud, M.K.; Nafees, M.; Ali, S.; Rizwan, M.; Bajwa, R.A.; Shakoor, M.B.; Arshad, M.U.; Chatha, S.A.S.; Deeba, F.; Murad, W.; et al. Drinking water quality status and contamination in Pakistan. *BioMed Res. Int.* **2017**, *2017*, 7908183. [CrossRef] [PubMed]
4. Shahab, A.; Shihua, Q.; Rashid, A.; Hasan, F.U.; Sohail, M.T. Evaluation of water quality for drinking and agricultural suitability in the lower Indus plain in Sindh province, Pakistan. *Pol. J. Environ. Stud.* **2016**, *25*, 2563–2574. [CrossRef]
5. Lanjwani, M.F.; Khuhawar, M.Y.; Jahangir Khuhawar, T.M.; Lanjwani, A.H.; Jagirani, M.S.; Kori, A.H.; Rind, I.K.; Khuhawar, A.H.; Muhammad Dodo, J. Risk assessment of heavy metals and salts for human and irrigation consumption of groundwater in Qambar city: A case study. *Geol. Ecol. Landsc.* **2020**, *4*, 23–39. [CrossRef]
6. Lanjwani, M.F.; Khuhawar, M.Y.; Khuhawar, T.M.J. Assessment of groundwater quality for drinking and irrigation uses in taluka Ratodero, district Larkana, Sindh, Pakistan. *Int. J. Environ. Anal. Chem.* **2022**, *102*, 4134–4157. [CrossRef]
7. Ahmed, J.; Wong, L.P.; Chua, Y.P.; Channa, N. Drinking water quality mapping using water quality index and geospatial analysis in primary schools of Pakistan. *Water* **2020**, *12*, 3382. [CrossRef]
8. Ahmed, J.; Wong, L.P.; Chua, Y.P.; Yasmin, A.; Channa, N.; VanDerslice, J.A. Estimation of hepatitis a virus infection prevalence through drinking water supply of primary schools of Sindh, Pakistan. *Hepat. Mon.* **2020**, *20*, e98412. [CrossRef]
9. Rind, I.K.; Khuhawar, M.Y.; Lanjwani, M.F.; Khuhawar, T.M.J.; Samtio, M.S.; Soomro, W.A.; Baloch, A.R. Spatial variability of hydrochemistry and health risk assessment of groundwater of taluka Hala, district Matiari, Sindh, Pakistan. *Arab. J. Geosci.* **2023**, *16*, 99. [CrossRef]
10. Podgorski, J.E.; Eqani, S.A.M.A.S.; Khanam, T.; Ullah, R.; Shen, H.; Berg, M. Extensive arsenic contamination in high-pH unconfined aquifers in the Indus Valley. *Sci. Adv.* **2017**, *3*, e1700935. [CrossRef]
11. Khuhawar, M.; Ursani, H.; Khuahwar, T.; Lanjwani, M.; Mahessar, A. Assessment of water quality of groundwater of Thar Desert, Sindh. *Pakistan J. Hydrogeol. Hydrol. Eng.* **2019**, *7*, 2.
12. Ullah, Z.; Talib, M.A.; Rashid, A.; Ghani, J.; Shahab, A.; Irfan, M.; Rauf, A.; Bawazeer, S.; Almarhoon, Z.M.; Mabkhot, Y.N. Hydrogeochemical investigation of elevated arsenic based on entropy modeling, in the aquifers of District Sanghar, Sindh, Pakistan. *Water* **2021**, *13*, 3477. [CrossRef]
13. Bashir, E.; Huda, S.N.-U.; Naseem, S.; Hamza, S.; Kaleem, M. Geochemistry and quality parameters of dug and tube well water of Khipro, District Sanghar, Sindh, Pakistan. *Appl. Water Sci.* **2017**, *7*, 1645–1655. [CrossRef]
14. Din, I.U.; Muhammad, S.; Faisal, S.; Rehman, I.U.; Ali, W. Heavy metal (loid) s contamination and potential risk assessment via groundwater consumption in the district of Hangu, Pakistan. *Environ. Sci. Pollut. Res.* **2023**, *30*, 33808–33818. [CrossRef]
15. Selmane, T.; Dougha, M.; Djerbouai, S.; Djemiat, D.; Lemouari, N. Groundwater quality evaluation based on water quality indices (WQI) using GIS: Maadher plain of Hodna, Northern Algeria. *Environ. Sci. Pollut. Res.* **2023**, *30*, 30087–30106. [CrossRef]
16. Kaur, T.; Bhardwaj, R.; Arora, S. Assessment of groundwater quality for drinking and irrigation purposes using hydrochemical studies in Malwa region, southwestern part of Punjab, India. *Appl. Water Sci.* **2017**, *7*, 3301–3316. [CrossRef]
17. Adimalla, N.; Dhakate, R.; Kasarla, A.; Taloor, A.K. Appraisal of groundwater quality for drinking and irrigation purposes in Central Telangana, India. *Groundw. Sustain. Dev.* **2020**, *10*, 100334. [CrossRef]
18. Li, P.; Wu, J.; Qian, H. Assessment of groundwater quality for irrigation purposes and identification of hydrogeochemical evolution mechanisms in Pengyang County, China. *Environ. Earth Sci.* **2013**, *69*, 2211–2225. [CrossRef]
19. Soleimani, H.; Nasri, O.; Ojaghi, B.; Pasalari, H.; Hosseini, M.; Hashemzadeh, B.; Kavosi, A.; Masoumi, S.; Radfard, M.; Adibzadeh, A.; et al. Data on drinking water quality using water quality index (WQI) and assessment of groundwater quality for irrigation purposes in Qorveh & Dehgolan, Kurdistan, Iran. *Data Brief* **2018**, *20*, 375–386. [PubMed]
20. P.B.O. Statistics. Available online: https://www.pbs.gov.pk/sites/default/files/tables/district_at_glance/Sanghar.pdf (accessed on 5 December 2023).
21. Kadri, I.B. *Petroleum Geology of Pakistan*; Pakistan Petroleum Limited: Karachi, Pakistan, 1995.
22. Baig, M.O.; Harris, N.; Ahmed, H.; Baig, M. Controls on reservoir diagenesis in the Lower Goru sandstone formation, Lower Indus Basin, Pakistan. *J. Petrol. Geol.* **2016**, *39*, 29–47. [CrossRef]
23. Ashraf, U.; Zhang, H.; Anees, A.; Mangi, H.N.; Ali, M.; Zhang, X.; Imraz, M.; Abbasi, S.S.; Abbas, A.; Ullah, Z.; et al. A core logging, machine learning and geostatistical modeling interactive approach for subsurface imaging of lenticular geobodies in a clastic depositional system, SE Pakistan. *Nat. Resour. Res.* **2021**, *30*, 2807–2830. [CrossRef]
24. Ashraf, U.; Zhu, P.; Yasin, Q.; Anees, A.; Imraz, M.; Mangi, H.N.; Shakeel, S. Classification of reservoir facies using well log and 3D seismic attributes for prospect evaluation and field development: A case study of Sawan gas field, Pakistan. *J. Pet. Sci. Eng.* **2019**, *175*, 338–351. [CrossRef]
25. Li, M.; Ma, H.; Pan, H.; Ashraf, U.; Jiang, R. Building a rock physics model for the formation evaluation of the Lower Goru sand reservoir of the Southern Indus Basin in Pakistan. *J. Petrol. Sci. Eng.* **2020**, *194*, 107461.

26. Ashraf, U.; Zhang, H.; Anees, A.; Ali, M.; Zhang, X.; Shakeel Abbasi, S.; Nasir Mangi, H. Controls on reservoir heterogeneity of a shallow-marine reservoir in Sawan Gas Field, SE Pakistan: Implications for reservoir quality prediction using acoustic impedance inversion. *Water* **2020**, *12*, 2972. [CrossRef]
27. Shakir, U.; Hussain, M.; Mahmood, M.F.; Amjad, M.R.; Mehood, W.A.Z.S.; Abideen, Z.U.; Tahir, A.R.; Barron, M. Prespect generation studies of Cretaceous sands, Kipro area by integrating seismic and well data. *Nucleous* **2020**, *57*, 67–75.
28. Shahab, A.; Qi, S.; Zaheer, M. Arsenic contamination, subsequent water toxicity, and associated public health risks in the lower Indus plain, Sindh province, Pakistan. *Environ. Sci. Pollut. Res.* **2019**, *26*, 30642–30662. [CrossRef] [PubMed]
29. Jacobs, H.L.; Gabrielson, I.N.; Horton, R.K.; Lyon, W.A.; Habbard, E.C.; McCallum, G.E. Water quality criteria-stream vs. effluents standards. *J. Water Pollut. Control. Fed.* **1965**, *37*, 292–315.
30. American Public Health Association (APHA); WEF (Water Environment Federation). *Standard Methods for the Examination of Water and Wastewater*, 22nd ed.; American Public Health Association: Washington, DC, USA; American Water Works Association: Denver, CO, USA; Water Environment Federation: Washington, DC, USA, 2012; ISBN 978-087553-013-0.
31. Greenberg, A.E.; Clesceri, L.S.; Eaton, A.D. Standard Methods for the Examination of Water and Wastewater. Available online: https://www.standardmethods.org/doi/book/10.2105/SMWW.2882 (accessed on 5 December 2023).
32. Baired, R.; Eaton, A.D.; Rice, E.W.; Bridgewater, L.; American Public Health Association (APHA); American Water Works Association, Water Environment Federation. *Standard Methods for the Examination of Water and Wastewater*; American Public Health Association: Washington, DC, USA, 2017.
33. Masood, A.; Aslam, M.; Pham, Q.B.; Khan, W.; Masood, S. Integrating water quality index, GIS and multivariate statistical techniques towards a better understanding of drinking water quality. *Environ. Sci. Pollut. Res.* **2022**, *29*, 26860–26876. [CrossRef] [PubMed]
34. US EPA. Risk Assessment for Carcinogenic Effects. Available online: https://www.epa.gov/fera/risk-assessment-carcinogenic-effects (accessed on 5 December 2023).
35. Backman, B.; Bodiš, D.; Lahermo, P.; Rapant, S.; Tarvainen, T. Applications of a groundwater contamination index in Finland and Slovakia. *Environ. Geol.* **1998**, *36*, 55–64. [CrossRef]
36. Zabin, S.A.; Foaad, M.; Al-Ghamdi, A.Y. Non-carcinogenic risk assessment of heavy metals and fluoride in some water wells in the Al-Baha Region, Saudi Arabia. *Hum. Ecol. Risk Assess.* **2008**, *14*, 1306–1317. [CrossRef]
37. Muhammad, S.; Usman, Q.A. Heavy metal contamination in water of Indus River and its tributaries, Northern Pakistan: Evaluation for potential risk and source apportionment. *Toxin Rev.* **2022**, *41*, 380–388. [CrossRef]
38. Moldovan, Z. Spectrophotometric determination of nitrite by its catalytic effect on the oxidation of Congo red with bromate. *Bull. Chem. Soc. Ethiop.* **2012**, *26*, 159–169. [CrossRef]
39. WHO. *Guidelines for Drinking-Water Quality: First Addendum to the Fourth Edition*; WHO: Geneva, Switzerland, 2017.
40. Abadin, H.; Llanos, F.; Stevens, Y.W. *Toxicological Profile for Lead*; US Department of Health and Human Services, Agency for Toxic Substances and Diseases Registry: Atlanta, GA, USA, 2007.
41. Ketata, M.; Gueddari, M.; Bouhlila, R. Use of geographical information system and water quality index to assess groundwater quality in El Khairat deep aquifer (Enfidha, Central East Tunisia). *Arab. J. Geosci.* **2012**, *5*, 1379–1390. [CrossRef]
42. Gibbs, R.J. Mechanisms controlling world water chemistry. *Science* **1970**, *170*, 1088–1090. [CrossRef] [PubMed]
43. Khan, S.; Cao, Q.; Zheng, Y.; Huang, Y.; Zhu, Y. Health risks of heavy metals in contaminated soils and food crops irrigated with wastewater in Beijing, China. *Environ. Pollut.* **2008**, *152*, 686–692. [CrossRef]
44. Mgbenu, C.N.; Egbueri, J.C. The hydrogeochemical signatures, quality indices and health risk assessment of water resources in Umunya district, southeast Nigeria. *Appl. Water Sci.* **2019**, *9*, 22. [CrossRef]
45. Datta, P.; Deb, D.; Tyagi, S. Stable isotope (18O) investigations on the processes controlling fluoride contamination of groundwater. *J. Contam. Hydrol.* **1996**, *24*, 85–96. [CrossRef]
46. Egbueri, J.C.; Mgbenu, C.N.; Chukwu, C.N. Investigating the hydrogeochemical processes and quality of water resources in Ojoto and environs using integrated classical methods. *Model. Earth Syst. Environ.* **2019**, *5*, 1443–1461. [CrossRef]
47. Jankowski, J.; Acworth, R.I. Impact of debris-flow deposits on hydrogeochemical processes and the developement of dryland salinity in the Yass River Catchment, New South Wales, Australia. *Hydrogeol. J.* **1997**, *5*, 71–88. [CrossRef]
48. Piper, A.M. A graphic procedure in the geochemical interpretation of water-analyses. *Eos Trans. Am. Geophys. Union* **1944**, *25*, 914–928.

water

MDPI

Article

Assessment of Hydrogeochemical Characteristics and Seawater Intrusion in Coastal Aquifers by Integrating Statistical and Graphical Techniques: Quaternary Aquifer, West Nile Delta, Egypt

Samia S. Hasan [1], Zenhom E. Salem [2] and Ahmed Sefelnasr [3,*]

[1] Desert Research Center, El Matareya, Cairo 11753, Egypt; samiasamir@drc.gov.eg
[2] Geology Department, Faculty of Science, Tanta University, Tanta 31527, Egypt; zenhomsalem@yahoo.com
[3] National Water and Energy Center, United Arab Emirates University, Al Ain 15551, United Arab Emirates
* Correspondence: ahmed.sefelnasr@uaeu.ac.ae

Abstract: The Quaternary aquifer in the western Nile Delta is threatened by seawater intrusion. Few studies have integrated diverse techniques for the assessment of seawater intrusion in this aquifer. The present study aims to determine the geochemical processes and impact of seawater intrusion on this aquifer. To accomplish this investigation, the integration of hydrogeochemical, statistical, multivariate statistical, and graphical tools were implemented on 75 groundwater samples and 5 soil samples. The physicochemical variables were analyzed using hierarchical cluster analysis (HCA), saturation index (SI), ionic ratios, ionic relationships, the seawater intrusion index (SWI) and the correlations among 16 hydrochemical parameters, to identify the influencing processes of groundwater quality in the study area. According to the statistical study, the groundwater is divided into four groups. Those are distributed, from north to south: Group1 (G1), Group2 (G2), Group4 (G4), and Group3 (G3). The samples of G1 and G2 are distinguished by Na–Cl chemical type. While G4 has two main ion associations, HCO_3–Ca–Mg and Cl–SO_4–Na, G3 is characterized by HCO_3–Cl–SO_4–Ca–Na type. The processes that affect the chemistry of the groundwater are the seawater intrusion, ion exchange, silicate and Ca-rich mineral weathering, and mineral deposition. G1 and G2 groups are primarily influenced by seawater incursion, evaporation, and the ion exchange mechanism. In addition, the weathering of silicate minerals has a substantial effect on G3 and G4 groups, resulting in the creation of carbonate minerals.

Keywords: coastal aquifers; seawater intrusion index; hierarchical cluster analysis; saturation index; ion exchange; GIS techniques

Citation: Hasan, S.S.; Salem, Z.E.; Sefelnasr, A. Assessment of Hydrogeochemical Characteristics and Seawater Intrusion in Coastal Aquifers by Integrating Statistical and Graphical Techniques: Quaternary Aquifer, West Nile Delta, Egypt. *Water* **2023**, *15*, 1803. https://doi.org/10.3390/w15101803

Academic Editors: Liangping Li and Guanxing Huang

Received: 29 March 2023
Revised: 5 May 2023
Accepted: 8 May 2023
Published: 9 May 2023

1. Introduction

Groundwater is a valuable natural resource that is essential for social and economic growth and human needs, especially in arid and semi-arid regions which suffer from shortage of precipitation and fresh surface water sources [1–3]. Aquifer rock material, climate conditions, and human activities in addition to seawater intrusion have a significant impact on groundwater quality and quantity in coastal aquifers [4–7].

Due to extreme use of the groundwater in coastal aquifers for industrial, agricultural, and urban expansion, seawater intrusion has come to be a global problem [8–11]. Climate change has raised atmospheric surface temperatures and melted ice caps, mountain glaciers, and polar ice sheets, which raises ocean and sea water levels, compounding the problem [12]. In addition, both the slope of the aquifer bed and the slope of the seaside influence the seawater intrusion; increasing the slope of the bed toward the sea will increase the interference [13]. Seawater intrusion hazard induces salinization of aquifers, reducing groundwater quality and causing a salinity hazard to the superimposed soils [14].

Globally, the coastal areas are occupied by about 20% of the world's population [12]. However, due to groundwater overexploitation to face the increasing water demand associated with population expansion and fast urban development, these locations are vulnerable to seawater intrusion. Future climate change and the resulting sea level rise are both predicted to worsen this vulnerability. Nevertheless, groundwater management becomes more complicated when all these elements come together.

Groundwater salinization threatens coastal fresh water supplies such as north Kuwait [15], Kish Island, Iran [16], Thriassion Plain and Eleusis Gulf, Greece [17], China [18], Spain [19], Bangladesh [20], Mexico [21], and others. Contamination by seawater has been studied in many coastal aquifers. The assessment of this phenomenon involves determining the spatial variation of physicochemical properties such as electrical conductivity (EC), total dissolved solids (TDS), seawater mixing and groundwater level, groundwater hydrochemical analysis, and geoelectrical methods such as resistivity and electromagnetic techniques [22–29]. Several studies have been conducted to simulate the seawater intrusion using different numerical techniques, including Sefelnasr and Sherif [30], Mabrouk et al. [31], Mabrouk et al. [32], Abd-Elhamid et al. [12], Abd-Elhamid et al. [13], Sarker et al. [20], Masoud et al. [33]. The Nile Delta aquifer is one of the largest fresh groundwater sources in the world. Researchers found that the seawater intrusion front in the Nile Delta aquifer moved inland more than 100 km from the Mediterranean shoreline [34–36].

Eventually, the complex process of seawater intrusion is influenced by, among others, shoreline geomorphology, hydrogeochemical reactions, biological processes, and aquifer dynamics. Saltwater intrusion, mostly in unconfined coastal aquifers, is the main cause of salinization [19]. In confined or semi-confined aquifers, complicated hydrogeological processes such as water–rock interaction and mixing with different water sources may impact groundwater quality and geochemical evolution [30].

Understanding the hydrochemistry is essential for determining the groundwater quality and its suitability for various applications, and therefore, determining sound management options for this resource as well as efficient mitigation measures when required. Diverse statistical and graphical techniques, such as pattern diagrams and GIS, can also be used to characterize seawater incursion. However, evaluation of the groundwater chemistry in the region based on integration of hydrogeochemical methods, statistical analysis, and advanced geostatistical techniques is still weak. Such integration provides the means of investigating such complex coastal aquifers, and the results may function as the foundation for effective groundwater management.

The purpose of this study is to evaluate the anthropogenic impacts and saltwater intrusion on the geochemical constituents and processes governing groundwater in the study area. To achieve this objective, groundwater samples of soil and water were collected and analyzed. For organizing groundwater samples, managing this volume of data, and providing accurate interpretations, graphical representations of the data were accompanied by multiple statistical approaches and multiple indices.

2. Materials and Methods

2.1. Study Area

The study area is a part of the Northwest Rosetta branch. It lies between 30°42′ to 31°28′ N and 29°45′ to 30°48′ E (Figure 1). It is bounded to the north by the Mediterranean Sea, to the east by the branch of Rosetta, and to the west by El Nubaryia Canal. The study area has an arid to semi-arid climate with an average annual temperature of 30.4 °C and an average annual humidity of 59% [37]. The rainfall ranges from 50 mm/a in the south to about 200 mm/a in the north. The evapotranspiration ranges from 1648 in the north to 1680 mm/y in the south [38]. Geomorphologically and hydrogeologically, the investigated area was subjected to different studies such as Allam et al. [39], Dawoud et al. [40], Morsy [41], El Fakharany and Hagran [42], and Salem et al. [43]. Young alluvial plains characterize the research region. It features a nearly elevated ground surface that ranges from −3 m (amsl) in the northwest to +6 m (amsl) in the southeast [41]. The area is

minerals in an aqueous solution and the state of saturation is determined using the hydrochemical program PHREEQC Interactive Program [58]. An SI < 0 means that the mineral's groundwater is undersaturated. An SI > 0 indicates that the groundwater is supersaturated with this mineral phase and thus incapable of dissolving more of the mineral.

2.4. Seawater Intrusion Quality Index (SWI)

To identify the threat of seawater intrusion, Tomaszkiewicz et al. [59] created the groundwater quality index for seawater intrusion (GQI_{SWI}) (Equation (1)). This model translates the Piper diagram information $GQI_{piper\ (mix)}$ (freshwater seawater mixing index of Piper diagram) and the seawater fraction index GQI_{fsea} to create a new two-stage numerical indicator for seawater intrusion [59].

$$GQI_{SWI} = \frac{GQI_{piper(mix)} + GQI_{fsea}}{2} \quad (1)$$

The first stage included the calculation of $GQI_{piper\ (mix)}$ as follows:

$$GQI_{Piper(mix)}(meq/l) = \left[\frac{\left(Ca^{+2} + Mg^{+2}\right)}{Total\ cations} + \frac{\left(HCO_3^-\right)}{Total\ anions} \right] \times 50 \quad (2)$$

The second stage involves computing the seawater fraction index GQI_{fsea} (Equation (3)) from the seawater fraction (f_{sea}) (Equation (4)). GQI_{fsea} was then used to classify the SWI. These values vary from 0 to 100, the fresher waters of which have lower values.

$$GQI_{f_{sea}} = (1 - f_{sea}) \times 100 \quad (3)$$

The following formula can be used to measure the seawater fraction (f_{sea}) [60]:

$$f_{sea} = \frac{m_{Cl(sample)} - m_{Cl(freshwater)}}{m_{Cl(seawater)} - m_{Cl(freshwater)}} \quad (4)$$

where $m_{Cl\ (sample)}$ is the Cl concentration of the sample, $m_{Cl\ (seawater)}$ is the Cl concentration of the Mediterranean Sea (603 meq/L), and $m_{Cl\ (freshwater)}$ represents the Cl concentration of the freshwater (0.48 meq/L).

The difference in hydrochemistry between seawater and freshwater was classified by concentrations of anions and cations. Indeed, the freshwater is defined by an abundance of calcium and bicarbonate, while the seawater is classified by its dominance of chloride and sodium [61]. The concentration of chloride is the base which determines the mixing of these two types of water, which is clarified by the conservative presence of anions [62].

3. Results and Discussions

3.1. Hydrochemical Characteristics and Spatial Distribution

The hydrochemical characteristics of groundwater are discussed with respect to the study area and according to the results of water samples chemical analysis. As shown in Table 2, the groundwater pH values range from slightly acidic (5.05) to slightly alkaline (8.7) averaging 7.1. The differences in pH values are primarily due to the aquifer–rock interaction and the distance from the sea [63]. Based on the classification of Winslow and Kister [64], 53%, 21%, 15%, and 11% of the groundwater samples are fresh, slightly, moderately, and very saline water, respectively (Table 3). The EC of groundwater in the study area ranges from 0.38 to 55.3 mS/cm with an average of 7.1 mS/cm (Table 2). TDS levels range from 190 to 27680 mg/L, averaging 3806.2 (Table 2). The Rosetta branch and many irrigation canals nearby produce fresh to slightly salty water in the southern and central study area. Near irrigation canals, Nile water seepage dilutes groundwater. Figure 2 shows that seawater incursion increases salinity northward, where the shore is 30 km away.

Table 3. Classification of water according to Total Dissolved Solids [64].

Water Class	Salt Concentration (mg/L)	No. of Wells	%
Fresh	>1000	40	53
Slightly saline	1000–3000	16	21
Moderately saline	3000–10,000	11	15
Very saline	10,000–35,000	8	11
Brine	<35,000	-	-

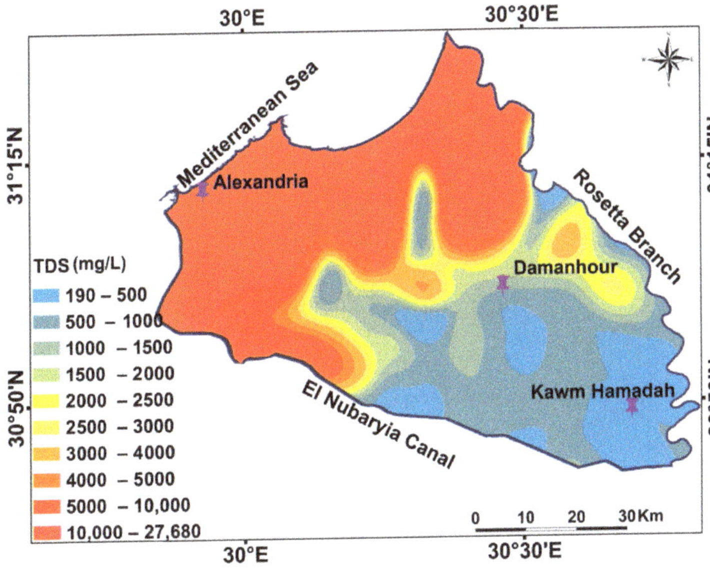

Figure 2. Spatial distribution of TDS across the study area.

As demonstrated in Figure 3a–c, Ca, Mg, and Na concentrations were lowest in the south and center regions and rose northward. It is largely seawater interference. The southern and central parts may have lower cations due to Nile water seepage and irrigation.

Figure 3d shows that the concentration of bicarbonate increases from south and northwest towards the central parts. Bicarbonates are introduced to natural water through recharging with meteoric water and interaction with sediments rich in either carbonate minerals or organic matter. The dissolution of $CaCO_3$ or the decay of organic matter increase in the presence of CO_2 leading to the formation of more soluble bicarbonate [65]. SO_4 and Cl spatial distributions showed that they have almost the same pattern. The lower values are located in the south and increase towards the north (Figure 3e,f). Such distribution patterns could be related to seawater intrusion effect.

Figure 4a–f shows Al, Mn, Ba, Fe, Sr, and Si spatial distribution maps. Al concentration increases southward and decreases northward (Figure 4a). Mn is highest in the northeast and declines westward to its minimum at El Nubaryia Canal (Figure 4b). Ba, Fe, and Sr have similar distributions. The concentration increases northeastward from the south (Figure 4c–e). Si levels are higher in central areas from northeastern to southwestern directions and lower along the western boundary (Figure 4f).

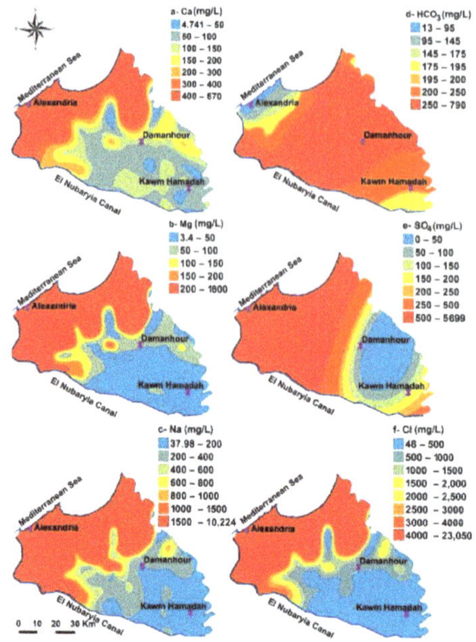

Figure 3. The spatial distribution maps of major ions.

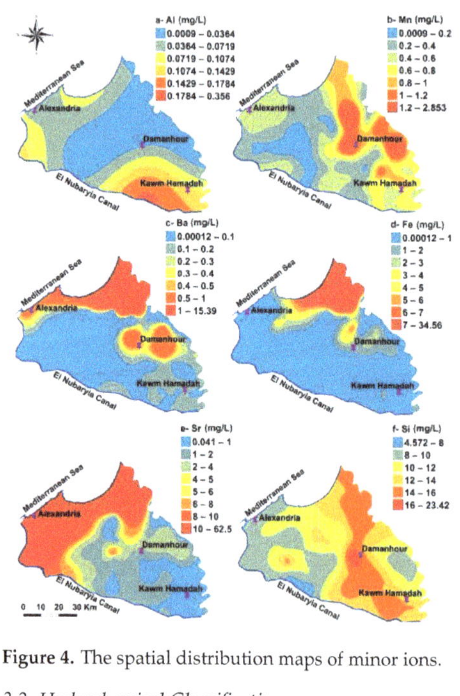

Figure 4. The spatial distribution maps of minor ions.

3.2. *Hydrochemical Classification*

3.2.1. Cluster Analysis

Based on the HCA procedures, four groups were identified. Figure 5a shows that G2, G3, and G4 are related, but G1 samples are distinct. This is primarily due to their occurrences in the south and center of the study area, which are relatively located far from the sea. On the other hand, G1 is located in the northern region near to the sea, thus, its chemical composition was significantly affected by the seawater.

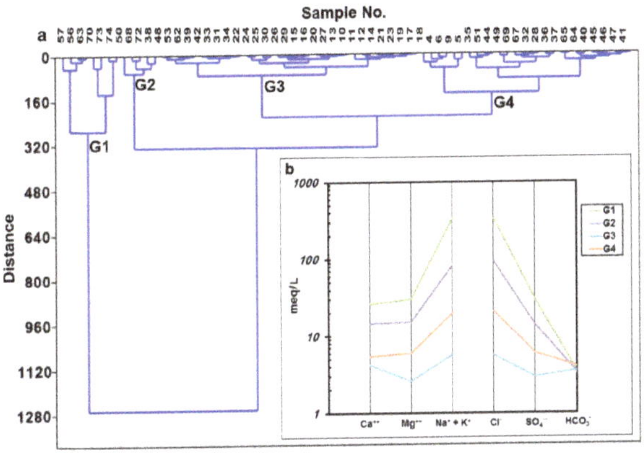

Figure 5. (**a**) Dendrogram shows the grouping of the groundwater samples into four groups depending on the major ion concentrations in milliequivalents per lite and (**b**) Scholler diagrams characterizing the chemical type for samples groups.

Table 4 shows the mean hydrochemical data values for each sample group to explain their features. Figure 6 depicts group dispersion by location.

Table 4. The average composition of the hydrochemical parameters of the four groundwater sample groups.

Parameter	G1	G2	G3	G4
TDS mg/L	22,686.75	6486.4	597.94	1829.1
Ca meq/L	26.09	14.27	4.24	5.37
Mg meq/L	30.12	14.97	2.57	5.94
Na meq/L	329.78	82.24	5.51	19.17
K meq/L	5.18	0.72	0.17	0.56
HCO_3 meq/L	3.42	187	3.49	3.95
SO_4 meq/L	29.71	13.94	2.92	5.93
Cl meq/L	332.59	92.92	5.6	20.91
Si mg/L	12.63	12.89	11.31	10.59
Al mg/L	0.09	0.03	0.01	0.00
Ba mg/L	0.12	3.11	0.13	0.24
Fe mg/L	0.25	10.01	0.55	1.47
Mn mg/L	0.49	1.55	0.49	0.88
Sr mg/L	0.80	28.01	1.97	6.14

Scholler's diagram is commonly used to correlate many of the chemical water compositions. The average hydrochemical data of the four groups (Table 4) are plotted on the semi-logarithmic diagrams (Figure 5b). The samples of G1, G2 and G4 have Na–Cl water type (Na > Ca > Mg > and Cl > SO_4 > HCO_3) with an average TDS values 22,686.7 mg/L, 6486.4 mg/L and 1829.1 mg/L, respectively. The spatial distribution of G1, G2 and G4 (Figure 6) indicated that these groups are influenced by seawater intrusion, while G3 samples have Na–Cl (Na > Ca > Mg and Cl > HCO_3 > SO_4) water type and lower concentrations of TDS (average 597.9 mg/L) and ions compared to the other groups. G3 is distributed in the southern part of the study area and its origin might be related to direct seepage from the irrigation canals.

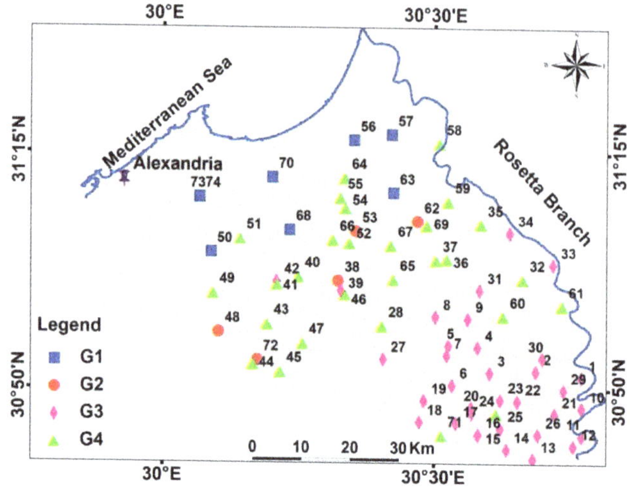

Figure 6. Spatial distribution map of the groundwater samples groups.

3.2.2. Piper's (Trilinear) Diagram

The collected samples were plotted in the trilinear diagram [66] (Figure 7) according to the HCA classification (Figure 5a). In most groundwater samples, Na is the major cation and Cl is the main anion. Generally, the samples of G1, G2 and G4 were plotted on sub-area 7 (Figure 7) indicating primary salinity character, where NaCl and Na_2SO_4 salts dominate the chemical properties. This is largely due to the impact of seawater intrusion and soil salinity. Most samples of G4, which occupied the southern parts, are located in sub-area 9 (Figure 7) indicating that none of the cation–anion pairs exceeds 50%. In the right triangle, samples of G3 and 4 show an increase in concentrations of either SO_4 or HCO_3 and decrease of Cl. In the lift triangle, due to ion exchange process, concentrations of Ca and Mg increase in groups 3 and 4 and Na decrease [67].

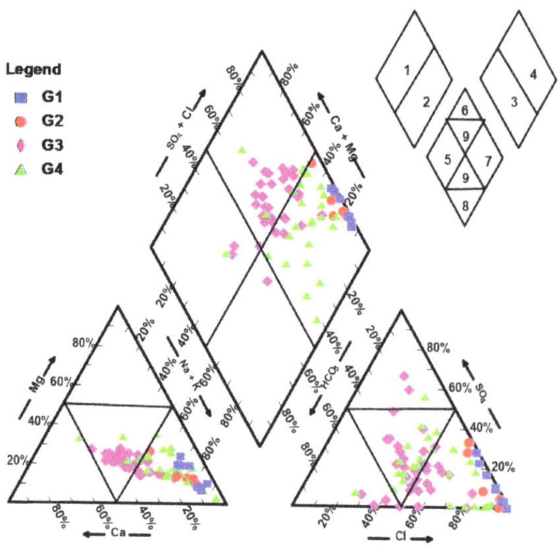

Figure 7. Piper diagrams show the hydrogeochemical characters of the studied groundwater. Samples are labeled according to their HCA groups shown in Figure 5a.

3.2.3. Statistical Ions Classification

Figure 8 shows sample group ion-association dendrograms. Na, Cl, and TDS in groundwater samples from groups G1 and G2 (Figure 8a,b) show seawater intrusion in the northern section of the research area (Figure 6). Assuming G3, HCO_3–Cl–SO_4–Ca–Na is the major ion association in the southern region, indicating freshwater origin and evaporation (Figure 8c). Figure 8d demonstrates ion-association in group four samples. HCO_3–Ca–Mg-minors and Cl–SO_4–Na showed that freshwater impacted by seawater intrusion and exchange processes is widespread. Al, Mn, Ba, Fe, Sr, Si, and NO_3 are linked in a short distance, suggesting a shared origin that may affect fertilizer or water–rock interactions (Figure 8).

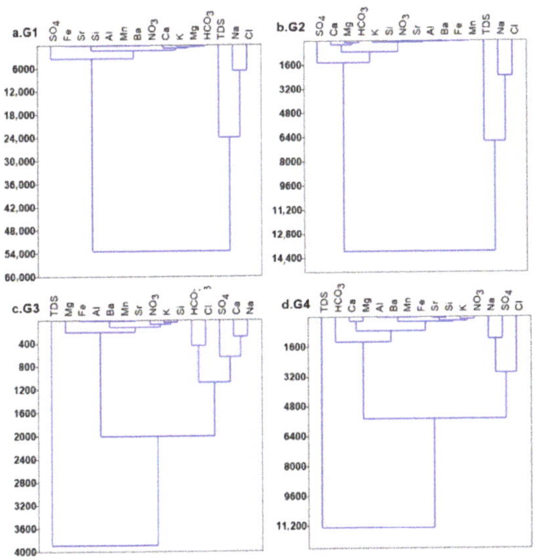

Figure 8. Dendrograms show the ion-associations in sample groups.

3.3. Saturation Index (SI)

Table 5 shows the statistical summary of dissolved mineral saturation indicators. The dissolved minerals were classified into five different groups: carbonate, sulphate, chlorides, oxides, and silicate minerals. Detected carbonate minerals include aragonite, calcite, dolomite, and witherite. Unlike groups 1, 2, and 4, group 3 is saturated with aragonite, calcite, and dolomite. All groups are undersaturation with respect to witherite. As the investigated aquifer is composed of clastic sediments, production of bicarbonate ion is due to the formation of CO_2 from decay of the organic matter (Equation (5)) [60]. Such conditions could lead to saturation conditions in the southern part where group 3 samples are located but the unsaturation conditions in groups 1, 2, and 4 might be related to carbonate minerals deposition where the increase in sulfate and chloride minerals due to seawater intrusion in the northern parts is accompanied by deposition of the carbonate minerals [36,68].

$$CO_2 + H_2O \rightarrow H_2CO_3 + H \rightarrow HCO_3^- + H \tag{5}$$

$$SO_4^{-2} + 2C + 2\,H_2O \rightarrow H_2S + 2\,HCO_3 \tag{6}$$

The studied groundwater is mostly unsaturated with respect to anhydrite and gypsum minerals, whereas most of G3, which is located in the southern part of the area, was saturated with respect to barite. Hydrochemical studies of major and trace elements made by Drevaliene et al. [69] showed that the probable barium source in groundwater is the dissolution of witherite. According to Table 5, all four groups of groundwater samples showed an undersaturation state with the dissolved chloride minerals (Halite and Sylvite).

Table 5. Dissolved minerals saturation indices.

	No.	Aragonite	Calcite	Dolomite	Witherite	Anhydrite	Gypsum	Barite	Halite
	Min	−1.7	−1.6	−2.7	−8.4	−1.3	−0.9	−2.5	−3.1
	Max	0.2	0.3	1.0	−3.6	−0.4	0.0	2.5	−2.2
G1	Av	−0.7	−0.6	−0.8	−5.5	−0.7	−0.4	−0.1	−2.7
	SI > 1%	12.5	25	25	0	0	0	57.2	0
	SI < 1%	87.5	75	75	100	100	100	42.8	100
	Min	−1.9	−1.7	−3.1	−8.6	−2.2	−1.9	−2.7	−4.2
	Max	1.1	1.2	2.5	−3.7	−0.8	−0.6	−0.7	−3.8
G2	Av	−0.5	−0.4	−0.5	−6.0	−1.4	−1.1	−1.5	−3.9
	SI > 1%	40	40	40	0	0	0	0	0
	SI < 1%	60	60	60	100	100	100	100	100
	Min	−1.1	−1.0	−1.9	−6.4	−3.8	−3.5	−4.2	−7.2
	Max	1.2	1.3	2.6	−1.8	−1.1	−0.7	0.7	−5.4
G3	Av	0.4	0.5	0.9	−2.9	−2.1	−1.8	−0.2	−6.3
	SI > 1%	84.90	87.90	87.9	0	0	0	63.7	0
	SI < 1%	15.10	12.10	12.1	100	100	100	36.3	100
	Min	−1.9	−1.7	−3.4	−8.0	−3.8	−3.5	−3.3	−7.1
	Max	0.8	0.9	1.7	−1.8	−1.2	−0.8	1.2	−4.4
G4	Av	−0.8	−0.6	−1.1	−5.2	−2.1	−1.8	−1.2	−5.4
	SI > 1%	22.6	25.9	22.6	0	0	0	25.9	0
	SI < 1%	77.4	74.1	77.4	100	100	100	74.1	100

	No.	Sylvite	Hematite	Pyrolusite	Albite	Anorthite	K-feldspar	Kaolinite	Quartz
	Min	−4.6	−2.7	−1.8	−4.4	−12.1	−4.3	−1.8	0.1
	Max	−3.7	16.5	2.9	0.1	−4.3	0.6	4.8	0.7
G1	Av	−4.0	10.6	0.7	−2.2	−7.7	−1.6	1.3	0.4
	SI > 1%	0	87.5	0	25	0	25	50	100
	SI < 1%	100	12.5	100	75	100	75	50	0
	Min	−5.7	1.7	−0.4	−2.9	−8.4	−2.9	1.1	0.1
	Max	−4.9	20.3	0.4	−1.7	−5.1	−1.1	1.7	0.5
G2	Av	−5.4	9.7	0.1	−2.5	−6.9	−2.2	1.4	0.3
	SI > 1%	0	100	0	0	0	0	100	100
	SI < 1%	100	0	100	100	100	100	0	0
	Min	−7.9	5.5	−0.2	−3.8	−6.6	−2.6	0.7	0.1
	Max	−6.4	20.3	2.1	0.1	−0.7	1.3	5.8	0.5
G3	Av	−7.1	15.0	1.1	−1.0	−2.2	0.1	3.6	0.3
	SI > 1%	0	100	0	5.8	0	79.5	100	100
	SI < 1%	100	0	100	94.2	100	20.5	0	0
	Min	−7.6	1.1	−0.6	−4.6	−9.6	−3.8	0.5	−0.1
	Max	−4.9	21.2	2.7	−0.6	−1.6	0.0	6.4	0.5
G4	Av	−6.4	11.4	0.5	−2.9	−6.5	−2.0	1.7	0.2
	SI > 1%	0	100	0	0	0	0	100	88
	SI < 1%	100	0	100	100	100	100	0	12

G1, G2, G3, and G4 are undersaturated with respect to pyrolusite but most of them are saturated with hematite (Table 5) due to the reductive dissolution of Fe and Mn. The dissolution of Fe and Mn oxides in the aquifer is at the cost of the oxidation of organic matter [70,71], and the bicarbonate will increase. The reaction Equations (7) and (8) are as follows:

$$CH_2O + 2MnO_2 + 3H^+ = 2Mn^{2+} + HCO_3^- + 2H_2O \quad (7)$$

$$CH_2O + 4Fe(OH)_3 + 7H^+ = 4Fe^{2+} + HCO_3^- + 10H_2O \quad (8)$$

The encountered dissolved silicate minerals are albite, anorthite, K-feldspar, kaolinite, and quartz. Groundwater samples are undersaturated with albite and anorthite. Most of G1, G2, and G4 are in the undersaturation state with K-feldspar but G3 is mostly saturated with this mineral. G2, G3, G4, and most of G1 are saturated with quartz and kaolinite (Table 5). Following Hosono et al. [72], incongruent dissolution reactions of silicate weathering in the study area can be hypothesized by following Equations (9) and (10):

$$2NaAlSi_3O_8 \text{ (Albite)} + 2CO_2 + 11H_2O \rightarrow Al2Si_2O_5(OH)_4 \text{ (Kaolinite)} + 2Na^+ + 2HCO_3^- + 4H_4SiO_4 \quad (9)$$

$$CaAl2Si_2O_8 \text{ (Anorthite)} + 2CO_2 + 3H_2O \rightarrow Al_2Si_2O_5(OH)_4 \text{ (Kaolinite)} + 2Ca2^+ + 2HCO_3^- \quad (10)$$

3.4. Hydrogeochemical Processes

Ions-TDS Relationships

- Gibbs diagram

Gibbs diagrams show the groundwater chemistry's main natural process. It shows how rock weathering, mineral precipitation, and evaporation affect groundwater chemistry [73]. High concentration of Ca^{2+} and HCO_3^- reveals rock–water interaction. On the other hand, the presence of high concentration of Na^+ and Cl^- indicates evaporation processes or seawater intrusion [74,75]. TDS concentrations are plotted against the weight ratios of Na/(Na + Ca) for cations (Figure 9a) and the weight ratios of Cl/(Cl + HCO_3) for anions (Figure 9b). The major processes governing groundwater chemistry are evaporation/seawater intrusion and rock interaction. Most G3 samples have dissolved HCO_3 and Ca via water-rock interaction. Evaporation and seawater intrusion supply Na and Cl to Groups 1, 2, 4, and soil water.

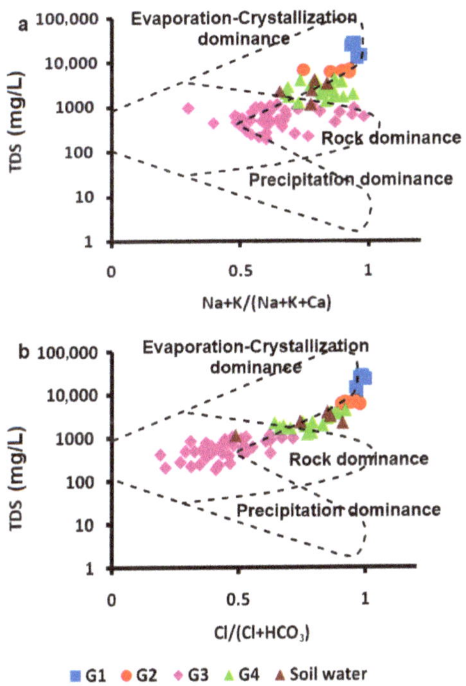

Figure 9. Gibbs diagrams illustrate (**a**) TDS concentrations are plotted against the weight ratios of Na/(Na + Ca) for cations and (**b**) the weight ratios of Cl/(Cl + HCO_3) for anions. Concentrations of TDS and ion are in mg/L.

- TDS vs. Ca, Mg, Na + K, Cl, HCO_3, and SO_4

TDS vs. Ca, Mg and Na + K relationship (Figure 10a–c) are very strong (r^2 = 0.9, 0.9, and 0.9, respectively, Table 6). For TDS relationships with Ca and Mg, samples were plotted under the mixing line which indicate ion-exchange process due to seawater intrusion, where Na replaces either Ca or/and Mg as shown in Equations (11) and (12). TDS vs. Cl relationship shows a very strong correlation (r^2 = 0.9, Table 6, Figure 10d). For the relationships between TDS and Na + K and Cl, most plotted groundwater samples are clustered along the irrigation water–seawater mixing line suggesting that the variance in concentrations of theses ions is due to the mixing between irrigation canal water and seawater. TDS vs. HCO_3 (Figure 10e) showed weak inverse relationship (r^2 = −0.1, Table 6), which means that freshwater seepage is not only the primary source of dissolved carbonates but may also be the reason for the weathering of silicate minerals. as shown in Equations (11) and (12) [76].

$$2Na^{+} + Ca - X_2 \rightarrow 2Na - X + Ca^{2+} \tag{11}$$

$$2Na^{+} + Mg - X_2 \rightarrow 2Na - X + Mg^{2+} \tag{12}$$

where x is an ion exchanger.

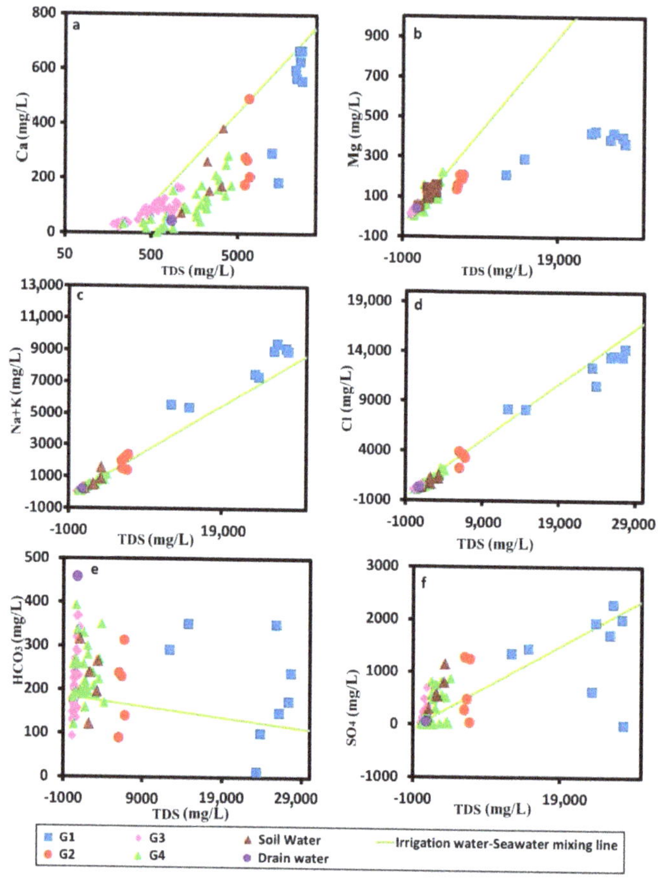

Figure 10. The relationships between Total Dissolved Solids (TDS) and the concentrations of (**a**) Ca, (**b**), Mg, (**c**) Na + Ca, (**d**) Cl, (**e**) HCO_3, and (**f**) SO_4.

Table 6. Correlation matrix between physico-chemical parameters in groundwater.

	TDS	EC	Ca	Mg	Na	K	HCO_3	SO_4	Cl	NO_3	Al	Ba	Fe	Mn	Si	Sr
TDS	1															
EC	0.9	1														
Ca	0.9	0.9	1													
Mg	0.9	0.9	0.9	1												
Na	0.9	0.9	0.8	0.9	1											
K	0.6	0.6	0.5	0.6	0.7	1										
HCO_3	0.1	0.1	0.2	0.1	0.1	0.1	1									
SO_4	0.6	0.6	0.5	0.7	0.6	0.7	0.0	1								
Cl	0.9	0.9	0.9	0.9	0.9	0.7	0.1	0.6	1							
NO_3	−0.1	−0.1	−0.1	0.0	−0.1	0.1	0.0	0.1	−0.1	1						
Al	−0.2	−0.2	−0.2	−0.3	−0.2	−0.1	−0.1	−0.1	−0.2	−0.1	1					
Ba	0.5	0.5	0.4	0.4	0.5	0.4	−0.1	0.5	0.4	−0.1	−0.1	1				
Fe	0.4	0.4	0.4	0.4	0.5	0.4	−0.1	0.5	0.4	−0.1	−0.1	0.9	1			
Mn	0.1	0.1	0.2	0.1	−0.1	−0.1	0.2	−0.1	0.1	−0.2	−0.1	0.1	0.1	1		
Si	0.1	0.1	0.1	0.1	0.1	0.0	−0.1	0.1	0.1	−0.2	0.3	0.1	0.1	0.2	1	
Sr	0.5	0.5	0.6	0.6	0.5	0.6	−0.1	0.6	0.5	0.1	−0.2	0.9	0.9	0.1	0.1	1

Note: values listed in red font have strong positive correlation, blue font have intermediate positive correlation, values listed in green font have weak correlation.

On the other hand, TDS–SO_4 (Figure 10f) have a strong relationship (r^2 = 0.7, Table 6). The origin of sulfates is not linked to the trend of surface water–seawater mixing but possibly related to soil origin, especially in low salinity water. The relationships between TDS vs. Ba, Fe and Sr (r^2 equals 0.5, 0.4, and 0.5, respectively, Table 6) showed intermediate correlation, on the other hand it showed weak relationships with Al, Mn, and Si (r^2 equals −0.2, 0.1, and 0.1 respectively, Table 6). Figure 10a–f shows that plotted soil water samples and indicates that the soil hydrochemical processes are very effective in the hydrochemical characteristics of the studied low salinity groundwater (G3 and G4) where it acts as an important origin of the dissolved ions.

3.5. Seawater Intrusion Quality Index (SWI)

In ArcGIS 10 framework, the calculated GQI_{SWI} results are transformed into a map using a kriging interpolation function. The GQI_{SWI} value varies from 17.8 to 82.3 as shown in Figure 11. According to Table 7, the groundwater of the study area is classified into saline, mixed, and fresh water. The coastal area in the northern parts has low values and reflects the salinization of groundwater.

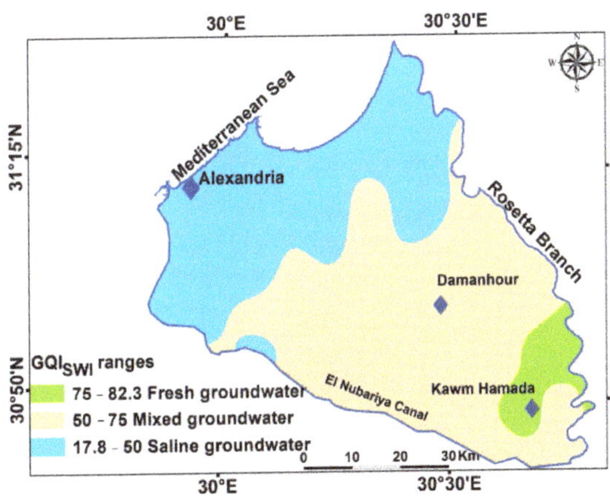

Figure 11. GQI_{SWI} in the study area.

Table 7. GQI_{SWI} ranges [77].

Water Type	GQI_{SWI} Based on Worldwide Literature			Typical GQI_{SWI}	
	Min	Max	Mean	Min	Max
Fresh water	73.5	90.1	82.7	75	100
Mixed groundwater	47.8	79.9	63.4	50	75
Saline groundwater	4.8	58.8	27.5	10	50
Seawater	3.1	9.2	5.8	0	10

4. Conclusions

In this work, the seawater intrusion and the processes governing the groundwater of the Quaternary aquifer of the Nile Delta were assessed using hydrochemical analysis in conjunction with several multivariate statistical methods and graphical approaches. It was concluded that groundwater chemistry is influenced by the evaporation process, seawater intrusion, ion exchange process, dissolution and weathering of Ca-rich minerals, and weathering of silicate minerals. The groundwater samples are divided into four groups in accordance with HCA, which were identified as G1, G2, G3, and G4. G1 and G2, which are found in the northern regions near to the Mediterranean Sea, are primarily controlled by the evaporation process and seawater intrusion. High Cl and Na concentrations and low HCO_3 concentrations mostly serve to demonstrate this. Additionally, seawater intrusion is joined by another process such as cation exchange where Na replaces Ca and/or Mg. On the other hand, in the case of G3 and G4, the weathering of silicate minerals led to the development of carbonate minerals. Most of the groundwater samples from the four groups included more calcium than HCO_3 and SO_4; this is mostly because calcium-rich minerals dissolve and weather.

Climate change and its parameters, sea-level rise, extensive groundwater pumping to meet various needs, land-use changes due to population growth and urbanization, changes in rainfall patterns, and reduced groundwater recharge all threaten saltwater intrusion coastal aquifers.

Since the coastal aquifer in the Nile Delta is responsive to climate change and groundwater extraction, sustainable water resource management is needed to protect it from saltwater intrusion. This research provides a useful assessment tool for seawater intrusion variability in coastal aquifers. The results can improve understanding of present and future saltwater intrusion in the investigated area and provide specific adaptive keys. As a response to the aquifer stresses, adaptation styles to any of the given hazards or challenges may be designed. Thus, the adaptation in the case of the Quaternary aquifer in the Nile Delta should be transformational, where changes in the fundamental attributes of a system (especially pumping) must be adopted. This may not preclude combining different adaptation strategies or using biodiversity and environmental services as part of an adaptation plan. Constant monitoring of the saline interface is essential for maintaining the aquifer's equilibrium and determining the most effective aquifer management strategy. An assessment study for the saltwater intrusion under different climate and anthropogenic activities is still needed for the aquifer.

Author Contributions: Conceptualization, S.S.H., Z.E.S. and A.S.; methodology, S.S.H. and Z.E.S.; software, S.S.H.; validation, S.S.H., Z.E.S. and A.S.; formal analysis, S.S.H., Z.E.S. and A.S.; investigation, S.S.H., Z.E.S. and A.S.; resources, S.S.H. and Z.E.S.; data curation, S.S.H., Z.E.S. and A.S.; writing—original draft preparation, S.S.H.; writing—review and editing, Z.E.S. and A.S.; visualization, S.S.H.; supervision, Z.E.S. and A.S.; project administration, Z.E.S. and A.S. All authors have read and agreed to the published version of the manuscript.

Funding: This research received no external funding.

Data Availability Statement: Not applicable.

Conflicts of Interest: The authors declare no conflict of interest.

References

1. Zhang, X.; Miao, J.; Hu, B.X.; Liu, H.; Zhang, H.; Ma, Z. Hydrogeochemical characterization and groundwater quality assessment in intruded coastal brine aquifers (Laizhou Bay, China). *Environ. Sci. Pollut. Res.* **2017**, *24*, 21073–21090. [CrossRef] [PubMed]
2. Makri, P.; Stathopoulou, E.; Hermides, D.; Kontakiotis, G.; Zarkogiannis, S.D.; Skilodimou, H.D.; Bathrellos, G.D.; Antonarakou, A.; Scoullos, M. The Environmental Impact of a Complex Hydrogeological System on Hydrocarbon-Pollutants' Natural Attenuation: The Case of the Coastal Aquifers in Eleusis, West Attica, Greece. *J. Mar. Sci. Eng.* **2020**, *8*, 1018. [CrossRef]
3. El Osta, M.; Masoud, M.; Alqarawy, A.; Elsayed, S.; Gad, M. Groundwater Suitability for Drinking and Irrigation Using Water Quality Indices and Multivariate Modeling in Makkah Al-Mukarramah Province, Saudi Arabia. *Water* **2022**, *14*, 483. [CrossRef]
4. Salem, Z.E.-S.; Abdelrahman, k.; Kováčiková, S.; Badran, O.M. Use of various statistical techniques to assess the vertical and lateral change in the groundwater chemistry of Quaternary aquifer in an irrigated highly populated area. *J. King Saud Univ. -Sci.* **2021**, *33*, 101556. [CrossRef]
5. Abu Salem, H.S.; Gemail, K.S.; Junakova, N.; Ibrahim, A.; Nosair, A.M. An Integrated Approach for Deciphering Hydrogeochemical Processes during Seawater Intrusion in Coastal Aquifers. *Water* **2022**, *14*, 1165. [CrossRef]
6. El Yousfi, Y.; Himi, M.; El Ouarghi, H.; Elgettafi, M.; Benyoussef, S.; Gueddari, H.; Aqnouy, M.; Salhi, A.; Alitane, A. Hydrogeochemical and statistical approach to characterize groundwater salinity in the Ghiss-Nekkor coastal aquifers in the Al Hoceima province, Morocco. *Groundw. Sustain. Dev.* **2022**, *19*, 100818. [CrossRef]
7. Makri, P.; Hermides, D.; Kontakiotis, G.; Zarkogiannis, S.D.; Besiou, E.; Janjuhah, H.T.; Antonarakou, A. Integrated Ecological Assessment of Heavily Polluted Sedimentary Basin within the Broader Industrialized Area of Thriassion Plain (Western Attica, Greece). *Water* **2022**, *14*, 382. [CrossRef]
8. Dibaj, M.; Javadi, A.A.; Akrami, M.; Ke, K.-Y.; Farmani, R.; Tan, Y.-C.; Chen, A.S. Modelling seawater intrusion in the Pingtung coastal aquifer in Taiwan, under the influence of sea-level rise and changing abstraction regime. *Hydrogeol. J.* **2020**, *28*, 2085–2103. [CrossRef]
9. Prusty, P.; Farooq, S.H. Seawater intrusion in the coastal aquifers of India—A review. *HydroResearch* **2020**, *3*, 61–74. [CrossRef]
10. Rajendiran, T.; Sabarathinam, C.; Chandrasekar, T.; Panda, B.; Mathivanan, M.; Nagappan, G.; Natesan, D.; Ghai, M.; Kumar Singh, D.; Alagappan, R. Geochemical variations due to salinization in groundwater along the southeast coast of India. *SN Appl. Sci.* **2021**, *3*, 581. [CrossRef]
11. Hajji, S.; Allouche, N.; Bouri, S.; Aljuaid, A.M.; Hachicha, W. Assessment of Seawater Intrusion in Coastal Aquifers Using Multivariate Statistical Analyses and Hydrochemical Facies Evolution-Based Model. *Int. J. Environ. Res. Public Health* **2021**, *19*, 155. [CrossRef] [PubMed]
12. Abd-Elhamid, H.F.; Abd-Elaty, I.; Hussain, M.S. Mitigation of seawater intrusion in coastal aquifers using coastal earth fill considering future sea level rise. *Environ. Sci. Pollut. Res.* **2020**, *27*, 23234–23245. [CrossRef] [PubMed]
13. Abd-Elhamid, H.F.; Abd-Elaty, I.; Sherif, M.M. Effects of Aquifer Bed Slope and Sea Level on Saltwater Intrusion in Coastal Aquifers. *Hydrology* **2019**, *7*, 5. [CrossRef]
14. Ding, Z.; Koriem, M.A.; Ibrahim, S.M.; Antar, A.S.; Ewis, M.A.; He, Z.; Kheir, A.M.S. Seawater intrusion impacts on groundwater and soil quality in the northern part of the Nile Delta, Egypt. *Environ. Earth Sci.* **2020**, *79*, 313. [CrossRef]
15. Yihdego, Y.; Al-Weshah, R.A. Assessment and Prediction of Saline Sea Water Transport in Groundwater Using 3-D Numerical Modelling. *Environ. Process.* **2017**, *4*, 49–73. [CrossRef]
16. Mahmoodzadeh, D.; Ketabchi, H.; Ataie-Ashtiani, B.; Simmons, C.T. Conceptualization of a fresh groundwater lens influenced by climate change: A modeling study of an arid-region island in the Persian Gulf, Iran. *J. Hydrol.* **2014**, *519*, 399–413. [CrossRef]
17. Hermides, D.; Makri, P.; Kontakiotis, G.; Antonarakou, A. Advances in the Coastal and Submarine Groundwater Processes: Controls and Environmental Impact on the Thriassion Plain and Eleusis Gulf (Attica, Greece). *J. Mar. Sci. Eng.* **2020**, *8*, 944. [CrossRef]
18. Han, G.; Liang, C.-Z.; Chung, T.-S.; Weber, M.; Staudt, C.; Maletzko, C. Combination of forward osmosis (FO) process with coagulation/flocculation (CF) for potential treatment of textile wastewater. *Water Res.* **2016**, *91*, 361–370. [CrossRef]
19. Llamas, M.R.; Martínez-Santos, P. Intensive Groundwater Use: Silent Revolution and Potential Source of Social Conflicts. *J. Water Resour. Plan. Manag.* **2005**, *131*, 337–341. [CrossRef]
20. Sarker, M.M.R.; Hermans, T.; Van Camp, M.; Hossain, D.; Islam, M.; Ahmed, N.; Bhuiyan, M.A.Q.; Karim, M.M.; Walraevens, K. Identifying the Major Hydrogeochemical Factors Governing Groundwater Chemistry in the Coastal Aquifers of Southwest Bangladesh Using Statistical Analysis. *Hydrology* **2022**, *9*, 20. [CrossRef]
21. Escolero, O.; Marin, L.E.; Domínguez-Mariani, E.; Torres-Onofre, S. Dynamic of the freshwater–saltwater interface in a karstic aquifer under extraordinary recharge action: The Merida Yucatan case study. *Environ. Geol.* **2007**, *51*, 719–723. [CrossRef]
22. Vengadesan, M.; Lakshmanan, E. Management of Coastal Groundwater Resources. In *Coastal Management*; Elsevier: Amsterdam, The Netherlands, 2019; pp. 383–397. ISBN 978-0-12-810473-6.
23. Torres Martínez, J.A.; Mora, A.; Ramos Leal, J.A.; Morán Ramírez, J.; Arango Galván, C.; Mahlknecht, J. Constraining a density-dependent flow model with the transient electromagnetic method in a coastal aquifer in Mexico to assess seawater intrusion. *Hydrogeol. J.* **2019**, *27*, 2955–2972. [CrossRef]
24. Senthilkumar, S.; Vinodh, K.; Johnson Babu, G.; Gowtham, B.; Arulprakasam, V. Integrated seawater intrusion study of coastal region of Thiruvallur district, Tamil Nadu, South India. *Appl. Water Sci.* **2019**, *9*, 124. [CrossRef]

25. Vann, S.; Puttiwongrak, A.; Suteerasak, T.; Koedsin, W. Delineation of Seawater Intrusion Using Geo-Electrical Survey in a Coastal Aquifer of Kamala Beach, Phuket, Thailand. *Water* **2020**, *12*, 506. [CrossRef]
26. Shin, J.; Hwang, S. A Borehole-Based Approach for Seawater Intrusion in Heterogeneous Coastal Aquifers, Eastern Part of Jeju Island, Korea. *Water* **2020**, *12*, 609. [CrossRef]
27. Hermans, T.; Paepen, M. Combined Inversion of Land and Marine Electrical Resistivity Tomography for Submarine Groundwater Discharge and Saltwater Intrusion Characterization. *Geophys. Res. Lett.* **2020**, *47*. [CrossRef]
28. Sae-Ju, J.; Chotpantarat, S.; Thitimakorn, T. Hydrochemical, geophysical and multivariate statistical investigation of the seawater intrusion in the coastal aquifer at Phetchaburi Province, Thailand. *J. Asian Earth Sci.* **2020**, *191*, 104165. [CrossRef]
29. Palacios, A.; Ledo, J.J.; Linde, N.; Luquot, L.; Bellmunt, F.; Folch, A.; Marcuello, A.; Queralt, P.; Pezard, P.A.; Martínez, L.; et al. Time-lapse cross-hole electrical resistivity tomography (CHERT) for monitoring seawater intrusion dynamics in a Mediterranean aquifer. *Hydrol. Earth Syst. Sci.* **2020**, *24*, 2121–2139. [CrossRef]
30. Sefelnasr, A.; Sherif, M. Impacts of Seawater Rise on Seawater Intrusion in the Nile Delta Aquifer, Egypt. *Groundwater* **2014**, *52*, 264–276. [CrossRef]
31. Mabrouk, M.; Jonoski, A.; HP Oude Essink, G.; Uhlenbrook, S. Impacts of Sea Level Rise and Groundwater Extraction Scenarios on Fresh Groundwater Resources in the Nile Delta Governorates, Egypt. *Water* **2018**, *10*, 1690. [CrossRef]
32. Mabrouk, M.; Jonoski, A.; Oude Essink, G.; Uhlenbrook, S. Assessing the Fresh–Saline Groundwater Distribution in the Nile Delta Aquifer Using a 3D Variable-Density Groundwater Flow Model. *Water* **2019**, *11*, 1946. [CrossRef]
33. Masoud, M.; El Osta, M.; Alqarawy, A.; Elsayed, S.; Gad, M. Evaluation of groundwater quality for agricultural under different conditions using water quality indices, partial least squares regression models, and GIS approaches. *Appl. Water Sci.* **2022**, *12*, 244. [CrossRef]
34. Sherif, M.; Sefelnasr, A.; Javadi, A. Incorporating the concept of equivalent freshwater head in successive horizontal simulations of seawater intrusion in the Nile Delta aquifer, Egypt. *J. Hydrol.* **2012**, *464–465*, 186–198. [CrossRef]
35. Salem, Z.E.; Al Temamy, A.M.; Salah, M.K.; Kassab, M. Origin and characteristics of brackish groundwater in Abu Madi coastal area, Northern Nile Delta, Egypt. *Estuar. Coast. Shelf Sci.* **2016**, *178*, 21–35. [CrossRef]
36. Salem, Z.E.-S.; Osman, O.M. Use of major ions to evaluate the hydrogeochemistry of groundwater influenced by reclamation and seawater intrusion, West Nile Delta, Egypt. *Environ. Sci. Pollut. Res.* **2017**, *24*, 3675–3704. [CrossRef]
37. El Deek, M.; Nessim, R. Organic pollutants in Abu Qir Bay. *J. High Inst. Public Health* **1995**, *25*, 221–226.
38. Armanuos, A.M.; Negm, A. Integrated Groundwater Modeling for Simulation of Saltwater Intrusion in the Nile Delta Aquifer, Egypt. In *Groundwater in the Nile Delta*; Negm, A.M., Ed.; The Handbook of Environmental Chemistry; Springer International Publishing: Cham, Switzerland, 2019; pp. 489–544. ISBN 978-3-319-94283-4.
39. Allam, A.R.; Saaf, E.-J.; Dawoud, M.A. Desalination of brackish groundwater in Egypt. *Desalination* **2003**, *152*, 19–26. [CrossRef]
40. Dawoud, M.A.; Darwish, M.M.; El-Kady, M.M. GIS-Based Groundwater Management Model for Western Nile Delta. *Water Resour. Manag.* **2005**, *19*, 585–604. [CrossRef]
41. Morsy, W. Environmental Management of Groundwater Resources in the Nile Delta Region. Ph.D. Thesis, Cairo University, Cairo, Egypt, 2009.
42. Awad, S.R.; El Fakharany, M.A.; Hagran, N.M. Environmental Impact on Water Resources at the Northwestern Part of the Nile Delta, Egypt. *J. Am. Sci.* **2015**, *11*, 1–11.
43. Salem, Z.E.; Sefelnasr, A.M.; Hasan, S.S. Assessment of groundwater vulnerability for pollution using DRASTIC Index, young alluvial plain, Western Nile Delta, Egypt. *Arab. J. Geosci.* **2019**, *12*, 727. [CrossRef]
44. USGS. *Earth Resources Observation And Science (EROS) Center Shuttle Radar Topography Mission (SRTM) 1 Arc-Second Global*; USGS: Reston, VA, USA, 2017. [CrossRef]
45. Salem, Z.E.; Hasan, S.S. Use of GALDIT model and HFE-Diagram to assess seawater intrusion vulnerability in West Nile Delta, Egypt. *Arab. J. Geosci.* **2021**, *14*, 1318. [CrossRef]
46. El-Arabi, N.E.; Morsy, W.S. Applying integrated ground- and surface-water management (case study: Nubaryia Basin, West Delta, Egypt). *J. Am. Sci.* **2013**, *9*, 43–53.
47. REGWA. *The Groundwater in a Study Area Southern of Wadi El Farigh*; REGWA: Cairo, Egypt, 1993; p. 74.
48. Domenico, P.A.; Schwartz, F.W. *Physical and Chemical Hydrogeology*, 2nd ed.; Wiley: Hoboken, NJ, USA, 1998; ISBN 978-0-471-59762-9.
49. Deutsch, W.J.; Siegel, R. *Groundwater Geochemistry: Fundamentals and Applications to Contamination*; CRC Press: Boca Raton, FL, USA, 1997; ISBN 978-1-00-306994-2.
50. HACH. *Chemical Procedures Explained*; Hach Technical Center for Applied Analytical Chemistry: Loveland, CO, USA, 1990.
51. US EPA. Method 200.7: Determination of Metals and Trace Elements in Water and Wastes by Inductively Coupled Plasma-Atomic Emission Spectrometry. Available online: https://www.epa.gov/esam/method-2007-determination-metals-and-trace-elements-water-and-wastes-inductively-coupled (accessed on 7 March 2023).
52. Okiongbo, K.S.; Douglas, R.K. Evaluation of major factors influencing the geochemistry of groundwater using graphical and multivariate statistical methods in Yenagoa city, Southern Nigeria. *Appl. Water Sci.* **2015**, *5*, 27–37. [CrossRef]
53. Alfy, M.E.; Lashin, A.; Faraj, T.; Alataway, A.; Tarawneh, Q.; Al-Bassam, A. Quantitative hydro-geophysical analysis of a complex structural karst aquifer in Eastern Saudi Arabia. *Sci. Rep.* **2019**, *9*, 2825. [CrossRef] [PubMed]

54. Dişli, E.; Gülyüz, N. Hydrogeochemical investigation of an epithermal mineralization bearing basin using multivariate statistical techniques and isotopic evidence of groundwater: Kestanelik Sub-Basin, Lapseki, Turkey. *Geochemistry* **2020**, *80*, 125661. [CrossRef]
55. Ren, X.; Li, P.; He, X.; Su, F.; Elumalai, V. Hydrogeochemical Processes Affecting Groundwater Chemistry in the Central Part of the Guanzhong Basin, China. *Arch. Environ. Contam. Toxicol.* **2021**, *80*, 74–91. [CrossRef]
56. Li, P.; Tian, R.; Liu, R. Solute Geochemistry and Multivariate Analysis of Water Quality in the Guohua Phosphorite Mine, Guizhou Province, China. *Expo Health* **2019**, *11*, 81–94. [CrossRef]
57. Wu, J.; Li, P.; Wang, D.; Ren, X.; Wei, M. Statistical and multivariate statistical techniques to trace the sources and affecting factors of groundwater pollution in a rapidly growing city on the Chinese Loess Plateau. *Hum. Ecol. Risk Assess. Int. J.* **2020**, *26*, 1603–1621. [CrossRef]
58. Charlton, S.R.; Parkhurst, D.L. *PHREEQCI—A Graphical User Interface to the Geochemical Model PHREEQC*; Fact Sheet; U.S. Geological Survey: Reston, VA, USA, 2002; Volume 031–032.
59. Tomaszkiewicz, M.; Abou Najm, M.; El-Fadel, M. Development of a groundwater quality index for seawater intrusion in coastal aquifers. *Environ. Model. Softw.* **2014**, *57*, 13–26. [CrossRef]
60. Appelo, C.A.J.; Postma, D. (Eds.) *Geochemistry, Groundwater and Pollution*, 2nd ed.; CRC Press: London, UK, 2005; ISBN 978-0-429-15232-0.
61. Kreitler, C.W. *Geochemical Techniques for Identifying Sources of Ground-Water Salinization*; CRC Press: Boca Raton, FL, USA, 1993; ISBN 978-1-56670-000-9.
62. Panteleit, B.; Hamer, K.; Kringel, R.; Kessels, W.; Schulz, H.D. Geochemical processes in the saltwater–freshwater transition zone: Comparing results of a sand tank experiment with field data. *Environ. Earth Sci* **2011**, *62*, 77–91. [CrossRef]
63. Sharaky, A.; Atta, S.; El Hassanein, A.; Khallaf, K. *Hydrogeochemistry of Groundwater in the Western Nile Delta Aquifers, Egypt*; Cairo University: Giza, Egypt, 2007; Volume 19, pp. 1–23.
64. Winslow, A.G.; Kister, L.R. *Saline-Water Resources of Texas*; U.S. Government Printing Office: Washington, DC, USA, 1956.
65. Abdel Baki, A. Hydrogeological and Hydrochemical Studies on the Area West of Rosetta Branch and South El Nasr Canal. Ph.D. Thesis, Ain Shams University, Cairo, Egypt, 1983.
66. Piper, A. A graphic procedure in the geochemical interpretation of water-analyses. *Eos Trans. Am. Geophys. Union* **1944**, *25*, 914–928. [CrossRef]
67. Sharaky, A.M.; El Hasanein, A.S.; Atta, S.A.; Khallaf, K.M. Nile and Groundwater Interaction in the Western Nile Delta, Egypt. In *The Nile Delta*; Negm, A.M., Ed.; The Handbook of Environmental Chemistry; Springer International Publishing: Cham, Switzerland, 2016; Volume 55, pp. 33–62. ISBN 978-3-319-56122-6.
68. Berkowitz, B.; Singurindy, O.; Lowell, R.P. Mixing-driven diagenesis and mineral deposition: $CaCO_3$ precipitation in salt water–fresh water mixing zones: Mixing-Driven Diagenesis. *Geophys. Res. Lett.* **2003**, *30*. [CrossRef]
69. Drevaliene, G.; Karro, E.; Mokrik, R.; Savitskaja, L. The origin of barium in the Cambrian–Vendian aquifer system, North Estonia. *Est. J. Earth Sci.* **2009**, *58*, 193–208. [CrossRef]
70. Rawson, J.; Siade, A.; Sun, J.; Neidhardt, H.; Berg, M.; Prommer, H. Quantifying Reactive Transport Processes Governing Arsenic Mobility after Injection of Reactive Organic Carbon into a Bengal Delta Aquifer. *Environ. Sci. Technol.* **2017**, *51*, 8471–8480. [CrossRef] [PubMed]
71. Biswas, A.; Gustafsson, J.P.; Neidhardt, H.; Halder, D.; Kundu, A.K.; Chatterjee, D.; Berner, Z.; Bhattacharya, P. Role of competing ions in the mobilization of arsenic in groundwater of Bengal Basin: Insight from surface complexation modeling. *Water Res.* **2014**, *55*, 30–39. [CrossRef]
72. Hosono, T.; Hossain, S.; Shimada, J. Hydrobiogeochemical evolution along the regional groundwater flow systems in volcanic aquifers in Kumamoto, Japan. *Environ. Earth Sci.* **2020**, *79*, 410. [CrossRef]
73. Marandi, A.; Shand, P. Groundwater chemistry and the Gibbs Diagram. *Appl. Geochem.* **2018**, *97*, 209–212. [CrossRef]
74. Tiwari, A.K.; De Maio, M.; Singh, P.K.; Singh, A.K. Hydrogeochemical characterization and groundwater quality assessment in a coal mining area, India. *Arab. J. Geosci.* **2016**, *9*, 177. [CrossRef]
75. Bouderbala, A.; Gharbi, B.Y. Hydrogeochemical characterization and groundwater quality assessment in the intensive agricultural zone of the Upper Cheliff plain, Algeria. *Environ. Earth Sci.* **2017**, *76*, 744. [CrossRef]
76. Shin, K.; Koh, D.-C.; Jung, H.; Lee, J. The Hydrogeochemical Characteristics of Groundwater Subjected to Seawater Intrusion in the Archipelago, Korea. *Water* **2020**, *12*, 1542. [CrossRef]
77. Hussien, R.A. Groundwater Quality Index Studies for Seawater Intrusion in Coastal Aquifer Ras Sudr, Egypt Using Geographic Information System. *Sciences* **2015**, *5*, 209–222.

MDPI AG
Grosspeteranlage 5
4052 Basel
Switzerland
Tel.: +41 61 683 77 34

Water Editorial Office
E-mail: water@mdpi.com
www.mdpi.com/journal/water

www.ingramcontent.com/pod-product-compliance
Lightning Source LLC
Chambersburg PA
CBHW061444210926
51850CB00008B/1614
9783725817986